Christian De Schryver
Editor

FPGA Based Accelerators for Financial Applications

Springer

Editor
Christian De Schryver
University of Kaiserslautern
Kaiserslautern, Germany

ISBN 978-3-319-15406-0   ISBN 978-3-319-15407-7 (eBook)
DOI 10.1007/978-3-319-15407-7

Library of Congress Control Number: 2015940116

Springer Cham Heidelberg New York Dordrecht London

Printed on acid-free paper

Springer International Publishing AG Switzerland is part of Springer Science+Business Media (www.springer.com)

Preface from the Editor

1 The Need for Reconfigurable Computing Systems in Finance

The finance sector is one of most prominent users of High Performance Computing (HPC) facilities. It is not only due to the aftermath of the financial crisis in 2008 that the computational demands have surged over the last years but due to increasing regulations (e.g., Basel III and Solvency II) and reporting requirements. Institutes are forced to deliver valuation and risk simulation results to internal risk management departments and external regulatory authorities frequently [2, 16, 17].

One important bottleneck in many investment and risk management calculations is the pricing of exotic derivatives in appropriate market models [2]. However, in many of these cases, no (semi)closed-form pricing formulas exist, and the evaluation is carried out by applying numerical approximations. In most cases, calculating those numbers for a complete portfolio can be very compute intensive and can last hours to days on state-of-the-art compute clusters with thousands of cores [17]. The increasing complexity of the underlying market models and financial products makes this situation even worse [2, 5, 6, 8]. In addition, the progress in online applications like news aggregation and analysis [9] and the competition in the field of low-latency and High-Frequency Trading (HFT) require new technologies to keep track with the operational and market demands.

Data centers and HPC in general are currently facing a massive energy problem [2, 3]. In particular, this also holds for financial applications: The energy needed for portfolio pricing is immense and lies in the range of several megawatts for a single average-sized institute today [17]. Already in 2008 the available power for Canary Wharf, the financial district of London, had to be limited to ensure a reliable supply for the Olympic Games in 2012 [15]. In addition, energy costs also force financial institutes to look into alternative ways of obtaining sufficient computational power at lower operating costs [16].

Two fundamental design principles for high-performance and energy-efficient computing appliances are the shifts to high data locality with minimum data movements and to heterogeneous computing platforms that integrate dedicated and specialized hardware accelerators. The performance of battery-driven mobile devices we experience today is grounded in these concepts. Nowadays, the need for heterogeneity is widely acknowledged in the HPC domain as well [2, 3]. Nevertheless, the vast majority of current data centers and in-house computing systems is still based on general-purpose Central Processing Units (CPUs), Graphics Processor Units (GPUs), or Intel Xeon Phi processors. The reason is that those architectures are tailored to providing a high flexibility on application level, but at the cost of low energy efficiency.

Dedicated Application Specific Integrated Circuit (ASIC) accelerator chips achieve the optimal performance and energy efficiency. However, ASICs come with some significant drawbacks regarding their use in supercomputing systems in general:

1. The Non-recurring Engineering (NRE) and fixed manufacturing costs for custom ASICs are in the range of several 100 million USD for state-of-the-art 28 nm processes [10]. This means that the cost per unit is enormous for low volume production and therefore economically unfeasible.
2. Manufactured ASICs are unalterably wired circuits and can therefore only provide the flexibility that has been incorporated into their architecture at design time. Changing their functionality or adding additional features beyond those capabilities would require a replacement of the hardware with updated versions.
3. The design effort and therefore also the Time to Market (TTM) is in the range of months to years for ASIC development. However, in particular in the finance domain, it can be necessary to implement new products or algorithms very fast. Designing a new ASIC for this is probably not viable.

In contrast to ASICs, reconfigurable devices like Field Programmable Gate Arrays (FPGAs) can be reprogrammed without limit and can change their functionality even while the system is running. Therefore, they are a very promising technology for integrating dedicated hardware accelerators in existing CPU- and GPU-based computing systems, resulting in so-called *High Performance Reconfigurable Computing (HPRC)* architectures [14].

FPGAs have already shown to outperform CPU- and GPU-only architectures with respect to speed and energy efficiency by far for financial applications [1, 2, 12]. First attempts to use reconfigurable technology in practice are made, for example, by J.P. Morgan [4] or Deutsche Bank [11].

However, the use of FPGAs still comes with a lot of challenges. For example, no standard design and integration flows exist up to now that make this technology available to software and algorithmic engineers right away. First approaches such as the Maxeler systems,[1] the MathWorks HDL Coder [13], the Altera OpenCL flow [7], or the Xilinx SDAccel approach [18] are moving into the right direction, but still require fundamental know-how about hardware design in order to end up

[1]http://www.maxeler.com

with powerful accelerator solutions. Hybrid devices like the recent Xilinx Zynq All Programmable system on chips (SoCs) combine standard CPU cores with a reconfigurable FPGA part and thus enable completely new system architectures also in the HPRC domain. This book summarizes the main ideas and concepts required for successfully integrating FPGAs into financial computing systems.

2 Intended Audience and Purpose of This Book

When I started my work as a researcher in the field of accelerating financial applications with FPGAs in 2010 at the University of Kaiserslautern, I found myself in a place where interdisciplinary collaboration between engineers and mathematicians was not only a buzzword, but had a long and lived tradition. It was not only established through informal cooperation projects between the departments and research groups within the university itself, but also materialized, for example, in the Center for Mathematical and Computational Modelling ($(CM)^2$). $(CM)^2$ is a research center funded by the German state Rhineland-Palatinate with the aim of showing that mathematics and computer science represent a technology that is essential to engineers and natural scientists and that will help advance progress in relevant areas.[2] I have carried out my first works as a member of the Microelectronic Systems Design Research Group headed by Prof. Norbert Wehn in the context of the very successful $(CM)^2$ project "Hardware assisted Acceleration for Monte Carlo Simulations in Financial Mathematics with a particular Emphasis on Option Pricing (HOPP)." As one outcome of $(CM)^2$, the Deutsche Forschungsgemeinschaft (DFG) has decided to implement a new research training group (RTG) 1932 titled "Stochastic Models for Innovations in the Engineering Sciences" at the University of Kaiserslautern for the period April 2014–September 2018 (see Preface from Prof. Ralf Korn, speaker of the RTG 1932).

In addition to the successful networking within the university, Kaiserslautern is a famous location for fruitful cooperations between companies and institutes in the fields of engineering and mathematics in general. Particularly active in the field of financial mathematics is the Fraunhofer Institute for Industrial Mathematics (ITWM),[3] a well-reputed application-oriented research institution with the mission of applying the latest mathematical findings from research to overcome practical challenges from industry. It is located only a short distance from the university campus.

Despite the beneficial circumstances, one of my first discoveries was that it was quite hard to get an overview about what is already going on in the field "accelerating financial applications with FPGAs." The reason is that we are entering a strongly interdisciplinary environment comprising hardware design,

[2] http://cmcm.uni-kl.de/en

[3] http://www.itwm.fraunhofer.de/en/departments/financial-mathematics.html

financial mathematics, computational stochastics, benchmarking, HPC, and software engineering. Although many particular topics had already been investigated in detail, their impact in the context of "accelerating financial applications with reconfigurable architectures" was not always obvious. In addition, up to now there is no accessible textbook available that covers all important aspects of using FPGAs for financial applications.

My main motivation to come up with this book is exactly to close this gap and to make it easier for readers to see the global picture required to identify the critical points from all cross-disciplinary viewpoints. The book summarizes the current challenges in finance and therefore justifies the needs for new computing concepts including FPGA-based accelerators, both for readers from finance business and research. It covers the most promising strategies for accelerating various financial applications known today and illustrates that real interdisciplinary approaches are crucial to come up with powerful and efficient computing systems for those in the end.

For people new to or particularly interested in this topic, the book summarizes the state-of-the-art work and therefore should act as a guide through all the various approaches and ideas. It helps readers from the academic domain to get an overview about possible research fields and points out those areas where further investigations are needed to make FPGAs accessible for people from practice. For practitioners, the book highlights the most important concepts and the latest findings from research and illustrates how those can help to identify and overcome bottlenecks in current systems. Quants and algorithmic developers will get insights into the technological effects that may limit their implementations in the end and how to overcome those. For managers and administrators in the Information Technology (IT) domain, the book gives answers about how to integrate FPGAs into existing systems and how to ensure flexibility and maintainability over time.

3 Outline and Organization of the Book

A big obstacle for researchers is the fact that it is generally very hard to get access to the real technological challenges that financial institutes are facing in daily business. My experience is that this information can only be obtained in face-to-face discussions with practitioners and will vastly differ from company to company. Chapter 1 by Desmettre and Korn therefore highlights the 10 biggest challenges in the finance business from a viewpoint of financial mathematics and risk management.

One particular computationally challenging task in finance is calibrating the market models against the market. Chapter 2 by Sayer and Wenzel outlines the calibration process and distills the most critical points in this process. Furthermore, it shows which steps in the calibration process are the main limiting factors and how they can be tackled to speed up the calibration process in general.

In Chap. 3, Delivorias motivates the use of FPGAs for pricing tasks by giving throughput numbers for CPU, GPU, and FPGA systems. He considers price paths generated in the Heston market model and compares the run time over all platforms.

Fairly comparing various platforms on application level is a nontrivial task, in particular when different algorithms are used. Chapter 4 by De Schryver and Noguiera introduces a generic benchmark approach together with appropriate metrics that can be used to characterize the performance and energy efficiency of (heterogeneous) systems independent of the underlying technology and implemented algorithm.

High-Level Synthesis (HLS) is currently moving into productive hardware designs and seems to be one of the most promising approaches to make FPGAs accessible to algorithm and software developers. In Chap. 5, Inggs, Fleming, Thomas, and Luk demonstrate the current performance of HLS for financial applications with an option pricing case study.

In addition to the design of the hardware accelerator architecture itself, its integration into existing computing system is a crucial point that needs to be solved. Chapter 6 by Sadri, De Schryver, and Wehn introduces the basics of Peripheral Component Interconnect Express (PCIe) and Advanced eXtensible Interface (AXI), two of the most advanced interfaces currently used in HPC and System on Chip (SoC) architectures. For the hybrid Xilinx Zynq device that comes with a CPU and an FPGA part it points out possible pitfalls and how they can be overcome whenever FPGAs need to be attached to existing host systems over PCIe.

Path-dependent options are particularly challenging for acceleration with dedicated architectures. The reason is that the payoff of those products needs to be evaluated at every considered point in time until the maturity. For American options, Varela, Brugger, Tang, Wehn, and Korn illustrate in Chap. 7 how a pricing system for path-dependent options can be efficiently implemented on a hybrid CPU/FPGA system.

One major benefit of FPGAs is their reconfigurability and therefore the flexibility they can provide once integrated into HPC computing systems. However, currently there is no standard methodology on how to exploit this reconfigurability efficiently at runtime. In Chap. 8, Brugger, De Schryver, and Wehn propose *HyPER*, a framework for efficient option pricer implementations on generic hybrid systems consisting of CPU and FPGA parts. They describe their approach in detail and show that HyPER is 3.4× faster and 36× more power efficient than a highly tuned software reference on an Intel Core i5 CPU.

While on CPUs and GPUs the hardware and therefore the available data types are fixed, FPGAs give complete freedom to the user about which precision and bit widths should be used in each stage of the architecture. This opens up a completely new degree of freedom and also heavily influences the costs of available algorithms whenever implemented on FPGAs. Chapter 9 by Omland, Hefter, Ritter, Brugger, De Schryver, Wehn, and Kostiuk outlines this issue and shows how so-called *mixed-precision* systems can be designed without losing any accuracy of the final computation results.

As introduced in Chap. 2, calibration is one of the compute intensive tasks in finance. Chapter 10 by Liu, Brugger, De Schryver, and Wehn introduces design concepts for accelerating this problem for the Heston model with an efficient accelerator for pricing vanilla options in hardware. It shows the complete algorithmic design space and exemplarily illustrates how to obtain efficient accelerator implementations from the actual problem level.

Fast methodologies and tools are mandatory for achieving high productivity whenever working with hardware accelerators in business. In Chap. 11, Becker, Mencer, Weston, and Gaydadjiev present the Maxeler data-flow approach and show how it can be applied to value-at-risk and low-latency trading in finance.

Kaiserslautern, Germany Christian De Schryver
15 Feb 2015

References

1. Brugger, C., de Schryver, C., Wehn, N.: HyPER: a runtime reconfigurable architecture for Monte Carlo option pricing in the Heston model. In: Proccedings of the 24th IEEE International Conference of Field Programmable Logic and Applications (FPL), Munich, pp. 1–8, Sept 2014
2. de Schryver, C.: Design methodologies for hardware accelerated heterogeneous computing systems. PhD thesis, University of Kaiserslautern (2014)
3. Duranton, M., Black-Schaffer, D., De Bosschere, K., Mabe, J.: The HiPEAC Vision for Advanced Computing in Horizon 2020 (2013) http://www.cs.ucy.ac.cy/courses/EPL605/Fall2014Files/HiPEAC-Roadmap-2013.pdf, last access: 2015-05-19
4. Feldman, M.: JP Morgan buys into FPGA supercomputing. http://www.hpcwire.com/2011/07/13/jp_morgan_buys_into_fpga_supercomputing/, July 2011. Last access 09 Feb 2015
5. Griebsch, S.A., Wystup, U.: On the valuation of Fader and discrete Barrier options in Heston's stochastic volatility model. Quant. Finance **11**(5), 693–709 (2011)
6. Heston, S.L.: A closed-form solution for options with stochastic volatility with applications to bond and currency options. Rev. Financ. Stud. **6**(2), 327 (1993)
7. Implementing FPGA design with the openCL standard. Technical report, Altera Corporation. http://www.altera.com/literature/wp/wp-01173-opencl.pdf, Nov 2011. Last access 05 Feb 2015
8. Lord, R., Koekkoek, R., van Dijk, D.: A comparison of biased simulation schemes for stochastic volatility models. Quant. Finance **10**(2), 177–194 (2010)
9. Mao, H., Wang, K., Ma, R., Gao, Y., Li, Y., Chen, K., Xie, D., Zhu, W., Wang, T., Wang, H.: An automatic news analysis and opinion sharing system for exchange rate analysis. In: Proceedings of the 2014 IEEE 11th International Conference on e-Business Engineering (ICEBE), Guangzhou, pp. 303–307, Nov 2014
10. Or-Bach, Z.: FPGA as ASIC alternative: past and future. http://www.monolithic3d.com/blog/fpga-as-asic-alternative-past-and-future, Apr 2014. Last access 13 Feb 2015
11. Schmerken, I.: Deutsche bank shaves trade latency down to 1.25 microseconds. http://www.advancedtrading.com/infrastructure/229300997, Mar 2011. Last access 09 Feb 2015
12. Sridharan, R., Cooke, G., Hill, K., Lam, H., George, A.: FPGA-based reconfigurable computing for pricing multi-asset Barrier options. In: Proceedings of Symposium on Application Accelerators in High-Performance Computing PDF (SAAHPC) (2012) Lemont, Illinois
13. The MathWorks, Inc.: HDL Coder. http://de.mathworks.com/products/hdl-coder. Last access 05 Feb 2015

14. Vanderbauwhede, W., Benkrid, K. (eds.): High-Performance Computing Using FPGAs. Springer, New York (2013)
15. Warren, P.: City business races the Games for power. The Guardian, May 2008
16. Weston, S., Marin, J.-T., Spooner, J., Pell, O., Mencer, O.: Accelerating the computation of Portfolios of tranched credit derivatives. In: IEEE Workshop on High Performance Computational Finance (WHPCF), New Orleans, pp. 1–8, Nov 2010
17. Weston, S., Spooner, J., Marin, J.-T., Pell, O., Mencer, O.: FPGAs speed the computation of complex credit derivatives. Xcell J. **74**, 18–25 (2011)
18. Xilinx Inc.: SDAccel development environment. http://www.xilinx.com/products/design-tools/sdx/sdaccel.html, Nov 2014. Last access 05 Feb 2015

Computing in Finance

Where Models and Applications Link Mathematics and Hardware Design

The table of contents of this book clearly indicates that the book is an interdisciplinary effort between engineers with a specialization in hardware design and mathematicians working in the area of financial mathematics.

Such a cooperation between engineers and mathematicians is a trademark for research done at the University of Kaiserslautern, the place related to most of the authors who contribute to this book. Many interdisciplinary research activities in recent years have benefitted from this approach, the most prominent one of them is the Research Training Group 1932 *Stochastic Models for Innovations in the Engineering Sciences* financed by the DFG, the German Research Foundation. The RTG considers four areas of application: production processes in fluids and non-wovens, multi-phase metals, high-performance concrete, and finally hardware design with applications in finance.

Mathematical modeling (and in particular stochastic modeling) is seen as the basis for innovations in engineering sciences. To ensure that this approach results in successful research, we have taken various innovative measures on the PhD level in the RTG 1932. Among them are:

- **PhD students attend all relevant lectures together**: This ensures that mathematics students can assist their counterparts from the engineering sciences to understand mathematics and vice versa when it comes to engineering talks.
- **Solid education in basics and advanced aspects**: Lecture series specially designed for the PhD students such as *Principles of Engineering* or *Principles of stochastic modeling* lift them quickly on the necessary theoretical level.
- **Joint language**: Via frequent meetings in the joint project, we urge the students to learn the scientific language of the partners. This is a key feature for true interdisciplinary research.

For this book, mainly the cooperation between financial mathematics, computational stochastics, and hardware design is essential. The corresponding contributions will highlight some advantages of these cooperations:

- Efficient use of modern hardware by mathematical algorithms that are implemented in adaptive ways
- Dealing with computational problems that do not only challenge the hardware, but that are truly relevant from the theoretical and the practical aspects of finance
- A mixed-precision approach that cares for the necessary accuracy required by theoretical numerics and at the same time considers the possible speedup

In total, this book is a proof that interdisciplinary research can yield breakthroughs that are possible as researchers have widened their scopes.

Kaiserslautern, Germany Ralf Korn
10 Dec 2014

Acknowledgements from the Editor

Coming up with a book about an emerging novel field of technology is not possible without valuable contributions from many different areas. Therefore, my warm thanks go to all people who have helped to make this book reality in the end.

First of all, I would like to thank all authors for their high-quality chapters and the time and effort each and everyone of them has invested to make this book as comprehensive as possible. I know from many discussions and talks with them throughout the writing process that it has always been a challenge to cover a large range of interesting aspects in this field and therefore provide a broad view on all the important aspects about using FPGAs in finance, but at the same time reduce the complexity to a level that allows readers without deep knowledge in this field to understand the big picture and the details.

I would also like to thank all involved reviewers for all their feedback that has significantly helped to write in a clear and precise wording and to make sure that we keep focused on the key points. In particular, I would like to thank Steffen de Schryver for his valuable suggestion on improving the language of parts of the book.

Very special thanks go to my boss Prof. Norbert Wehn for giving me the opportunity to assemble this book and his continuous and constructive support during the creation phase. Without his dedicated encouragement, this book would never have happened.

A considerable amount of content included in this book has been investigated in the context of funded research and industry projects. I would like to thank the German state Rhineland-Palatinate for funding the (CM)^2[4] at the University of Kaiserslautern. My thanks also go to the Deutsche Forschungsgemeinschaft (DFG) for supporting the RTG GRK 1932 "Stochastic Models for Innovations in the Engineering Sciences,"[5] and its speaker Prof. Ralf Korn for contributing a preface to this book. Furthermore, I give thanks to the German Federal Ministry of Education and

[4]http://cmcm.uni-kl.de/en

[5]http://www.mathematik.uni-kl.de/en/research/dfg-rtn1932

Research for sponsoring the project "Energieeffiziente Simulationsbeschleunigung für Risikomessung und -management"[6] under grant number 01LY1202D.

I would like to give thanks to Hemachandirane Sarumathi from SPi Global for coordinating the typesetting of this book. Last but not least, I would like to thank Charles "Chuck" Glaser and Jessica Lauffer from Springer and my family for their patience and continuous support over the last months.

Kaiserslautern, Germany Christian De Schryver
15 Feb 2015

---

[6]http://www.esr-projekt.de/en

Contents

Chapter 1
10 Computational Challenges in Finance

Sascha Desmettre and Ralf Korn

Abstract With the growing use of both highly developed mathematical models and complicated derivative products at financial markets, the demand for high computational power and its efficient use via fast algorithms and sophisticated hard- and software concepts became a hot topic in mathematics and computer science. The combination of the necessity to use numerical methods such as Monte Carlo simulation, of the demand for a high accuracy of the resulting prices and risk measures, of online availability of prices, and the need for repeatedly performing those calculations for different input parameters as a kind of sensitivity analysis emphasizes this even more.

In this survey, we describe the mathematical background of some of the most challenging computational tasks in financial mathematics. Among the examples are the pricing of exotic options by Monte Carlo methods, the calibration problem to obtain the input parameters for financial market models, and various risk management and measurement tasks.

1.1 Financial Markets and Models as Sources for Computationally Challenging Problems

With the growing use of both highly developed mathematical models and complicated derivative products at financial markets, the demand for high computational power and its efficient use via fast algorithms and sophisticated hard- and software concepts became a hot topic in mathematics and computer science. The combination of the necessity to use numerical methods such as Monte Carlo (MC) simulations, of the demand for a high accuracy of the resulting prices and risk measures, of online availability of prices, and the need for repeatedly performing those calculations for different input parameters as a kind of sensitivity analysis emphasizes this even more.

S. Desmettre (✉) • R. Korn
Department of Mathematics, TU Kaiserslautern, 67663 Kaiserslautern, Germany
e-mail: desmettre@mathematik.uni-kl.de; korn@mathematik.uni-kl.de

C. De Schryver (ed.), *FPGA Based Accelerators for Financial Applications*,
DOI 10.1007/978-3-319-15407-7_1

In this survey, we describe the mathematical background of some of the most challenging computational tasks in financial mathematics. Among the examples are the pricing of exotic options by MC methods, the calibration problem to obtain the input parameters for financial market models, and various risk management and measurement tasks.

We will start by introducing the basic building blocks of stochastic processes such as the Brownian motion and stochastic differential equations, present some popular stock price models, and give a short survey on options and their pricing. This will then be followed by a survey on option pricing via the MC method and a detailed description of different aspects of risk management.

1.2 Modeling Stock Prices and Further Stochastic Processes in Finance

Stock price movements in time as reported in price charts always show a very irregular, non-smooth behavior. The irregular fluctuation seems to dominate a clear tendency of the evolution of the stock price over time. The appropriate mathematical setting is that of diffusion processes, especially that of the Brownian Motion (BM).

A one-dimensional BM $W(t)$ is defined as a stochastic process with continuous path (i.e. it admits continuous realizations as a function of time) and

- $W(0) = 0$ almost surely,
- Stationary increments with $W(t) - W(s) \sim \mathcal{N}(0, t-s)$, $t > s \geq 0$,
- Independent increments, i.e. $W(t) - W(s)$ is independent of $W(u) - W(r)$ for $t > s \geq u > r \geq 0$.

A d-dimensional BM consists of a vector $W(t) = (W_1(t), \ldots, W_d(t))$ of independent one-dimensional BMs $W_i(t)$. A correlated d-dimensional BM is again a vector of one-dimensional BMs $Z_i(t)$, but with

$$Corr(Z_i(t), Z_j(t)) = \rho_{ij}$$

for a given correlation matrix $\boldsymbol{\rho}$.

A simulated path of a one-dimensional BM, i.e. a realization of the BM $W(t)$, $t \in [0, 1]$ is given in Fig. 1.1. It exhibits the main characteristics of the BM, in particular its non-differentiability as a function of time.

In this survey, we will consider a general diffusion type model for the evolution of stock prices, interest rates or additional processes that influence those prices. The corresponding modeling tool that we are using are Stochastic Differential Equations

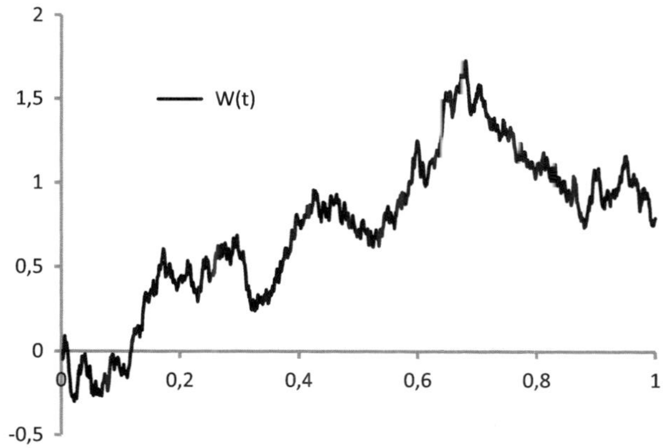

Fig. 1.1 A path of a Brownian motion

(SDEs) (see [11] for a standard reference on SDEs). In particular, we assume that the price process $S(t)$ of d stocks and an additional, m-dimensional state process $Y(t)$ are given by the SDE

$$\begin{aligned} dS(t) &= \mu(t,S(t),Y(t))\,dt + \sigma(t,S(t),Y(t))\,dW(t),\ S(0) = s, \\ dY(t) &= \kappa(t,Y(t))\,dt + \nu(t,Y(t))\,dW(t),\ Y(0) = y. \end{aligned}$$

Here, we assume that the coefficient functions μ, σ, κ, ν satisfy appropriate conditions for existence and uniqueness of a solution of the SDE. Such conditions can be found in [11]. Sufficient (but not necessary) conditions are e.g. the affine linearity of the coefficient functions or suitable Lipschitz and growth conditions. Further, $W(t)$ is a k-dimensional BM.

The most popular special case of those models is the Black-Scholes (BS) model where the stock price does not depend on the state process $Y(t)$ (or where formally the state process Y is a constant). We assume that we have $d = k = 1$ and that the stock price satisfies the SDE

$$dS(t) = S(t)(\mu dt + \sigma dW(t)),\ S(0) = s \tag{1.1}$$

for given constants μ, σ and a positive initial price of s. By the variation of constants formula (see e.g. [13], Theorem 2.54) there exists a unique (strong) solution to the SDE (1.1) given by the *geometric BM*

$$S(t) = s\exp\left(\left(\mu - \frac{1}{2}\sigma^2\right)t + \sigma W(t)\right).$$

As the logarithm of $S(t)$ is normally distributed, we speak of a *log-normal model*. In this case, we further have

$$\mathbb{E}(S(t)) = s\exp(\mu t).$$

Multi-dimensional generalizations of this example are available for linear coefficient functions $\mu(.), \sigma(.)$ without dependence on the state process $Y(t)$.

A popular example for a stock price model with dependence on an underlying state process is the Stochastic Volatility (SV) model of Heston (for short: *Heston model*, see [10]). There, we have one stock price and an additional state process $\nu(t)$ that is called the *volatility*. They are given by

$$dS(t) = S(t)\left(\mu dt + \sqrt{\nu(t)}dW^S(t)\right), \; S(0) = s,$$

$$d\nu(t) = \kappa(\theta - \nu(t))\,dt + \sigma\sqrt{\nu(t)}dW^\nu(t), \nu(0) = s$$

with arbitrary constants μ, σ and positive constants κ, θ. Further, we assume

$$corr\left(W^S(t), W^\nu(t)\right) = \rho$$

for a given constant $\rho \in [-1, 1]$ for the two one-dimensional Brownian motions W^S and W^ν. A particular aspect of the volatility process $\nu(t)$ is that it is non-negative, but can attain the value zero if we have

$$2\theta\kappa \leq \sigma^2.$$

The Heston model is one of the benchmark models in the finance industry that will also appear in further contributions to this book. One of its particular challenges is that the corresponding SDE does not admit an explicit solution. Thus, it can only be handled by simulation and discretization methods, a fact that is responsible for many computational issues raised in this book.

1.3 Principles of Option Pricing

Options are derivative securities as their future payments depend on the performance of one or more underlying stock prices. They come in many ways, plain and simple, and complicated, with many strange features when it comes to determine the actual final payment that their owner receives. As they are a characteristic product of modern investment banking, calculating their prices in an efficient and accurate way is a key task in financial mathematics.

The most popular example of an option is the European call option on a stock. It gives its owner the right (but not the obligation!) to buy one unit of the stock at a predefined future time T (the *maturity*) for an already agreed price of K (the *strike*). As the owner will only buy it when the price of the underlying at maturity $S(T)$ is above the strike, the European call option is identified with the random payment of

$$H = (S(T) - K)^+$$

at time T.

One of the reasons for the popularity of the European call option is that it admits an explicit pricing formula in the BS model, the BS formula

$$c(t,S(t)) = S(t)\,\Phi\left(\frac{\ln\left(\frac{S(t)}{K}\right)+\left(r+\frac{1}{2}\sigma^2\right)(T-t)}{\sigma\sqrt{T-t}}\right)$$
$$-Ke^{-r(T-t)}\,\Phi\left(\frac{\ln\left(\frac{S(t)}{K}\right)+\left(r-\frac{1}{2}\sigma^2\right)(T-t)}{\sigma\sqrt{T-t}}\right)$$

where $\Phi(.)$ denotes the cumulative distribution function of the standard normal distribution. This formula which goes back to [2] is one of the cornerstones of modern financial mathematics. Its importance in both theory and application is also emphasized by the fact that Myron Scholes and Robert C. Merton were awarded the Nobel Prize in Economics in 1997 for their work related to the BS formula.

The most striking fact of the BS formula is that the stock price drift μ, i.e. the parameter that determines the expected value of $S(t)$, does not enter the valuation formula of the European call option. This is no coincidence, but a consequence of a deep theoretical result. To formulate it, we introduce a riskless investment opportunity, the so-called money market account with price evolution $M(t)$ given by

$$M(t) = e^{rt},$$

i.e. the evolution of the value of one unit of money invested at time $t = 0$ that continuously earns interest payments at rate r.

The financial market made up of this money market account and the stock price of the BS model is called the BS market. There, we have the following result:

Theorem 1 (Option price in the BS model). *The price X_H of an option given by a final payment H with $\mathbb{E}\left(H^b\right) < \infty$ for some $b \geq 1$ is uniquely determined by*

$$X_H = \tilde{\mathbb{E}}\left(e^{-rT}H\right),$$

where the expectation is taken with respect to the unique probability measure Q under which the discounted stock $\tilde{S}(t) = S(t)/M(t)$ is a martingale. In particular, for the purpose of option pricing, we can assume that we have

$$\mu = r.$$

The reason for this very nice result is the completeness of the BS market, i.e. the fact that every (sufficiently integrable) final payoff H of an option can be created in a synthetic way by following an appropriate trading strategy in the money market account and the stock (see [13] for the full argumentation and the concept of completeness and replication).

In market models where the state process $Y(t)$ has a non-vanishing stochastic component that is not contained in the ones of the stock price, one does not have such a nice result as in the BS setting. However, even there, we can assume that for the purpose of option pricing we can model the stock prices in such a way that their discounted components $\tilde{S}_i(t) = S_i(t)/M(t)$ are martingales. In particular, now and in the following we directly assume that we only consider probability measures P such that we have

$$S_i(0) = \mathbb{E}(S_i(t)/M(t)),\ i = 1,\ldots,d.$$

Thus, all trade-able assets are assumed to have the same expected value for their relative increase in this artificial market. We therefore speak of *risk-neutral valuation.*

As we have now seen, calculating an option price boils down to calculating an expected value of a function or a functional of an underlying stochastic process. For simplicity, we thus assume that the underlying (possibly multi-dimensional) stock price process is given as the unique solution of the SDE

$$dS(t) = \mu(t,S(t))\,dt + \sigma(t,S(t))\,dW(t),\ S(0) = s$$

with $W(t)$ a d-dimensional BM, $s = (s_1,\ldots,s_n)'$, and μ,σ being functions satisfying appropriate conditions such that the above SDE possesses a unique (strong) solution. Further, for the moment, we consider a function

$$f : \mathbb{R}^m \to \mathbb{R}$$

which is non-negative (or polynomially bounded). Then, we can define the (conditional) expectation

$$\mathcal{V}(t,s) = \mathbb{E}^{(t,s)}\left(e^{-r(T-t)} f(S(T))\right)$$

for a given starting time $t \in [0,T]$ at which we have $S(t) = s$. Of course, we can also replace the function f in the two preceding equations by a functional F that can depend on the whole path of the stock price. However, then the conditional expectation at time t above is in general not determined by only starting in (t,x). Depending on the functional's form one needs more information of the stock price performance before time t to completely describe the current value of the corresponding option via an appropriate expectation.

However, in any case, to compute this expectation is indeed our main task. There are various methods for computing it. Examples are:

- *Direct calculation of the integral*

$$\mathbb{E}^{(0,s)}\left(e^{-rT} f(S(T))\right) = e^{-rT} \int_{\mathbb{R}^n} f(x) h(x) \, dx$$

 if the density $h(.)$ of $S(T)$ (conditioned on $S(0) = s$) is explicitly known.
- *Approximation of the price process* $S(t), t \in [0, T]$, by simpler processes – such as binomial trees – $S^{(n)}(t), t \in [0, T]$, and then calculating the corresponding expectation

$$\mathbb{E}_n^{(0,s^{(n)})}\left(f\left(S^{(n)}(T)\right)\right)$$

 in the simpler model as an approximation for the original one (see [15] for a survey on binomial tree methods applied to option pricing).
- *Solution of the partial differential equation* for the conditional expectation $\mathcal{V}(t,s)$ that corresponds to the stock price dynamics. For notational simplicity, we only state it in the one-dimensional case as

$$\mathcal{V}_t(t,s) + \frac{1}{2}\sigma(t,s)^2 \mathcal{V}_{ss}(t,s) + \mu(t,s)\mathcal{V}_s(t,s) - r(t,s)\mathcal{V}(t,s) = 0,$$
$$\mathcal{V}(T,s) = f(T,s)$$

 For more complicated option prices depending on the state process $Y(t)$, we also obtain derivatives with respect to y and mixed derivatives with respect to t, s, y.
- Calculating the expectation via *MC simulation*, i.e. simulating the final payoff H of an option N times and then estimating the option price via

$$\mathbb{E}\left(e^{-rT} H\right) \approx e^{-rT} \frac{1}{N} \sum_{i=1}^{N} H^{(i)}$$

 where the $H^{(i)}$ are independent copies of H.

We will in the following restrict ourselves to the last method, the MC method. The main reason for this decision is that it is the most flexible of all the methods presented, and it suffers the most from heavy computations, as the number N of simulation runs usually has to be very large.

Before doing so, we will present options with more complicated payoffs than a European call option, so called *exotic options*. Unfortunately, only under very special and restrictive assumptions, there exist explicit formulae for the prices of such exotic options. Typically, one needs numerical methods to price them. Some popular examples are:

- Options with payoffs depending on multiple stocks such as *basket options* with a payoff given by

$$H_{basket} = \left(\frac{1}{d} \sum_{j=1}^{d} S_j(T) - K \right)^+$$

- Options with payoffs depending on either a finite number of stock prices at different times $t_j \in [0,T]$ such as *discrete Asian options* given by e.g.

$$H_{disc.\ Asian\ call} = \left(\frac{1}{d} \sum_{j=1}^{d} S(t_j) - K \right)^+$$

or a continuous average of stock prices such as *continuous Asian options* given by

$$H_{cont.\ Asian\ call} = \left(\frac{1}{T} \int_0^T S(t)\, dt - K \right)^+$$

- Barrier options that coincide with plain European put or call options as long as certain barrier conditions are either satisfied on $[0,T]$ or are violated such as e.g. a *knock-out-double-barrier call option* with a payoff given by

$$H_{dbkoc} = (S(T) - K)^+ 1_{\{B_1 < S(t) < B_2\}}$$

for constants $0 \leq B_1 < B_2 \leq \infty$
- Options with local and global bounds on payoffs such as *locally and globally capped and floored cliquet options* given by

$$H_{cliquet} = \max\left\{ F, \min\left\{ C, \sum_{j=1}^{d} \max\left\{ F_j, \min\left\{ C_j, \frac{S(t_j) - S(t_{j-1})}{S(t_{j-1})} \right\}\right\}\right\}\right\}$$

for different time instants $0 \leq t_0 < t_1 < \ldots < t_d \leq T$ and constants $F < C, F_j < C_j$

All of those exotic options are tailored to the needs of special customers or markets. As an example, cliquet options are an essential ingredient of modern pension insurance products.

At the end of this section, we will formulate our first computational challenge:

Computational challenge 1: Find a universal framework/method for an efficient calculation of prices of exotic options.

An obvious candidate is the Monte Carlo (MC) method which we are going to present in the next section.

1.4 Monte Carlo Methods for Pricing Exotic Options

MC methods are amongst the simplest methods to compute expectations (and thus also option prices) and are on the other hand a standard example of a method that causes a big computing load when applied in a naive way. Even more, we will show by an example of a simple barrier option that a naive application of the MC method will lead to a completely wrong result that even pretends to be of a high accuracy.

Given that we can generate random numbers which are distributed as the considered real-valued, integrable random variable H, the standard MC method to calculate the expectation $\mathbb{E}(H)$ consists of two steps:

1. Generate N independent, identically distributed copies H_i of H.
2. Estimate $\mu \triangleq \mathbb{E}(H)$ by

$$\hat{\mu}_N = \frac{1}{N}\sum_{i=1}^{N} H_i.$$

Due to the linearity of the expectation the MC estimator $\hat{\mu}_N$ is unbiased. Further, the convergence of the standard MC method is ensured by the strong law of large numbers. One obtains an approximate confidence interval of level $1-\alpha$ for μ as (see e.g. [14], Chapter 3)

$$\left[\frac{1}{N}\sum_{i=1}^{N} H_i - z_{1-\alpha/2}\frac{\sigma}{\sqrt{N}}, \frac{1}{N}\sum_{i=1}^{N} H_i + z_{1-\alpha/2}\frac{\sigma}{\sqrt{N}}\right].$$

Here, $z_{1-\alpha/2}$ is the $(1-\alpha/2)$-quantile of the standard normal distribution and σ is defined via

$$\sigma^2 \triangleq Var(H).$$

If σ^2 is unknown (which is the typical situation) then it will be estimated by

$$\hat{\sigma}_N^2 = \frac{1}{N-1}\sum_{i=1}^{N} (H_i - \hat{\mu}_N)^2.$$

σ is then replaced by $\hat{\sigma}_N$ in the MC estimator of the confidence interval for μ. In both cases, the message is that – measured in terms of the length of the confidence interval – the **accuracy of the unbiased MC method is of the order** $O\left(1/\sqrt{N}\right)$

This in particular means that we need to increase the number of simulations of H_i by a factor 100 if we want to increase the accuracy of the MC estimator for μ by one order. Thus, we have in fact a very slow rate of convergence.

Looking at the ingredients in the MC method we already see the first challenge of an efficient and robust implementation:

Computational challenge 2: Find an appropriate Random Number Generator (RNG) to simulate the final payments H of an exotic option.

Here, the decision problem is crucial with respect to both performance and accuracy. Of course, the (typically deterministic) RNG should mimic the distribution underlying H as good as possible. Further, as the biggest computational advantage of the MC method is the possibility for parallelization, the RNG should allow a simple way of parallel simulation of independent random numbers.

The standard method here is to choose a suitable RNG that produces good random numbers that are uniformly distributed on $(0,1)$ and to use the inverse transformation method for getting the right distribution. I.e. let U_i be the ith random number which is uniformly distributed on $(0,1)$, let F be the desired distribution function of H. Then

$$H_i \triangleq F^{-1}(U_i)$$

has the desired distribution. This method mostly works, in particular in our diffusion process setting which is mainly dominated by the use of the normal distribution. Thus, for the normal distribution one only has to decide between the use of the classical Box-Muller transform or an approximate inverse transformation (see [14], Chapter 2). While the approximate inverse transformation method preserves a good grid structure of the original uniformly distributed random numbers, the Box-Muller transform ensures that even extreme values outside the interval $[-8,8]$ can occur which is not the case for the approximate inverse method. Having made the decision about the appropriate transformation method, it still remains to find a good generator for the uniformly distributed random numbers U_i. Here, there is an enormous choice. As parallelization is one of the major advantages, the suitability for parallelization is a major issue for deciding on the RNG. Thus, the Mersenne Twister is a favorable choice (see [14], Chapter 2 and [16]).

For a simple standard option with a final payment of $H = f(S(T))$ (such as a European call option) in the Black-Scholes setting, we only have to simulate independent standard normally distributed random variables $Z_i, i = 1, \ldots, N$, to obtain

$$H^{(i)} = f\left(se^{\left(r-\frac{1}{2}\sigma^2\right)T+\sigma\sqrt{T}Z_i}\right).$$

However, things become more involved when one either cannot generate the price process $S(t)$ exactly or when one can only simulate a suitably discretized version of the payoff functional.

For the first case, one has to use a discretization scheme for the simulation of the stock price (see [12] for a standard reference on the numerical solution of SDE). The most basic such scheme is the Euler-Maruyama scheme (EMS). To illustrate it, we apply it to a one dimensional SDE

$$dS(t) = \mu(S(t))\,dt + \sigma(S(t))\,dW(t), S(0) = s_0.$$

Then, for a step size of $\Delta = T/n > 0$, the discretized process $S^{(\Delta)}(t)$ generated by the EMS is defined by

$$\begin{aligned} S^{(\Delta)}(0) &\triangleq s_0, \\ S^{(\Delta)}(k\Delta) &\triangleq S^{(\Delta)}((k-1)\Delta) + \mu\left(S^{(\Delta)}((k-1)\Delta)\right)\Delta \\ &\quad + \sigma\left(S^{(\Delta)}((k-1)\Delta)\right)\Delta W_k, \;\; k = 1,\ldots,n. \end{aligned}$$

Here, $\Delta W_k, k = 1,\ldots,n$, is a sequence of independent, $\mathcal{N}(0,\Delta)$-distributed random variables. Between two consecutive discretization points, we obtain the values of $S^{(\Delta)}(t)$ by linear interpolation. The EMS can easily be generalized to a multi-dimensional setting.

If we now replace the original process $S(t)$ by $S^{(\Delta)}(t)$ in the standard MC approach, then we obtain

$$\hat{\mu}_{N,\Delta} \triangleq \frac{1}{N}\sum_{i=1}^{N} f\left(S_i^{(\Delta)}(T)\right) \xrightarrow{a.s.} \mathbb{E}\left(f\left(S^{(\Delta)}(T)\right)\right) \text{ for } N \to \infty$$

In particular, this application of the MC that uses the discretized process leads to a biased result. The accuracy of the MC method can then no longer be measured by the variance of the estimator. We have to consider the Mean Squared Error (MSE) to judge the accuracy instead, i.e.

$$\begin{aligned} MSE(\hat{\mu}_{N,\Delta}) &\triangleq \mathbb{E}\left[\left(\hat{\mu}_{N,\Delta} - \mathbb{E}(f(S(T)))\right)^2\right] \\ &= Var\left(\hat{\mu}_{N,\Delta}\right) + \left(\mathbb{E}(f(S(T))) - \mathbb{E}\left(f\left(S^{(\Delta)}(T)\right)\right)\right)^2 \end{aligned}$$

Thus, the MSE consists of two parts, the MC variance and the so-called discretization bias. We consider this bias a bit more detailed by looking at the convergence behavior of the EMS: Given suitable assumptions on the coefficient functions μ, σ, we have weak convergence of the MSE of order 1 (see e.g. [12]). More precisely, for μ, σ being four times continuously differentiable we have

$$\left|\mathbb{E}(f(S(T))) - \mathbb{E}\left(f\left(S^{(\Delta)}(T)\right)\right)\right| \leq C_f \Delta$$

for four times differentiable and polynomially bounded functions f and a suitable constant C_f not depending on Δ.

With regard to the MSE it is optimal to choose the discretization step size $\Delta = T/n$ and the number of MC simulations N in such a way that both components of the MSE are of the same order. So, given that we have weak convergence of order 1 for the EMS then an MSE of order $\varepsilon^2 = 1/n^2$ can be obtained by the choices of

$$\Delta = O(1/n), \quad N = n^2$$

which lead to an order of $O\left(n^3\right)$ measured in the random numbers simulated in total. As this leads to a high computational effort for pricing an option by the standard MC method, we can formulate another computational challenge:

Computational challenge 3: Find a modification of the standard MC method that has an effort of less than $O(n^3)$ for pricing an option including path simulation.

There are some methods now available that can overcome this challenge. Among them are weak extrapolation, the statistical Romberg method and in particular the multi-level MC method which will also play a prominent role in further contributions to this book (see e.g. [14] for a survey on the three mentioned methods).

However, unfortunately, the assumptions on f are typically not satisfied for option type payoffs (simply consider all the examples given in the last section). Further, the assumptions on the coefficients of the price process are not satisfied for e.g. the Heston model.

Thus, in typical situations, although we know the order of the MC variance, we cannot say a lot about the actual accuracy of the MC estimator. This problem will be illustrated by the second case mentioned above where we have to consider the MSE as a measure for accuracy of the MC method, the case where the payoff functional can only be simulated approximately. Let therefore $f(S)$ be a functional of the path of the stock price $S(t)$, $t \in [0,T]$ and $\hat{\mu}_{N,\Delta}$ be a MC estimator based on N simulated stock price paths with a discretization step size for the payoff functional of Δ. Then, we obtain a similar decomposition of the MSE

$$\begin{aligned} MSE(\hat{\mu}_{N,\Delta}) &= \mathbb{E}\left(\hat{\mu}_{N,\Delta} - \mathbb{E}(f(S))\right)^2 \\ &= \mathbb{E}\left(\left(\hat{\mu}_{N,\Delta} - \mathbb{E}(f(S;\Delta))\right)^2\right) + \mathbb{E}\left(\left(\mathbb{E}(f(S)) - \mathbb{E}(f(S;\Delta))\right)^2\right) \\ &= Var\left(\hat{\mu}_{N,\Delta}\right) + bias(\Delta) \end{aligned}$$

where now the bias is caused by the discretization of the payoff functional.

To illustrate the dependence of the accuracy of the MC method on the bias, we look at the problem of computing the price of a one-sided down-and-out barrier call option with a payoff functional given by

$$f\left(S\left(t\right);t\in\left[0,T\right]\right)=\left(S\left(T\right)-K\right)^{-}1_{S(t)>B\ \forall t\in[0,T]}.$$

As the one-sided down-and-out barrier call option possesses an explicit valuation formula in the BS model (see e.g. [13], Chapter 4), it serves well to illustrate the effects of different choices of the discretization parameter $\Delta=1/m$ and the number of MC replications N.

As input parameters we consider the choice of

$$T=1,\ r=0,\ \sigma=0.1,\ S\left(0\right)=K=100,\ B=95.$$

We first fix the number of discretization steps m to 10, i.e. we have $\Delta=0.1$. As we then only check the knock-out condition at 10 time points, the corresponding MC estimator (at least asymptotically for large N) overestimates the true value of the barrier option. This is underlined in Fig. 1.2 where the 95 %-confidence intervals do not contain the true value of the barrier option. This, however is not surprising as in this case the sequence of MC estimators converges to the price of the discrete down-and-out call given by the final payoff

$$f\left(S;N,\Delta\right)=\left(S\left(T\right)-K\right)^{+}1_{S(i\Delta T/N)>B\ \forall i=1,\ldots,m}.$$

As a contrast, we now fix the number $N=100{,}000$ of simulated stock price paths and consider a varying number of discretization points m in Fig. 1.3. As can be seen from the nearly identical length of the confidence intervals for varying m, the variance of the MC estimator is estimated consistently. Considering the differences

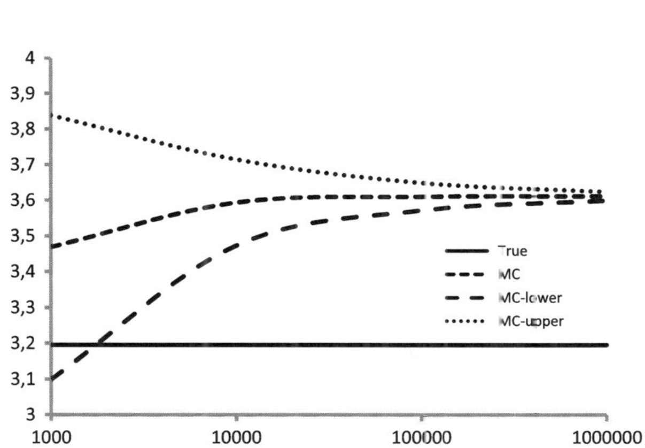

Fig. 1.2 MC estimators with 95 %-confidence intervals for the price of a barrier option with fixed time discretization 0, 1 and varying number N of simulated stock price paths

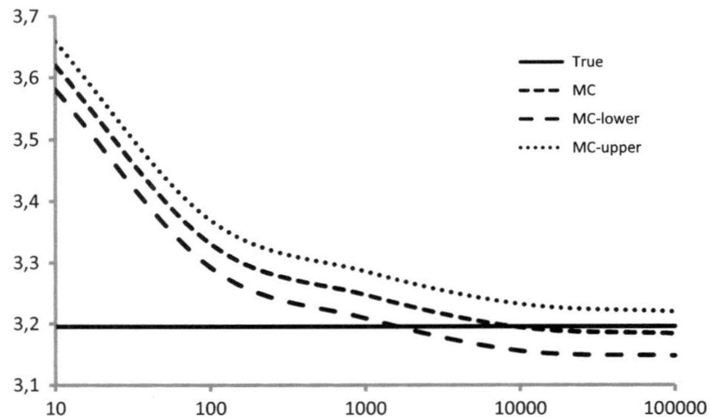

Fig. 1.3 MC estimators with 95 %-confidence intervals for the price of a barrier option with varying time discretization $1/m$ for 100,000 stock price paths

of the bias of the different MC estimators from the true value, one can conjecture that the bias behaves as $O(1/\sqrt{m})$, and thus converges at the same speed as the unbiased MC estimator. This example highlights that the order of the convergence of the bias is the critical aspect for the MC method in such a situation. Fortunately, in the case of the barrier options, there are theoretical results by Gobet (see e.g. [9]) that prove the above conjecture of a discretization bias of order 0.5. There are also good modifications of the crude MC method above that produce an unbiased estimator (such as the Brownian bridge method (see e.g. Chapter 5 of [14])), but the effects demonstrated above are similar for other types of exotic options. And moreover, there are not too many results on the bias of the MC estimator for calculating the price of an exotic option.

Thus, in calculating the prices of exotic options by MC methods, we face another computational challenge:

Computational challenge 4: Develop an efficient algorithm to estimate the order of the discretization bias when calculating the price of an exotic option with path dependence by the MC method.

A possibly simple first suggestion is to perform an iterative search in the following way:

1. Start with a rough discretization (i.e. a small number m) and increase the number N of MC simulation runs until the resulting (estimated) variance is below the order of the desired size of the MSE.

2. Increase the number of discretization points by a factor 10 and repeat calculating the corresponding MC estimation with the final N from Step 1 10 times. Take the average over the 10 calculations as an estimator for the option price.
3. Repeat Step 2 until the estimator for the option price is no longer significantly changing between two consecutive steps.

Of course, this is only a kind of simple cooking recipe that leaves a lot of space for improvement. One can also try to estimate the order of the discretization bias from looking at its behavior as a function of the varying step size $1/(10^k m)$.

In any case, not knowing the discretization bias increases the computational effort enormously, if one wants to obtain a trustable option price by the MC method. So, any strategy, may it be more based on algorithmic improvements or on an efficient hardware/software concept, will be a great step forward.

1.5 Risk Measurement and Management

The notion of risk is ubiquitous in finance, a fact that is also underlined by the intensive use of such terms as market risk, liquidity risk, credit risk, operational risk, model risk, just to mention the most popular names. As measuring and managing risk is one of the central tasks in finance, we will also highlight some corresponding computational challenges in different areas of risk.

1.5.1 Loss Distributions and Risk Measures

While we have concentrated on the pricing of single derivative contracts in the preceding sections, we will now consider a whole bunch of financial instruments, a so-called **portfolio** of financial positions. This can be the whole book of a bank or of one of its departments, a collection of stocks or of risky loans. Further, we will not price the portfolio (this would just be the sum of the single prices), but will instead consider the sum of the risks that are inherent in the different single positions simultaneously. What interests us is the potential change, particularly the losses, of the total value of the portfolio over a future time period.

The appropriate concepts for measuring the risk of such a portfolio of financial assets are those of the **loss function** and of **risk measures**. In our presentation, we will be quite brief and refer the reader for more details to the corresponding sections in [17] and [14].

We denote the **value** at time s of the portfolio under consideration by $V(s)$ and assume that the random variable $V(s)$ is observable at time s. Further, we assume that the composition of the portfolio does not change over the period we are looking at.

For a time horizon of Δ the **portfolio loss** over the period $[s, s+\Delta]$ is given by

$$L_{[s,s+\Delta]} \triangleq -(V(s+\Delta) - V(s)) .$$

Note that we have changed the sign for considering the differences of the future and the current portfolio value. This is because we are concerned with the possibilities of big losses only. Gains do not play a big role in risk measurement, although they are the main aim of performing the business of a company in general.

Typical time horizons that occur in practice are 1 or 10 days or even a year. As $L_{[s,s+\Delta]}$ is not known at time s it is considered to be a random variable. Its distribution is called the **(portfolio) loss distribution**. We do not distinguish between the *conditional loss* and *unconditional loss* in the following as our objective are computational challenges. We always assume that we perform our computations based on the maximum information available at the time of computation.

As in [17] we will work in units of the fixed time horizon Δ, introduce the notation $V_t \triangleq V(t\Delta)$, and rewrite the loss function as

$$L_{t+1} \triangleq L_{[t\Delta,(t+1)\Delta]} = -(V_{t+1} - V_t) . \tag{1.2}$$

Fixing the time t, the distribution of the loss function $L \triangleq L_{t+1}$ for $\ell \in \mathbb{R}$ (conditional on time t) is introduced using a simplified notation as

$$F_L(\ell) \triangleq P(L \leq \ell).$$

With the distribution of the loss function, we are ready to introduce so-called **risk measures**. Their main purpose is stated by Föllmer and Schied in [7] as:

> ...a risk measure is viewed as a capital requirement: We are looking for the minimal amount of capital which, if added to the position and invested in a risk-free manner, makes the position acceptable.

For completeness, we state:

A **risk measure** ρ is a real-valued mapping defined on the space of random variables (*risks*).

To bring this somewhat meaningless, mathematical definition closer to the above intention, there exists a huge discussion in the literature on reasonable additional requirements that a good risk measure should satisfy (see e.g. [7, 14, 17]).

As this discussion is beyond the scope of this survey, we restrict ourselves to the introduction of two popular examples of risk measures: The one which is mainly used in banks and has become an industry standard is the *value-at-risk*.

The **value-at-risk** of level α (VaR_α) is the α-quantile of the loss distribution of the portfolio:

$$VaR_\alpha(L) \triangleq \inf\{\ell \in \mathbb{R} | P(L > \ell) \leq 1 - \alpha\} = \inf\{\ell \in \mathbb{R} | F_L(\ell) \geq \alpha\},$$

where α is a high percentage such as 95 %, 99 % or 99.5 %.

By its nature as a quantile, values of VaR_α have an understandable meaning, a fact that makes it very popular in a wide range of applications, mainly for the measurement of market risks, but also in the areas of credit risk and operational risk management. VaR_α is not necessarily sub-additive, i.e. the $VaR_\alpha(X+Y) > VaR_\alpha(X) + VaR_\alpha(Y)$ for two different risks X, Y is possible. This feature is the basis for most of the criticism of using value-at-risk as a risk measure. Furthermore, as a quantile, VaR_α does not say anything about the actual losses above it.

A risk measure that does not suffer from these two drawbacks (compare e.g. [1]), and, which is therefore also popular in applications, is the *conditional value-at-risk*:

The **conditional value-at-risk** (or average value-at-risk) is defined as

$$CVaR_\alpha(L) \triangleq \frac{1}{1-\alpha} \int_\alpha^1 VaR_\gamma(L) \mathrm{d}\gamma.$$

If the probability distribution of L has no atoms, then the $CVaR_\alpha$ has the interpretation as the expected losses above the value-at-risk, i.e. it then coincides with the **expected shortfall** or **tail conditional expectation** defined by

$$TCE_\alpha(L) \triangleq \mathbb{E}(L | L \leq VaR_\alpha(L)).$$

As the conditional value-at-risk is the value at risk integrated w.r.t. the confidence level, both notions do not differ remarkably from the computational point of view. Thus, we will focus on the value-at-risk below.

However, as typically the portfolio value V and thus by (1.2) the loss function L depend on a d-dimensional vector of market prices for a very large dimension d, the loss function will depend on the market prices of maybe thousands of different derivative securities. This directly leads us to the first obvious computational challenge of risk management:

Computational challenge 5: Find an efficient way to evaluate the *loss function* of large portfolios to allow for a fast computation of the *value-at-risk.*

1.5.2 Standard Methods for Market Risk Quantification

The importance of the quantification of market risks is e.g. underlined by the popular JPMorgan's Risk Metrics document (see [18]) from the practitioners site or by the reports of the Commission for the Supervision of the Financial Sector (CSSF) (see [19]) from the regulatory point of view. This has the particular consequence that every bank and insurance company have to calculate risk measures, of course for different horizons. While for a bank, risk measures are calculated typically for a horizon of 1–10 days, insurance companies typically look at the horizon of a year.

To make a huge portfolio numerically tractable, one introduces so-called **risk factors** that can explain (most of) the variations of the loss function and ideally reduce the dimension of the problem by a huge amount. They can be log-returns of stocks, indices or economic indicators or a combination of them. A classical method for performing such a model reduction and to find risk factors is a principal component analysis of the returns of the underlying positions.

We do not go further here, but simply assume that the portfolio value is modeled by a so-called **risk mapping**, i.e. for a d-dimensional random vector $\mathbf{Z}_t \triangleq (Z_{t,1}, \ldots, Z_{t,d})'$ of risk factors we have the representation

$$V_t = f(t, \mathbf{Z}_t) \tag{1.3}$$

for some measurable function $f : \mathbb{R}_+ \times \mathbb{R}^d \to \mathbb{R}$. Of course, this representation is only useful if the risk factors $\mathbf{Z}_t$ are observable at time t, which we assume from now on. By introducing the **risk factor changes** $(\mathbf{X}_t)_{t \in \mathbb{N}}$ by $\mathbf{X}_t \triangleq \mathbf{Z}_t - \mathbf{Z}_{t-1}$ the portfolio loss can be written as

$$L_{t+1}(X_{t+1}) = -\left(f(t+1, \mathbf{Z}_t + \mathbf{X}_{t+1}) - f(t, \mathbf{Z}_t)\right) \tag{1.4}$$

highlighting that the loss is completely determined by the risk factor changes.

In what follows we will discuss some standard methods used in the financial industry for estimating the value-at-risk.

1.5.2.1 The Variance-Covariance Method

The variance-covariance method is some crude, first-order approximation. Its basis is the assumption that risk factor changes X_{t+1} follow a multivariate normal distribution, i.e.

$$X_{t+1} \sim \mathcal{N}_d(\mu, \Sigma)$$

where μ is the mean vector and Σ the covariance matrix of the distribution.

The second fundamental assumption is that f is differentiable, so that we can consider a first-order approximation L^{lin}_{t+1} of the loss in (1.4) of the form

$$L^{lin}_{t+1}(X_{t+1}) \triangleq -\left(f(t, \mathbf{Z}_t) + \sum_{i=1}^{d} f_{z_i}(t, \mathbf{Z}_t) X_{t+1,i}\right). \quad (1.5)$$

As the portfolio value $f(t, \mathbf{Z}_t)$ and the relevant partial derivatives $f_{z_i}(t, \mathbf{Z}_t)$ are known at time t, the linearized loss function has the form of

$$L^{lin}_{t+1}(X_{t+1}) = -(c_t + \mathbf{b}_t' \mathbf{X}_{t+1}) \quad (1.6)$$

for some constant c_t and a constant vector $\mathbf{b}_t$ which are known to us at time t. The main advantage of the above two assumptions is that the linear function (1.6) of $\mathbf{X}_{t+1}$ preserves the normal distribution and we obtain

$$L^{lin}_{t+1}(X_{t+1}) \sim \mathcal{N}\left(-c_t - \mathbf{b}_t' \mu, \mathbf{b}_t' \Sigma \mathbf{b}_t\right).$$

This yields the following explicit formula:

The **value-at-risk of the linearized loss** corresponding to the confidence level α is given by

$$VaR_\alpha(L^{lin}_{t+1}) = -c_t - \mathbf{b}_t' \mu + \sqrt{\mathbf{b}_t' \Sigma \mathbf{b}_t}\, \Phi^{-1}(\alpha), \quad (1.7)$$

where Φ denotes the standard normal distribution function and $\Phi^{-1}(\alpha)$ is the α-quantile of Φ.

To apply the value-at-risk of the linearized loss to market data, we still need to estimate the mean vector μ and the covariance matrix Σ based on the historical risk factor changes $\mathbf{X}_{t-n+1}, \dots, \mathbf{X}_t$ which can be accomplished using standard estimation procedures (compare Section 3.1.2 in [17]).

Remark 1. The formulation of the variance-covariance method based on the first-order approximation L_{t+1}^{lin} in (1.5) of the loss is often referred to as the *Delta-approximation* in analogy to the naming of the first partial derivative with respect to underlying prices in option trading.

Remark 2. Another popular version of the variance-covariance method is the *Delta-Gamma-approximation* which is based on a second-order approximation of the loss function in order to capture the non-linear structure of portfolios that contain a high percentage of options. However, the general advantages and weaknesses of these methods are similar. We therefore do not repeat our analysis for the Delta-Gamma-approximation here.

Merits and Weaknesses of the Method

The main advantage of the variance-covariance method is that it yields an explicit formula for the value-at-risk of the linearized losses as given by (1.7). However, this closed-form solution is only obtained using two crucial simplifications:

1. Linearization (in case of the Delta-approximation) or even a second order approximation (in case of the Delta-Gamma-approximation) is in the fewest cases a good approximation of the risk mapping as given in (1.3), in particular when the portfolio contains many complex derivatives.
2. Empirical examinations suggest that the distribution of financial risk factor returns is leptokurtic and fat-tailed compared to the Gaussian distribution. Thus the assumption of normally distributed risk factor changes is questionable and the value-at-risk of the linearized losses (1.7) is likely to underestimate the true losses.

1.5.2.2 Historical Simulation

Historical simulation is also a very popular method in the financial industry. It is based on the simple idea that instead of making a model assumption for the risk factor changes, one simply relies on the **empirical distribution** of the already observed past data $\mathbf{X}_{t-n+1}, \ldots, \mathbf{X}_t$. We then evaluate our portfolio loss function for each of those data points and obtain a set of synthetic losses that would have occurred if we hold our portfolio on the past days $t-1, t-2, \ldots, t-n$:

$$\left\{\tilde{L}_s(\mathbf{X}_s) : s = t-n+1, \ldots, t\right\}. \tag{1.8}$$

Based on these historically simulated loss data, one now estimates the value-at-risk by the corresponding *empirical quantile*, i.e. the quantile of the just obtained historical empirical loss distribution:

Let $\tilde{L}_{n,n} \leq \ldots \leq \tilde{L}_{1,n}$ be the ordered sequence of the values of the historical losses in (1.8). Then, the **estimator for the value-at-risk obtained by historical simulation** is given by

$$VaR_\alpha(\tilde{L}_s) \triangleq \tilde{L}_{[n(1-\alpha)],n},$$

where $[n(1-\alpha)]$ denotes the largest integer not exceeding $n(1-\alpha)$.

Merits and Weaknesses of the Method

Besides being a very easy method, a convincing argument of historical simulation is its independence on distributional assumptions. We only use data that have already appeared, no speculative ones.

From the theoretical point of view, however, we have to assume stationarity of the risk factor changes over time which is also quite a restrictive assumption. And even more, we can be almost sure that we have not yet seen the worst case of losses in the past. The dependence of the method on reliable data is another aspect that can cause problems and can lead to a weak estimator for the value-at-risk.

1.5.2.3 The Monte Carlo Method

A method that overcomes the need for linearization and the normal assumption in the variance-covariance method and that does not rely on historical data is the Monte Carlo (MC) method. Of course, we still need an assumption for the distribution of the future risk factor changes.

Given that we have made our choice of this distribution, the MC method only differs to the historical simulation by the fact that we now simulate our data, i.e. we **simulate** independent identically distributed random future risk factor changes $\tilde{\mathbf{X}}_{t+1}^{(1)},\ldots,\tilde{\mathbf{X}}_{t+1}^{(M)}$, and then compute the corresponding portfolio losses

$$\left\{\tilde{L}_{t+1}(\tilde{\mathbf{X}}_{t+1}^{(i)}) : i=1,\ldots,M\right\}. \tag{1.9}$$

As in the case of the historical simulation, by taking the relevant quantile of the empirical distribution of the simulated losses we can estimate the value-at risk:

The **MC estimator for the value-at-risk** is given by

$$VaR_\alpha(\tilde{L}_{t+1}) \triangleq \inf\{\ell \in \mathbb{R} | \tilde{F}_{t+1}(\ell) \geq \alpha\},$$

(continued)

(continued)

where the empirical distribution function $\tilde{F}_{t+1}(\ell)$ is given by

$$\tilde{F}_{t+1}(\ell) \triangleq \frac{1}{M}\sum_{i=1}^{M} I_{\{\tilde{L}_{t+1}(\tilde{\mathbf{X}}_{t+1}^{(i)})\leq \ell\}}.$$

Remark 3 (Some aspects of the MC method).

(i) Of course, the crucial modeling aspect is the choice of the distribution for the risk factor changes and the calibration of this distribution to historical risk factor change data $\mathbf{X}_{t-n+1},\ldots,\mathbf{X}_t$. This can be a computational challenging problem itself (compare also Sect. 1.6.1 and the chapter by Sayer and Wenzel in this book).

(ii) The above simulation to generate the risk factor changes is often named the *outer simulation*. Depending on the complexity of the derivatives included in the portfolio, we will need an *inner simulation* in order to evaluate the loss function of the risk factor changes. This means, we have to perform MC simulations to calculate the future values of options in each run of the outer simulation. As this is also an aspect of the historical simulation, we postpone this for the moment and assume that the simulated realizations of the loss distribution given by (1.9) are available.

Merits and Weaknesses of the Method

Of course, the quality of the MC method depends heavily on the choice of an appropriate distribution for the risk factor changes. On the up side, we are not limited to normal distributions anymore. A further good aspect is the possibility to generate as many loss values as one wants by simply choosing a huge value M of simulation runs. This is a clear advantage over the historical simulation where data are limited.

As there is no simplification to evaluate the portfolio, each simulation run will possibly need a huge computational effort, in particular if complicated options are held. On the other hand, this evaluation is then exact given the risk factor changes, which is a clear advantage compared to the variance-covariance method.

1.5.2.4 Challenges When Determining Market Risks

The Choice of a Suitable Risk Mapping

The above three methods have the main problem in common that it is not clear at all how to determine the appropriate risk factors yielding an accurate approximation of the actual loss. On top of that, their dimension can still be remarkably high. This is a modeling issue and is closely connected to the choice of the function f in (1.3). As already indicated, performing a principal component analysis (compare e.g. [3]) can lead to a smaller number of risk factors which explain the major parts of the market risks. However, the question if the postulated risk factors approximate the actual loss well enough then remains still an issue and translates into the problem of the appropriate choice of the input for the principal component analysis.

The different approaches we explained above each have their own advantages and drawbacks. While the Delta-approximation is usually not accurate enough if the portfolio contains non-linear securities/derivatives, the Delta-Gamma-approximation already performs much better than the Delta-approximation. However, the resulting approximation of the loss function only has a known distribution if we stick to normally distributed risk factors. The most accurate results can be achieved by the MC method but at the cost of a high computational complexity compared to the other methods. The trade-off therein consists of balancing out accuracy and computability. Further, we sometimes have to choose between accuracy and a fast computation which can be achieved via a smart approximation of the loss function (especially with regard to the values of the derivatives in the portfolio). And in the end, the applicability of all methods highly depends on the structure of the portfolio at hand. Also, the availability of computing power can play an important role on the decision for the method to use. Thus, a (computational) challenge when determining market risks is the choice of the appropriate value-at-risk computation method.

(Computational) challenge 6: Given the structure of the portfolio and of the computing framework, find an appropriate algorithm to decide on the adequate method for the computation of the value-at-risk.

Nested Simulation

As already pointed out, in both the historical simulation and in the MC method we have to evaluate the portfolio in its full complexity. This computational challenge carries to extremes, when the portfolio contains a lot of complex derivatives, for which no closed-form price representation is available. In such a case, we will need an *inner MC simulation* in addition to the outer one to compute the realized losses.

To formalize this, assume for notational convenience that the time horizon Δ is fixed, that time $t+1$ corresponds to time $t+\Delta$, and that the risk mapping $f : \mathbb{R}_+ \times \mathbb{R}^d \to \mathbb{R}$ corresponds to a portfolio of derivatives with payoff functions $H_1, \ldots, H_K$ with maturities $T_1, \ldots, T_K$. From our main result Theorem 1.3 we know that the fair time-t price of a derivative is given by the discounted conditional expectation of its payoff function under the risk neutral measure Q (we here assume that our market satisfies the assumptions of Theorem 1.3). Thus, the risk mapping f at time $t+\Delta$ is given by

$$f(t+\Delta, \mathbf{Z}_t + \tilde{\mathbf{X}}_{t+\Delta}^{(i)}) = \sum_{k=1}^{K} \tilde{\mathbb{E}}\left[e^{-r(T_k-(t+\Delta))} H_k | \tilde{\mathbf{X}}_{t+\Delta}^{(i)}\right], \tag{1.10}$$

where $\tilde{\mathbb{E}}(.)$ denotes the expectation under the risk neutral measure Q. For standard derivatives like European calls or puts the conditional expectations in (1.10) can be computed in closed-form (compare again Theorem 1.3). For complex derivatives, however, they have to be determined via MC simulation. This then causes an **inner simulation** as follows that has to be performed for **each** (!!!) realization of the outer simulation:

Inner MC simulation for complex derivatives in the portfolio:

1. Generate N independent realizations $H_k^{(1)}, \ldots, H_k^{(N)}$ of the $k = 1, \ldots, K$ (complex) payoffs given $\tilde{\mathbf{X}}_{t+\Delta}^{(i)}$.
2. Estimate the discounted conditional expectation of the payoff functions by

$$\tilde{\mathbb{E}}\left[e^{-r(T_k-(t+\Delta))} H_k | \tilde{\mathbf{X}}_{t+\Delta}^{(i)}\right] \approx \frac{1}{N} e^{-r(T_k-(t+\Delta))} \sum_{j=1}^{N} H_k^{(j)}$$

for $k = 1, \ldots, K$.

Remark 4 (Important!).

(i) The amount of simulation work in the presence of the need for an inner simulation is enormous as the inner simulations have to be redone for each run of the outer simulation. A possible challenge is to find a framework for reusing the simulations in the inner loop for each new outer simulation. A possibility could be to perform the inner simulations only a certain times and then setting up something as an interpolation polynomial for the price of the derivatives as a function of the risk factors.

(ii) Note further, that for notational simplicity, we have assumed that each derivative in the inner simulation requires the same number N of simulation paths to achieve a desired accuracy for the MC price calculation. This, however, heavily depends on the similarity of the derivatives and the volatility of the underlyings. If the variety of option types in the portfolio is large, substantial savings can be obtained by having a good concept to choose the appropriate number of inner simulation runs per option type.

(iii) As a minor issue, note that the input for the option pricing typically has to be the price of the underlying(s) at time $t + \Delta$ or even more, the paths of the price of the underlying(s) up to time $t + \Delta$. This input has to be reconstructed from the risk factor changes.

(iv) Finally, the biggest issue is the load balance between inner and outer simulation. Given only a limited computing time and capacity, one needs a well-balanced strategy. Highly accurate derivative prices in the inner simulation lead to an accurate evaluation of the loss function (of course, conditioned on the correctness of the chosen model for the distribution of the risk factor changes). On the other hand, they cause a big computational effort which then results in the possibility of performing only a few outer simulation runs. This then leads to a poor estimate of the value-at-risk. A high number of outer simulation runs however only allows for a very rough estimation of the derivative prices on the inner run, again a non-desirable effect.

The foregoing remark points in the direction of the probably most important computational challenge of risk management:

Computational challenge 7: Find an appropriate concept for balancing the workload between the inner and outer MC simulation for the determination of the value-at-risk of complex portfolios and design an efficient algorithm that ensures sufficient precision of both the derivative prices in the inner simulation and the MC estimator for the value-at-risk on the outer simulation.

1.5.3 Liquidity Risk

Besides the measurement of market risks, another important strand of risk management is the measurement of liquidity risk. We understand thereby liquidity risk as the risk not to be able to obtain needed means of payment or to obtain them only at increased costs. In this article we will put emphasis on liquidity risks which arise in the fund management sector. Fund management focuses in particular on calling risk (liquidity risk on the liabilities side) which is the risk of unexpectedly high claims or claims ahead of schedule as for instance the redemption of shares in a fund. **Liquidity risk in fund management** has gained importance in recent times which manifests itself in European Union guidelines that require appropriate liquidity risk management processes for UCITS (=Undertakings for Collective Investment in Transferable Securities) and AIFMs (=Alternative Investment Funds Managers); compare therefore [20] and [21].

One approach which covers these liquidity risk regulations is to calculate the **peaks over threshold (POT) quantile of the redemptions** of mutual funds. It is well-known (compare e.g. [6]) that the excess distribution can be approximated by the **generalized Pareto distribution (GPD)** from a certain threshold u. This fact is due to the famous theorem of Pickands, Balkema and de Haan, on which we will give a short mathematical excursion: Define the excess distribution of a real-valued random variable X with a distribution function F as

$$F_u(y) \triangleq P(X - u \leq y | X > u) \text{ , where } 0 \leq y < x_F - u,$$

for a fixed right endpoint x_F

$$x_F \triangleq \sup\{x \in \mathbb{R} : F(x) < 1\} \leq \infty, \text{ where } u < x_F.$$

Then we have the following:

Theorem 2 (Pickands, Balkema, de Haan). *There exists an appropriate function* $\beta(u)$ *such that*

$$\lim_{u \uparrow x_F} \sup_{0 < x < x_F - u} \left| F_u(x) - G_{\xi, \beta(u)(x)} \right| = 0,$$

where

$$G_{\xi,\beta}(x) = \begin{cases} 1 - \left(1 + \frac{\xi x}{\beta}\right)^{-1/\xi} & \xi \neq 0, \\ 1 - e^{-x/\beta} & \xi = 0, \end{cases} \quad , x \in D(\xi, \beta) = \begin{cases} [0, \infty) & \xi \geq 0, \\ [0, -\beta/\xi] & \xi < 0, \end{cases}$$

is the generalized Pareto distribution (GPD) with shape parameter $\xi \in \mathbb{R}$ *and scale parameter* $\beta > 0$.

As a consequence, the excess distribution can be approximated in a similar way by a suitable generalized Pareto distribution as the distribution of a sum can be approximated by the normal distribution. The quantile of the excess distribution then gives a liquidity reserve which is not exceeded by a certain probability p and is called POT quantile. The POT quantile is also referred to as **liquidity-at-risk** and was applied by [22] for the banking sector. Desmettre and Deege [5] then adapted it to the mutual funds sector and provided a thorough backtesting analysis.

The p-quantile of the excess distribution, i.e. the liquidity-at-risk, is given as

$$LaR_p \triangleq u + \frac{\hat{\beta}}{\hat{\xi}} \left(\left(\frac{n}{N_u} (1 - p) \right)^{-\hat{\xi}} - 1 \right), \tag{1.11}$$

where N_u is the number of exceedances over the threshold u, n is the sample size, $\hat{\xi}$ is an estimator for the shape parameter and $\hat{\beta}$ is an estimator for the scale parameter of the generalized Pareto distribution.

Thus in order to calculate the liquidity-at-risk, it is necessary to estimate the threshold parameter u, the shape parameter ξ and the scale parameter β of the GPD. The estimation of shape and scale parameter can be achieved using standard maximum likelihood estimators; a procedure for the **estimation of the threshold parameter** u and also its detailed derivation is also given in [5] and is the time-consuming part when computing the liquidity-at-risk as given by (1.11). In what follows we sketch the calibration method and explain how it leads to a computational challenge.

Using well-known properties of the generalized Pareto distribution $G_{\xi,\beta}$, we can conclude that the estimator $\hat{\xi}$ of the scale parameter ξ of the excess distribution (which is approximated by a suitable GPD) is approximately invariant under shifts in the threshold parameter u. Thus a procedure for the determination of the threshold parameter u is given by

> Choose the first threshold parameter $u > 0$ such that the estimator $\hat{\xi}$ of the shape parameter ξ of the corresponding GPD is approximately invariant under shifts in the threshold parameter $u > 0$.

The implementation of this method can be sketched as follows (see also [5]):

1. Sort the available data by ascending order and keep a certain percentage of the data.
2. Start with u being the lowest possible threshold and increase it up to the value for which at least k percent of the original data are left. With increasing threshold u truncate the data at the threshold u.
3. Estimate the unknown parameters ξ and β of the GPD by their maximum likelihood estimators for every u from 2.
4. For each u, calculate a suitable deviation measure of the corresponding maximum likelihood estimators $\hat{\xi}(i)_{i=1,\dots,K}$ within a sliding interval.
5. The appropriate threshold u is determined as the threshold which lies in the middle of the interval with the lowest deviation measure. Take the number of exceedances N_u corresponding to this u and the sample size n.
6. The estimates $\hat{\xi}$ and $\hat{\beta}$ are the maximum likelihood estimates which correspond to the threshold u.

The computational challenge now arises when we look at typical data sets. Often, fund redemption data is available over a quite long time horizon such that a time series of a single share class can contain thousands of data points. Moreover, management companies will typically face a large portfolio of share classes which can have a dimension of several hundreds. Combining these two facts we see that a fund manager will have to determine a large amount of estimates for the shape and scale parameter of the Generalized Pareto distribution in order to calibrate the threshold parameter u (compare steps 1–4 of the above algorithm) for a daily liquidity risk management process of her portfolio. Therefore it is important to have a grip on a fast calibration of the threshold and our next computational challenge can be formulated as

Computational challenge 8: Speed up the calibration of the threshold parameter u for a fast computation of the liquidity-at-risk.

1.5.4 Intraday Simulation and Calculation

Up to now we considered daily or yearly time horizons Δ. Nowadays in practice, the demand for so called *intraday calculations and calibrations* is growing, i.e. we face time horizons $\Delta \ll 1$ day and in the extreme the time horizon can have the dimension of a few hours or even 15 and 30 min which represents the time horizon of intraday returns. Especially within times of crises it may be of use to be able to recalibrate all corresponding risk measures of portfolios in order to have as much information as possible. This will allow fund managers to take well-founded decisions. For a concise overview of intraday market risk we refer to [8].

The recalibration and recalculation of the risk measures typically involves a reevaluation of the actual portfolio value as we have for instance seen within the nested simulations of the MC value-at-risk method. Therefore the *intraday evaluation of large portfolios* is also of importance. Summarizing our considerations above we face the computational challenge

Computational challenge 9: Speed up the calculation and calibration of the risk management process of financial firms such that intraday calculations become feasible.

1.6 Further Aspects of Computationally Challenging Problems in Financial Markets

Besides the optimization of MC methods and of risk management calculations, there are various other computational issues in financial mathematics. We will mention only three more, two of them are very important from a practical point of view, the other has big consequences for designing an efficient hardware/software concept:

1.6.1 Calibration: How to Get the Parameters?

Every financial market model needs input parameters as otherwise we cannot calculate any option price or, more general, cannot perform any type of calculation. To highlight the main approach at the derivatives markets to obtain the necessary parameters we consider the BS model. There, the riskless interest rate r can (in principle) be observed at the market. The volatility σ however has to be determined in a suitable way. There are in principle two ways,

- A classical maximum likelihood estimation (or any other conventional estimation technique) based on past stock prices using the fact that the logarithmic differences (i.e. $\ln(S(t_i)/S(t_{i-1})), \ln(S(t_{i-1})/S(t_{i-2})), \ldots$) are independent,
- A calibration approach, i.e. the determination of the parameter σ_{imp} which minimizes the squared differences between model and market prices of traded options.

As the second approach is the one chosen at the derivatives markets, we describe it a little bit more detailed. Let us for simplicity assume that at a derivatives market we are currently observing only the prices c_{K_i,T_i} of n call options that are characterized by their strikes K_i and their (times to) maturities T_i. The calibration task now consists of solving

$$\min_{\sigma>0} \sum_{i=1}^{n} \left(c_{K_i,T_i} - c\left(0, S(0); \sigma, K_i, T_i\right)\right)^2$$

where $c(0, S(0); K, T)$ denotes the BS formula with volatility $\sigma > 0$, strike K and maturity T. Of course, one can also use a weighted sum as the performance measure to care for the fact that some of these options are more liquidly traded than others.

Note that calibration typically is a highly non-linear optimization problem that even gets more involved if more parameters have to be calibrated. We also recognize the importance of having closed pricing formulae in calibration. If the theoretical prices have to be calculated by a numerical method (say the MC method) then the computational effort per iteration step in solving the calibration problem increases dramatically.

For a much more complicated calibration problem we refer to the work by Sayer and Wenzel in this book.

1.6.2 Money Counts: How Accurate Do We Want Our Prices?

The financial markets are known for their requirement of extremely accurate price calculations. However, especially in the MC framework, a huge requirement for accurate prices increases the computational effort dramatically. It is therefore worth to point out that high accuracy is worthless if the parameter uncertainty (i.e. the error in the input parameters), the algorithmic error (such as the order of (weak) convergence of the MC method) or the model error (i.e. the error caused by using an idealized model for simulation that will certainly not exactly mimic the real world price dynamics) are of a higher order than the accuracy of the performed computations.

On the other hand, by using a sparse number format, one can speed up the computations and reduce storage capacity by quite a factor. It is therefore challenging to find a good concept for a variable treatment of precision requirements.

For an innovative suggestion of a mixed precision multi-level MC framework we refer to the work by Omland, Hefter and Ritter in this book.

1.6.3 Data Maintenance and Access

All mentioned computational methods in this article have in common that they can only be efficiently executed once the data is available and *ready to use*. A good many times, the data access takes as much time as the computations themselves. In general, the corresponding data like market parameters or information about the composition of derivatives and portfolios are stored in large data bases whose maintenance can be time-consuming; for an overview on the design and maintenance of database systems we refer to the textbook of Connolly and Begg [4].

In that regard it is also very useful to thoroughly review the computations that have to be done and to do them in a clever way; for instance a smart approximation of the loss function where feasible may already tremendously accelerate the value-at-risk computations. We thus conclude with the computational challenge

Computational challenge 10: Maintain an efficient data storage and provide an efficient data access.

Acknowledgements Both authors are grateful to the Deutsche Forschungsgemeinschaft for funding within the Research Training Group 1932 "Stochastic Models for Innovations in the Engineering Sciences". Sascha Desmettre wishes to thank Matthias Deege from IPConcept (Luxemburg) S.A. and Heiko Reiss for useful discussions about the practical challenges of derivative pricing and risk management.

References

1. Acerbi, C., Tasche, D.: On the coherence of expected shortfall. J. Bank. Financ. **26**(7), 1487–1503 (2002)
2. Black, F., Scholes, M.: The pricing of options and corporate liabilities. J. Political Econ. **81**, 637–654 (1973)

3. Brummelhuis, R., Cordoba, A., Quintanilla, M., Seco, L.: Principal component value at risk. Math. Financ. **12**(1):23–43 (2002)
4. Connolly, T.M., Begg, C.E.: Database Systems: A Practical Approach to Design, Implementation and Management. Prentice Hall, Upper Saddle River (2005)
5. Desmettre, S., Deege, M.: Liquidity at risk for mutual funds. Preprint available at SSRN. http://ssrn.com/abstract=2440720 (2014)
6. Embrechts, P., Klüppelberg, C., Mikosh, T.: Modeling Extremal Events. Springer, Berlin (1997)
7. Föllmer, H., Schied, A.: Stochastic Finance: An Introduction in Discrete Time. de Gruyter, Berlin (2002)
8. Giot, P.: Market risk models for intraday data. Eur. J. Financ. **11**(4) 309–324 (2005)
9. Gobet, E.: Advanced Monte Carlo methods for barrier and related exotic options. In: Ciarlet, P., Bensoussan, A., Zhang, Q. (eds.) Mathematical Modeling and Numerical Methods in Finance. Handbook of Numerical Analysis, vol. 15, pp. 497–528. Elsevier, Amsterdam (2009)
10. Heston, S.L.: A closed-form solution for options with stochastic volatility with applications to bond and currency options. Rev. Financ. Stud. **6**(2), 327–343 (1993)
11. Karatzas, I., Shreve, S.E.: Brownian Motion and Stochastic Calculus, 2nd edn. Springer, Berlin (1991)
12. Kloeden, P.E., Platen, E.: Numerical Solution of Stochastic Differential Equations. Springer, Berlin (1999)
13. Korn, R., Korn, E.: Option Pricing and Portfolio Optimization. Graduate Studies in Mathematics, vol. 31. AMS, Providence (2001)
14. Korn, R., Korn, E., Kroisandt, G.: Monte Carlo Methods and Models in Finance and Insurance. Chapman & Hall/CRC Financial Mathematics Series. CRC, London (2010)
15. Korn, R., Müller, S.: Binomial trees in option pricing – history, practical applications and recent developments. In: Devroye, L., Karasozen, B., Kohler, M., Korn, R. (eds.) Recent Developments in Applied Probability and Statistics, pp. 119–138. Springer, Berlin (2010)
16. Matsumoto, M., Nishimura, T.: Mersenne Twister: A 623-dimensionally equidistributed uniform pseudo-random number generator. ACM Trans. Model. Comput. Simul. **8**, 3–30 (1998)
17. McNeil, A.J., Frey, R., Embrechts, P.: Quantitative Risk Management. Princeton University Press, Princeton/Oxford (2005)
18. RiskMetrics: RiskMetrics Technical Document, 4th edn. J.P. Morgan/Reuters, New York (1996)
19. The Commission for the Supervision of the Financial Sector: CSSF Circular 11/512 (2011)
20. The European Commission: Commission Directive 2010/43/EU. Official Journal of the European Union (2010)
21. The European Parliament and the Council of the European Union: Directive 2011/61/EU of the European Parliament and of the Council. Official Journal of the European Union (2011)
22. Zeranski, S.: Liquidity at Risk – Quantifizierung extremer Zahlungsstromrisiken. Risikomanager **11**(1), 4–9 (2006)

Chapter 2
From Model to Application: Calibration to Market Data

Tilman Sayer and Jörg Wenzel

Abstract We present the procedure of model calibration within the scope of financial applications. We discuss several models that are used to describe the movement of financial underlyings and state closed or semi-closed pricing formulas for basic financial instruments. Furthermore, we explain how these are used in a general calibration procedure with the purpose to determine sensible model parameters. Finally, we gather typical numerical issues that often arise in the context of calibration and that have to be handled with care.

2.1 Introduction

This contribution focuses on one of the most time consuming areas of financial mathematics, namely the calibration of a financial model to observed market data. In fact, calibration ensures the applicability of complex models and is a necessary requirement for accurate pricing and a thorough risk management as demanded by regulatory agencies.

In Sect. 2.2, we will abstractly explain the concepts behind model calibration, present different objective functions and focus on relevant instruments and optimization algorithms. Section 2.3 presents several financial equity and interest rate models often employed in applications, where we focus on model prices of instruments that are used for calibration. We conclude the contribution in Sect. 2.4 by amassing the most challenging numerical parts for calibration purposes.

T. Sayer (✉) • J. Wenzel
Department of Financial Mathematics, Fraunhofer ITWM, Fraunhofer-Platz 1, 67663 Kaiserslautern, Germany
e-mail: tilman.sayer@itwm.fraunhofer.de; joerg.wenzel@itwm.fraunhofer.de

C. De Schryver (ed.), *FPGA Based Accelerators for Financial Applications*,
DOI 10.1007/978-3-319-15407-7_2

2.2 Model Calibration: A General View

When valuing financial derivatives, like options for example, certain assumptions on the development of the underlying financial instruments, like for instance the stock price or some interest rates, have to be made. Usually, these assumptions give rise to a mathematical model, which in our case typically is a stochastic differential equation, describing the future random movement of the underlying. Then, the model can be used to derive pricing formulas for financial derivatives, or to simulate possible future evolutions of the financial market.

In most cases, the model will depend on a set of *model parameters* that are not directly observable from the market. We denote this set by $\mathcal{M}$. The set of parameters directly observable from the market, i.e. the *market parameters*, such as spot prices or interest rates, is denoted by $\mathcal{O}$. Finally, there is a third parameter set $\mathcal{P}$, entering a pricing formula, which contains parameters of the financial product, also referred to as *product parameters*. For instance, for a European call on a stock we have $\mathcal{P} = \{K,T\}$, where K is the *strike* and T the *maturity* of the call. For $t = 0$, we have $\mathcal{O} = \{S_0,r\}$ with today's price of the underlying S_0 and *interest rate* r.

While the meaning of market and product parameters generally is quite clear, the model parameters might not even have a financial meaning at all. Still, in order to use the model, sensible parameter values have to be found such that the model mirrors market reality. We will call this process *model calibration*.

Let f denote the type of a financial instrument, for example $f = call$ for calls. Combined with a particular choice of market and product parameters, this completely defines the specific product of type f. Throughout this contribution, the corresponding *market price* of this product is denoted by

$$X_f^{\text{market}}(\mathcal{O},\mathcal{P}) .$$

This price can generally be obtained as *market quote* from recent trades on an exchange. In fact, we assume to observe N such prices, that result for different product parameters $\mathcal{P}_i$, and we will refer to them as

$$X_f^{\text{market}}(\mathcal{O},\mathcal{P}_i) , \quad i = 1,\ldots,N .$$

Accordingly, let

$$X_f(\mathcal{M},\mathcal{O},\mathcal{P})$$

be the *model price* of a particular product of type f. Hence, for given model and market parameters,

$$X_f(\mathcal{M},\mathcal{O},\mathcal{P}_i)$$

refers to the model price, calculated for the product parameter set $\mathcal{P}_i$.

Since $\mathcal{O}$ can be observed and $\mathcal{P}_i$ is stipulated in the contract of the product, the purpose of calibration is to determine $\mathcal{M}$ such that the resulting model prices are as close as possible to the observable market prices, hence

$$X_f(\mathcal{M},\mathcal{O},\mathcal{P}_i) = X_f^{\text{market}}(\mathcal{O},\mathcal{P}_i)$$

simultaneously for all i. If the model reflects reality, model and market prices coincide. However, since a model always is only an approximation of reality, equality cannot be expected in general and the fit might not be perfect.

Therefore, we can only expect the distance

$$\text{dist}\left(\left(X_f(\mathcal{M},\mathcal{O},\mathcal{P}_i)\right)_{i=1,\ldots,N}, \left(X_f^{\text{market}}(\mathcal{O},\mathcal{P}_i)\right)_{i=1,\ldots,N}\right) \tag{2.1}$$

to be small, where $\text{dist}(\cdot,\cdot)$ is a suitable distance or *objective function* on $\mathbb{R}^N$, which will be discussed in more detail later on. Finally, we are left with the optimization problem to find a particular set $\mathcal{M}$ such that the given distance is small.

The remainder of this section deals with possible choices of an objective function, focuses on financial instruments that can be used for calibration and briefly discusses types of optimization algorithms.

2.2.1 *Objective Function*

Obviously, the primary objective of calibration is to find an optimal fit between market and model prices. However, optimality cannot be guaranteed and the choice of a particular distance function is quite arbitrary.

In the following, we state the most common objective functions detailing Eq. (2.1).

Mean squared error: A first obvious choice to specify the distance function is to use the "smoothest" norm on $\mathbb{R}^N$ which is the L_2-norm. The minimization problem then corresponds to the common least-squares problem and reads as

$$\min_{\mathcal{M}} \sum_{i=1}^{N} \left(X_f(\mathcal{M},\mathcal{O},\mathcal{P}_i) - X_f^{\text{market}}(\mathcal{O},\mathcal{P}_i)\right)^2 .$$

Instead of the L_2-norm any other L_p-norm with $1 \leq p < \infty$ can also be used. Larger values of p put more emphasis on larger deviations.

Relative errors: This type of objective function emphasizes out-of-the-money products. It is given by

$$\min_{\mathcal{M}} \sum_{i=1}^{N} \left(\frac{X_f(\mathcal{M},\mathcal{O},\mathcal{P}_i) - X_f^{\text{market}}(\mathcal{O},\mathcal{P}_i)}{X_f^{\text{market}}(\mathcal{O},\mathcal{P}_i)}\right)^2 .$$

Logarithmic errors: Similar to relative errors, logarithmic errors sometimes are advantageous, compare for example [14]. Here, the objective function is given as

$$\min_{\mathcal{M}} \sum_{i=1}^{N} \left(\log\left(X_f\left(\mathcal{M},\mathcal{O},\mathcal{P}_i\right)\right) - \log\left(X_f^{\text{market}}\left(\mathcal{O},\mathcal{P}_i\right)\right)\right)^2 .$$

It is important to note, that depending on the particular objective function, the resulting model parameters may vary, see for example [14]. So, in applications, one can incorporate further desirable properties. Mostly these features deal with the stability of the calibration, as detailed in the following list.

- If market conditions change, market prices change. Ideally, the model is designed in such a way that it can reflect these changes. However at certain times a re-calibration of the model parameters has to be performed in order to capture new market beliefs. If the change in the market prices is rather small, we expect the change in the parameters to be relatively small as well, i.e. we want the calibration to be continuous and smooth in the market prices.
- If market conditions remain constant, still, as time goes by, prices of financial derivatives change due to the decreasing time to maturity. Calibrating on the progressed prices, however, should then not distort the resulting model parameters. This is what we call time homogeneity. The model as well as the calibration is required to price derivatives consistently with respect to time.

Prices of illiquid products with a wide bid-ask spread are less reliable and often might change even randomly. These prices imply different calibration results. This opposes the stability conditions above.

The consequential disparity of information is often incorporated into the calibration by assigning different weights $\omega_i, i = 1,\dots,N$ and $\sum_{i=1}^{N} \omega_i = 1$ to the influence of the market prices depending on their importance, i.e. the influence of the prices might be stressed or damped. In applications, the weights are often set in such a way that they correspond to the liquidity of the product. Since frequently traded instruments are more liquid, hence contain more reliable information, they might be of higher importance.

For example, in the case of equity options, setting the weights to the Vega of the product, i.e. to the derivative of the option price with respect to volatility, emphasizes at-the-money products.

Even using weights, calibration problems typically are *ill-posed* in the sense of Hadamard, see for instance [11]. Usually, the surface of the objective function is very rugged. It is not convex and does not feature any regular shapes. Often there exist several local minima, separated by relatively high ridges, what makes the calibration process even more difficult. Further, small changes in market prices typically lead to quite different model parameters.

In general, a stabilizing functional that incorporates prior knowledge will smooth the minimization problem to some extent. In our case, the prior knowledge, or *penalty term*, can be seen as the expectation of the market participants, who usually assume that the model parameters remain relatively constant over time. In terms of the common least-squares formulation, for $\mathcal{M} = \{p_1,\dots,p_M\}$, the minimization problem changes to

$$\min_{\mathcal{M}} \sum_{i=1}^{N} \omega_i \left(X_f(\mathcal{M}, \mathcal{O}, \mathcal{P}_i) - X_f^{\text{market}}(\mathcal{O}, \mathcal{P}_i) \right)^2 - \alpha \sum_{j=1}^{M} (p_j - \bar{p}_j)^2 \,, \qquad (2.2)$$

where $\bar{p}_j$ are previously calibrated model parameters, if such parameters are available.

In (2.2), the multiplier α has to be chosen with care. If too small, the problem is still unstable, if too large, the dependence on prior data is too intense and we are in a situation of a self-fulfilling prophecy. For a broader view on ill-posed problems and possible regularization methods, we refer to [11] or [21].

Although model prices rarely match observed market prices exactly, they should at least lie within the corresponding bid-ask spread. In some cases, this relaxation is very reasonable because, not only due to noisy data, accuracy in the context of model calibration quickly might be spurious. In [17] Kalman filters are used to incorporate bid and ask prices into the calibration.

A further possibility to smooth calibration is to consider functionals of the product prices. These can for example be the time value of an equity option or the implied volatility, which we will discuss in Sect. 2.3.1.1.

2.2.2 Relevant Instruments

The selection of financial instruments for calibration is an important task and has to be done carefully.

First of all, the chosen instruments must allow to determine model prices fast and accurately for different model parameter sets. Note that, even though the time needed for calculating a single model price might be almost negligible, calculating the prices for a whole set of instruments and for several sets of model parameters is very time consuming. Moreover, especially when valuing derivatives for trading, calibration is done on demand for each trade rather than only occasionally, compare for instance [14].

For equity and foreign exchange markets this requirement leads to considering plain vanilla European exercise feature options almost exclusively. In interest rate markets the corresponding instruments are options on the forward rate, so called *caps* and *floors*. Although slightly more complicated, options on the swap rate, so called *swaptions*, are also computable in closed-form, and hence are used for calibration too.

In general, more complicated products can be relevant as well. In particular, for *multi-asset models* correlation parameters must be calibrated. In this case, basket options can be used.

However, the pricing of more complex products often requires advanced numerical schemes like flexible Monte Carlo methods or finite difference schemes. But due to the slow convergence of Monte Carlo simulations and the strong dependence between grid size and pricing accuracy of finite difference schemes, products that require advanced numerical methods are generally not usable for calibration.

In a nutshell: *"Therefore, closed pricing formulae are the basis of a convenient model calibration."*, see [16, Remark 5.30].

2.2.3 Optimization: Local Versus Global

Obviously, minimizing the objective function is a non-linear problem. Due to the rugged shape of the objective function, both local and global optimization algorithms should be applied.

Local or deterministic algorithms like the *downhill simplex algorithm* are simple and fast. Usually, in each step these algorithms determine the direction in which the objective function has its steepest descend. However, they will terminate in a local minimum and should be restarted with different initial parameters again.
These algorithms are most useful if a strong expectation about the model parameters is available, such as if previously calibrated parameters exist. In this case it makes sense to also use these parameters as starting values for the optimization.

Global or stochastic algorithms like *adaptive simulated annealing* do not depend on the initial parameter set, since they randomly sweep the search space. Obviously, these algorithms are able to continuously improve the value of the objective function, but this comes with a high computational cost.
This type of algorithms should in particular be used for an initial calibration or when the market situation changed drastically.

Further, it is important to note that for many practical applications a perfect fit is not really necessary. Rather, stable, more reliable values are usually preferred. Inaccuracies in pricing financial derivatives come from many sources and calibration of model parameters is just one of them. Others include the appropriateness of the model and the use of approximation formulas.

2.3 Modeling Financial Markets

So far, we have introduced market prices of financial instruments, possible objective functions and optimization algorithms. As the final ingredient used for calibration, we now focus on financial equity and interest rate models and present closed and semi-closed pricing formulas for relevant products.

2.3.1 Equity Models

We first introduce the popular Black-Scholes, highlight two of its weaknesses and give a flexible generalization of it.

2.3.1.1 The Black-Scholes Model

Without doubt, the most famous equity model is the Black-Scholes, see [4], where the stock price S_t is modeled as a *geometric Brownian motion*. For the sake of simplicity, we will introduce the model directly under the *risk-neutral measure* and refer to Chap. 1 by Desmettre and Korn for a brief discussion on risk-neutral valuation and to [15] for a detailed explanation. Thus, the model reads as

$$\mathrm{d}S_t = rS_t\,\mathrm{d}t + \sigma S_t\,\mathrm{d}W_t\,, \quad S_0 = s \geq 0\,,$$

with interest rate r, constant volatility σ and Brownian motion W_t. In this model, the price of a European exercise feature call with strike K and maturity T at time $t = 0$ can be calculated as

$$X_{call} = S_0 \Phi\left(d_+\right) - K \exp\left(-rT\right) \Phi\left(d_-\right),$$

where $\Phi(\cdot)$ denotes the *cumulative distribution function* of the standard normal distribution and

$$d_\pm = \frac{\log\left(S_0\right) - \log\left(K\right) + \left(r \pm \sigma^2/2\right)T}{\sigma\sqrt{T}}\,.$$

Note that, for convenience, we suppress the dependence on $\mathcal{M}$, $\mathcal{O}$, and $\mathcal{P}$ here and in the following when it is clear from the context.

Due to the simple form of the model and since all parameters except the volatility are observable from market data, this model is the fundamental model for various financial applications.

However, empirical findings suggest that the Black-Scholes does not explain reality in a satisfactory way. A detailed view on these findings, which are also called *stylized facts*, is beyond the scope of this contribution. Instead we only briefly discuss the most important ones.

Volatility clustering refers to an effect, frequently observed in historical price data. Figure 2.1 presents historical daily logarithmic returns of the German stock index DAX from January 2008 to December 2012. One can observe phases with relatively high or low volatility. If, as the Black-Scholes assumes, data was distributed normally, these clusters would not occur.

Skew and kurtosis both are measures of a probability distribution and characterize the shape of a distribution. For normal data, the skew equals zero and the

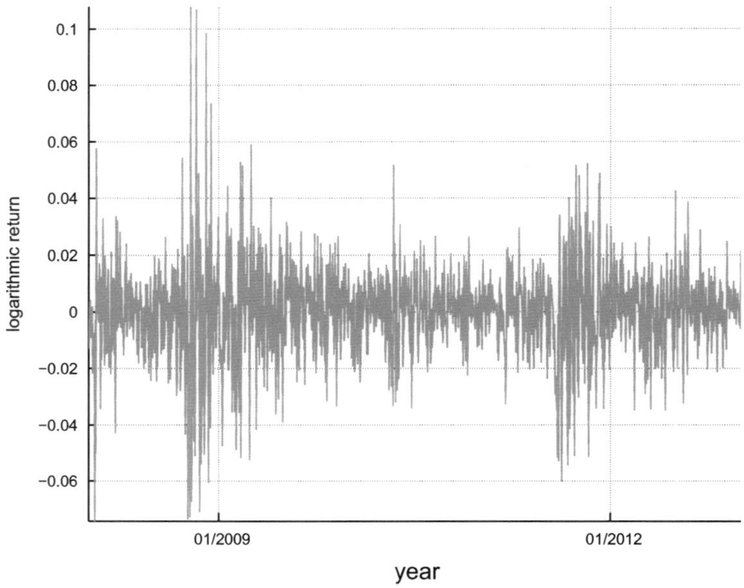

Fig. 2.1 Historical daily logarithmic returns of the DAX between January 2008 and December 2012

kurtosis equals three. Yet, in real applications, the empirical skew of logarithmic returns typically is negative, meaning that the left tail of the empirical density function is fatter than the right one. Additionally, the empirical kurtosis of the corresponding data is larger than three, which means that we face distributions with higher peaks and fatter tails, compared to the normal distribution.
Figure 2.2 shows the empirical distribution of logarithmic returns of the DAX index for daily observations between January 2008 and December 2012. Here, the solid line corresponds to the density function of the fitted normal distribution. As can be seen, in comparison with the normal distribution, we obtain a higher peak as well as fatter tails.

While these phenomena are directly inferred from asset prices, there is a further inconsistency between the Black-Scholes and the market regarding option prices. The usage of a constant volatility in the Black-Scholes model market implies that the price of each traded European exercise feature call on the same stock, should result from applying the same value of σ. In general, this assumption is heavily violated throughout many financial markets.

Looking at this issue the other way round implies that we can find a particular volatility value for each observed call. In fact, since the call price is monotone in σ, this value is unique and is called *implied volatility*.

Figure 2.3 shows the implied volatility surface obtained from European exercise feature calls on the stock of Allianz SE for different maturities and strikes quoted on December 14th, 2011. On that date, the closing stock price was €71.53.

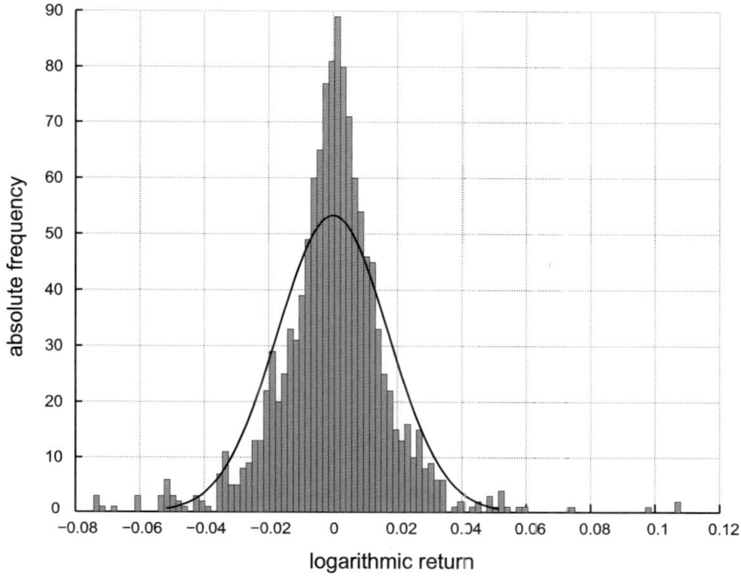

Fig. 2.2 Empirical distribution of daily logarithmic returns of the DAX from January 2008 to December 2012

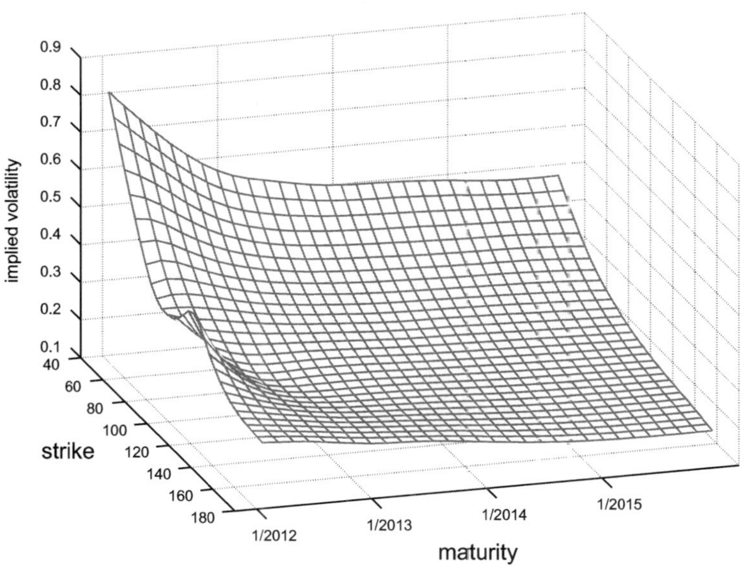

Fig. 2.3 Implied volatility surface on December 14th, 2011 from calls on Allianz

The shape of the surface is very common in equity markets and is not at all flat, as the Black-Scholes would imply. In fact, for fixed maturity, the graph is typically U-shaped, reminiscent of a smile, hence practitioners often call it the *volatility smile* or *smirk*.

All in all, there is a variety of empirical findings that cannot be matched by the Black-Scholes. To overcome these drawbacks, we introduce a more complex model, which is detailed in [1] and forms a superset of many renowned equity models.

2.3.1.2 A Generalization

Here, we present a quite substantial generalization of the Black-Scholes, which offers more stochastic processes and free parameters and is thus able to capture more market effects. Under the risk-neutral measure it reads as

$$
\begin{aligned}
\mathrm{d}S_t &= (r - \lambda\xi)\, S_t\, \mathrm{d}t + \sqrt{v_t} S_t\, \mathrm{d}W_t + J_t S_t\, \mathrm{d}N_t\,, \quad S_0 = s \geq 0\,,\\
\mathrm{d}v_t &= \kappa\,(\theta - v_t)\, \mathrm{d}t + \sigma\sqrt{v_t}\, \mathrm{d}\tilde{W}_t\,, \quad v_0 = v \geq 0\,,
\end{aligned}
$$

where the *variance process* v_t is modeled as stochastic process, a so called *Cox-Ingersoll-Ross* or *CIR* process. In fact, v_t is a *mean-reverting* stochastic process with *long-term mean* $\theta \geq 0$ and *speed of mean-reversion* $\kappa \geq 0$ while $\sigma \geq 0$ is its *volatility*.

The variance process cannot be written in an explicit form but its distribution, the *non-central chi-squared distribution*, is known. Further, the process is almost surely non-negative and finite. If additionally the *Feller condition*

$$
2\kappa\theta \geq \sigma^2
$$

is satisfied, the process is strictly positive, i.e. the probability that v_t is greater than zero equals one.

As before, r is the risk-free interest rate. Further, the Brownian motions W_t and $\tilde{W}_t$ are correlated with ρ, hence $\rho \in [-1, 1]$. In fact, typically ρ is close to -1 for equity data.

The process N_t is an independent *Poisson process* with jump intensity $\lambda \geq 0$. The jump sizes J_t are identically and independently distributed according to the log-normal distribution, i.e.

$$
\log(1 + J_t) \sim \mathcal{N}\left(\log(1 + \xi) - \frac{\psi^2}{2}, \psi^2\right)
$$

for constant ξ and ψ. In this model, today's price of a European exercise feature call can be determined as

$$
X_{call} = S_0 Q_1 - K \exp(-rT)\, Q_2 \tag{2.3}
$$

with

$$Q_1 = \frac{1}{2} + \frac{1}{\pi} \frac{\exp(-rT)}{S_0} \int_0^\infty \mathfrak{Re}\left(\frac{\varphi(u-i)}{iu\exp(iu\log(K))}\right) du,$$

$$Q_2 = \frac{1}{2} + \frac{1}{\pi} \int_0^\infty \mathfrak{Re}\left(\frac{\varphi(u)}{iu\exp(iu\log(K))}\right) du.$$

Here, $\mathfrak{Re}(\cdot)$ and $\varphi(\cdot)$ respectively denote the real part of a complex number and the *characteristic function* of the logarithmic stock price, which is given as

$$\begin{aligned}\varphi(u) &= \exp\left[iu\left(\log(S_0) + (r - \lambda\xi)T\right)\right] \\ &\times \exp\left[\frac{\kappa\theta}{\sigma^2}\left((\kappa - iu\rho\sigma - d)T + 2\log\left(\frac{1-g\exp(-dT)}{1-g}\right)\right)\right] \\ &\times \exp\left[\frac{v_t(\kappa - iu\rho\sigma - d)(1-\exp(-dT))}{\sigma^2(1-g\exp(-dT))}\right] \\ &\times \exp\left[\lambda T\left((1+\xi)^{iu}\exp\left(\frac{-\psi^2 u(i+u)}{2}\right) - 1\right)\right],\end{aligned}$$

with

$$d = \sqrt{(iu\rho\sigma - \kappa)^2 + \sigma^2(u^2 + iu)} \text{ and } \qquad g = \frac{\kappa - iu\rho\sigma - d}{\kappa - iu\rho\sigma + d},$$

compare for instance [1] or [20].

For European puts, a slightly different formula can be derived as well. In this contribution, however, we will solely focus on calls.

As already indicated, the model forms a superset of several famous equity models, which we will present shortly especially in respect of their model parameters.

The model of Bakshi, Cao and Chen, see [1] has eight free parameters, that is $\mathcal{M} = \{\kappa, \theta, \sigma, v, \rho, \lambda, \xi, \psi\}$, and it is also called *Bates*, see [2]. Due to its complexity, it obviously can describe many effects seen in reality and might fit the implied volatility surface quite well. However, due to the number of parameters, the search space is very large and the calibration might take too long. Besides, overfitting might occur, i.e. the model inadvertently ascribes too much importance to random noise or noisy data.

The model of Heston, see [13] is obtained for $\lambda = 0$, i.e. in the case of no jumps. Here we obtain $\mathcal{M} = \{\kappa, \theta, \sigma, v, \rho\}$. The resulting model is one of the most popular ones and it is widely applied among practitioners. Because of its popularity, there also exist several extensions. These include for example time-dependent parameters or versions for more than one stock, see for instance [19] or [9].

One of the disadvantages of the Heston is that typically prices of short maturity out-of-the-money options cannot be matched properly. This is due to the fact that the processes involved in the model are continuous and cannot move so steeply.

The model of Merton, see [18] results for constant variance $v_t = v$ and free parameters $\mathcal{M} = \{v, \lambda, \xi, \psi\}$. Due to the jumps, the model is able to capture extreme implied volatilities for options that are out-of-the-money.

The model of Black and Scholes, see [4] only has one free parameter, namely the constant variance and is not able to capture realistic market effects.

Having the model chosen according to the requirements of the application, model prices of European exercise feature calls can be determined via Eq. (2.3) and applied for model calibration.

2.3.2 *Interest Rate Models*

The main difference of interest rate models compared to equity models is the additional time scale. While for equity or foreign exchange models the underlying assets have a potentially unbounded life time, in interest rate models the underlyings are interest rates or bonds that mature. Therefore, we have to model functions of time that change over time.

This makes the modeling of interest rates and the respective model calibration a much more cumbersome task. In particular, the correlation between interest rates with different maturities has a high effect on pricing and thus has to be considered as well.

Since we only briefly discuss the main concepts needed in our context here, we refer to [5] for further information on interest rate instruments and relevant models in interest rate markets. Note that for simplicity, we neglect basis spreads and do not focus on overnight indexed swaps and their use for discounting.

2.3.2.1 Relevant Products

The most basic building blocks of the interest rate market are *zero bonds*. These guarantee a payment of the nominal, which we here consider to be € 1 at maturity T. We denote today's market price of a zero bond by

$$X_{zero\ bond}^{\text{market}}(T) .$$

The bond prices translate one-to-one to *zero rates* $r_z(T)$ for a given maturity T via

$$X_{zero\ bond}^{\text{market}}(T) = \exp(-r_z(T)T) .$$

Coupon bearing bonds can be seen as a suitable sum of zero bonds with different nominals and maturities.

Forward rate agreements fix an interest rate for a zero bond starting at a future time T with maturity $S > T$. The price of such an instrument determines the *forward rate* $r_f(T,S)$. The forward rates can in fact be calculated using observable zero rates as

$$r_f(T,S) = \frac{r_z(S)S - r_z(T)T}{S-T}.$$

A call (put) on a forward rate is called a *caplet* (*floorlet*). These instruments correspond to calls and puts in the equity market where here the underlying is the forward rate. Series of caplets or floorlets with different maturities and equal strike are traded as *caps* or *floors*. Their prices are the sum of the individual caplet or floorlet prices.

A *swap* exchanges a series of cash flows determined by the forward rates against a series of cash flows determined by a fixed coupon rate c. The fixed coupon rate making the swap a fair trade, i.e. a zero price trade at time T, is called the *swap rate* $r_s(T)$. This rate is a non-linear deterministic function of the zero rates for the different payment times of the cash flows.

An option on the swap rate is called a *swaption*. As well as caps and floors, swaptions are also frequently traded. Contrary to the prices of caplets or floorlets, swaption prices carry information about the correlation of forward rates, since a swaption is an option on a portfolio of forward rates.

The *instantaneous forward rate*

$$r_f(T) = -\frac{\partial \ln\left(X_{zero\ bond}^{\text{market}}(T)\right)}{\partial T}$$

is an artificial instrument and represents the interest rate paid for an infinitesimal small time step at time T. Employing this rate, the zero as well as the forward rates can be determined via

$$r_z(T) = \frac{1}{T}\int_0^T r_f(u)\,\mathrm{d}u,$$

$$r_f(T,S) = \frac{1}{S-T}\int_T^S r_f(u)\,\mathrm{d}u.$$

The collection of all observable zero rates can be completed to a function on $[0,\infty)$ using suitable inter- and extrapolation methods. The resulting function $r_z(\cdot)$ is called the *term structure of interest rates*. By construction, nearly all used interest rate models automatically calibrate to the current term structure of interest rates. In fact, this means that the zero bond prices for all traded maturities are fitted by the model by default.

The calibration procedure uses caps, floors, and swaption data. Since caps and floors depend on maturity and strike, here we fit a *price surface*, as already discussed for the equity markets. However, swaptions are characterized by maturity and strike of the corresponding option as well as by the maturity of the underlying swap. Therefore, instead of a surface, we are facing a whole cube of price data.

For the sake of simplicity, we call

$$D(T) = X^{\text{market}}_{zero\ bond}(T)$$

the *discount factor* for maturity T. Furthermore we focus on caps and swaptions rather than on floors.

2.3.2.2 Black '76 Model

The model of Black, see [3], which is usually called *Black '76*, is less an interest rate model, but a market convention when quoting prices of caps and swaptions. It assumes the underlying forward rate of a contract, i.e. the forward rate for caplets or the swap rate for swaptions, to follow a geometric Brownian motion with volatility parameter σ. Thus, the option price of the respective instrument can be derived via the Black-Scholes formula.

In this sense, today's price of a caplet on the forward rate from T to $S > T$ with strike c is given by

$$X_{caplet} = (S-T)D(S)\left(r_f(T,S)\Phi(d_+) - c\Phi(d_-)\right),$$

where

$$d_\pm = \frac{\log\left(r_f(T,S)\right) - \log(c) \pm \sigma^2 T/2}{\sigma\sqrt{T}}.$$

Similarly, the price of an option with maturity T on a swap paying a fixed coupon c at times S_i for $i = 1,\dots,N$ and $T < S_1 < S_2 < \cdots < S_n = S$ is given by

$$X_{swaption} = \sum_{i=1}^{n} (S_i - S_{i-1}) D(S_i) \left(r_s(T)\Phi(d_+) - c\Phi(d_-)\right),$$

where

$$d_\pm = \frac{\log(r_s(T)) - \log(c) \pm \sigma^2 T/2}{\sigma\sqrt{T}}.$$

Note, that here the forward rate $r_f(T,S)$ is simply replaced by the swap rate $r_s(T)$.

Since, as in the Black-Scholes case, there is just one model parameter, namely the volatility σ, the model is very appealing. However, it can be shown that in any interest rate model that models the whole interest rate curve, not both, forward and swap rates, can be log-normally distributed. Therefore the formulas given above are inconsistent for any reasonable interest rate model.

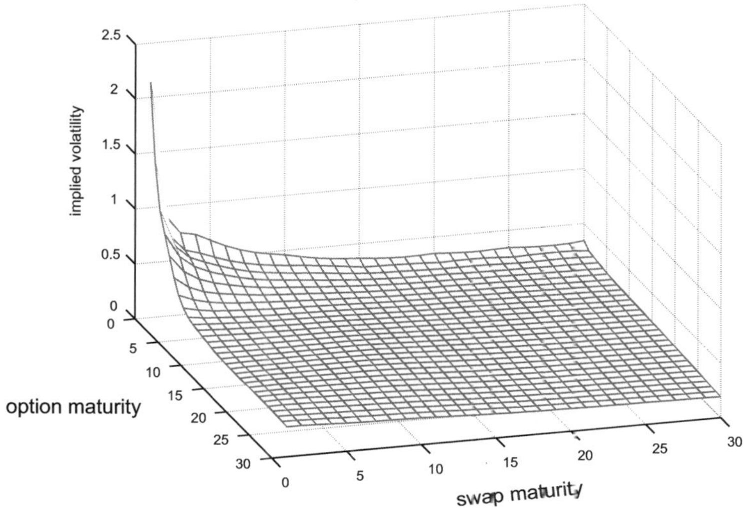

Fig. 2.4 EUR swaption volatility surface on October 30th, 2014

Despite this fact, they have become standard in practice, as prices of caps and swaptions are usually quoted in terms of their *implied Black '76 volatility*. This means, the corresponding formula is inverted to find the implied volatility. For swaptions, the resulting volatility data is called the *implied volatility cube*. Usually quotes are given only for at-the-money swaptions, i.e. swaptions whose strikes equal the current swap rate. Additionally, for each maturity of the option and each maturity of the swap, a volatility smile is quoted, listing the difference to the at-the-money swaption volatility for several deviations of the strike from the current swap rate. An at-the-money swaption volatility surface is shown in Fig. 2.4.

Caps and floors are quoted by a single implied Black '76 volatility, which, when entered in each individual caplet formula, yields the correct cap price as sum of the caplet prices. Note that when constructing a consistent caplet volatility surface the same caplets can enter several caps. Therefore the caplet surface must be bootstrapped from short maturities to long maturities in such a way that caplets entering different caps use the same implied volatility in all caps.

2.3.2.3 Models of Hull and White

The most simple interest rate models describe the short rate as a single stochastic differential equation. All other rates can be directly derived from it. However, this implies full correlation between rates with different maturities. Here we consider the *one-factor Hull-White short rate model*, which is given by

$$\mathrm{d}r_t = (\theta(t) - \kappa r_t)\,\mathrm{d}t + \sigma\,\mathrm{d}W_t\,, \quad r_0 = r_f(0)\,,$$

with constant *mean reversion* κ and *volatility* σ. The time-dependent parameter $\theta(\cdot)$ is determined by the initial term structure of zero rates $r_z(\cdot)$ as well as by the model parameters $\mathcal{M} = \{\kappa, \sigma\}$.

In this model, today's price of a caplet on the forward rate from T to $S > T$ with strike c is given by

$$X_{caplet} = -(1 + c(S - T))D(S)\Phi(-d_+) + D(T)\Phi(-d_-), \tag{2.4}$$

where

$$d_\pm = \frac{\log(D(S)) + \log((1 + c(S - T))) + \left(r_z(T) \pm \Sigma(T,S)^2/2\right)T}{\Sigma(T,S)\sqrt{T}}, \tag{2.5}$$

$$\Sigma(T,S) = \sigma(S - T)B(S - T, \kappa)\sqrt{B(T, 2\kappa)}, \tag{2.6}$$

$$B(T, \kappa) = \frac{1 - \exp(-\kappa T)}{\kappa T}. \tag{2.7}$$

We now determine the price of a swaption on the swap rate paying a fixed coupon c at times $T < S_1 < S_2 < \cdots < S_n = S$, where T is the maturity of the option. Let c_i be the fixed cash flow paid at time S_i, that is $c_i = c(S_i - S_{i-1})$ for $i = 1, \ldots, n-1$ and $c_n = 1 + c(S_n - S_{n-1})$. The swaption price in the one-factor Hull-White is given by

$$X_{swaption} = \sum_{i=1}^{n} c_i \left(D(t_i)\Phi(-d_{+,i}) + X_i(r^*)D(t)\Phi(-d_{-,i})\right),$$

with

$$d_{\pm,i} = \frac{\log(D(S_i)) - \log(X_i(r^*)) + \left(r_z(T) \pm \Sigma(T,S_i)^2/2\right)T}{\Sigma(T,S_i)\sqrt{T}},$$

$$A(T,S) = \frac{D(S)}{D(T)} \exp\left(r_f(T)(S - T)B(S - T, \kappa) - \Sigma(T,S)^2/2T\right),$$

$$X_i(r) = A(T,S_i)\exp(-(S_i - T)B(T,\kappa)r),$$

and $\Sigma(\cdot,\cdot)$ and $B(\cdot,\cdot)$ given in Eqs. (2.6) and (2.7). Moreover, r^* is chosen such that

$$\sum_{i=1}^{n} c_i X_i(r^*) = 1. \tag{2.8}$$

As in the case of caplets, we recover a formula that is quite similar to the Black-Scholes formula. In fact, the option on the swap rate can be written as sum of options with strike $X_i(r^*)$. Thus, calibrating the model to swaption prices adds no additional complexity but solving Eq. (2.8). Fortunately, it can be shown, that the left-hand-side of Eq. (2.8) is monotone in r^* and its solution is unique. Therefore a plain bisection search can be used to obtain the root.

The derivation of the formula above exhibits the most severe shortcoming of one-factor models. Since, regardless of the random state of the economy at time T, either all expected $D(S_i)$ are larger than the strike $X_i(r^*)$ or none, the option on the sum of the cash flows c_i can be written as a sum of options on the individual c_i. This is very unrealistic and is due to the fact that short and long term interest rates move in common.

Therefore, multi-factor interest rate models are preferred to one-factor ones. These models can decorrelate long and short term interest rates as well as alter the curvature of the interest rate curve.

A *two-factor Hull-White* is given by the equations

$$r(t) = x_1(t) + x_2(t) + \theta(t)\,,$$
$$\mathrm{d}x_i(t) = -\kappa_i x_i(t)\,\mathrm{d}t + \sigma_i \mathrm{d}W_i(t)\,,\quad x_i(0) = 0\,,$$

with the constant *mean reversion* parameters κ_i and *volatilities* σ_i for $i = 1,2$. Moreover the Brownian motions $W_1(\cdot)$ and $W_2(\cdot)$ are correlated with *correlation* parameter ρ. Again, the time dependent parameter $\theta(\cdot)$ is determined by the initial term structure of the zero rates, hence $\mathcal{M} = \{\kappa_1, \kappa_2, \sigma_1, \sigma_2, \rho\}$.

Compared to the one-factor model, the price of a caplet solely changes in the volatility function $\Sigma(\cdot,\cdot)$. Hence, the price can be determined using Eq. (2.4), where $d_\pm$ is given in Eq. (2.5),

$$\Sigma(T,S) = \frac{\sqrt{C_{11}(T,S) + 2C_{12}(T,S) + C_{22}(T,S)}}{S-T}\,,$$
$$C_{ii}(T,S) = \sigma_i^2 B(S-T,\kappa_i)^2 B(T,2\kappa_i)\,,$$
$$C_{12}(T,S) = \rho\sigma_1\sigma_2 B(S-T,\kappa_1) B(S-T,\kappa_2) B(T,\kappa_1+\kappa_2)$$

and $B(\cdot,\cdot)$ is given in Eq. (2.7).

The price of a swaption becomes a little more involved. Again, let c_i be the cash flow paid at time S_i, that is $c_i = c(S_i - S_{i-1})$ for $i = 1,\ldots,n-1$ and $c_n = 1 + c(S_n - S_{n-1})$. Then,

$$X_{swaption} = \frac{D(T)}{\sqrt{2\pi}} \int_{-\infty}^{+\infty} \exp\left(-\frac{x_1^2}{2}\right) I(x_1)\,\mathrm{d}x_1\,, \tag{2.9}$$

where

$$I(x_1) = \Phi(-h_0(x_1,x_2^*))$$
$$- \sum_{i=1}^{n} c_i \lambda(S_i) \exp(x_1 g_1(S_i))\, \Phi(-h_0(x_1,x_2^*) + h_1(S_i))\,.$$

Moreover, for fixed x_1, x_2^* is the solution of

$$\sum_{i=1}^{n} c_i \lambda(S_i) \exp(x_1 g_2(S_i) + x_2^* g_3(S_i)) = 1 . \tag{2.10}$$

As before, for fixed x_1, it can be shown that the left hand side of Eq. (2.10) is monotone in x_2^* and the solution can be found easily. Further, $h_0(\cdot,\cdot)$ is linear in its arguments and the remaining functions $\lambda(\cdot)$, $h_1(\cdot)$, $g_1(\cdot)$, $g_2(\cdot)$, $g_3(\cdot)$ only depend on model parameters.

The main problem in the calculation of the swaption price remains the calculation of the integral, since Eq. (2.10) has to be solved for each x_1 where the integrand is evaluated. Figure 2.5 shows the integrand for an at-the-money swaption with maturity $T = 1$ on a swap with $S = 3$ calculated using several model parameters, which are given in Table 2.1.

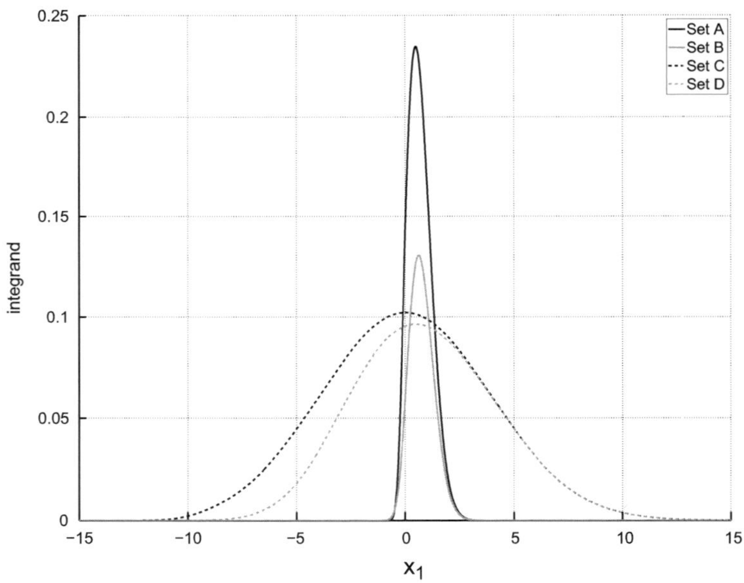

Fig. 2.5 Differently scaled integrands arising when computing the price of a swaption in the two-factor Hull-White model

Table 2.1 Model parameters corresponding to Fig. 2.5

$\mathcal{M}$	σ_1	σ_2	κ_1	κ_2	ρ
Set A	4.0	2.0	0.05	0.01	−0.75
Set B	4.0	2.0	0.05	0.01	0.75
Set C	1.0	0.5	0.50	0.75	−0.75
Set D	1.0	0.5	0.50	0.75	0.75

2.3.2.4 SABR Model

As in the Black-Scholes for equities, the single most advantage of the Black '76 is the dependence on only observable parameters as well as on a single volatility. However, it turned out that this is grossly inadequate due to the smile effect observed in the interest rate market.

Intuitively, introducing a deterministic volatility parameter that depends on the strike and the maturity might solve this issue. But the resulting model as well as its calibration is very unstable.

Another model with far better properties, which can be used to determine a single volatility to plug into the Black '76 formula, is the *SABR*, where the initial letters stand for **S**tochastic **A**lpha, **B**eta and **R**ho. It was introduced in [12] and models an interest rate F, i.e. a forward or swap rate for a given maturity, jointly with its corresponding stochastic volatility as

$$
\begin{aligned}
\mathrm{d}F_t &= \alpha_t F_t^{\beta}\,\mathrm{d}W_t\,, \quad F_0 = f\,,\\
\mathrm{d}\alpha_t &= \nu\alpha_t\,\mathrm{d}\tilde{W}_t\,, \quad \alpha_0 = \alpha\,,
\end{aligned}
$$

where the Brownian motions W_t and $\tilde{W}_t$ are correlated with ρ, f is the currently observed forward or swap rate, α is the initial volatility, and β determines the distribution of the rate process.

For a given strike and maturity, there exist approximations that directly translate the calibrated parameters into the corresponding Black '76 volatility.

The calibration of the model differs from the previous models since β typically is either fixed in advance or is determined approximatively for a fixed maturity T from the at-the-money Black '76 volatilities. After fixing β the remaining parameters are calibrated for a fixed maturity time slice to incorporate the smile effect.

2.4 Numerical Challenges

So far, we have seen several different models and their closed pricing formulas for the most important products used for calibration. In this section we want to extract numerical problems and bottlenecks that deserve most attention.

When calibrating a model to market data, the objective function has to be called for each particular set of model parameters. This subsequently results in the calculation of the model prices for all relevant products. Here a set of roughly about 200 products is not uncommon in financial applications. So, in total, it is quite obvious that most of the computation time is spent when computing the product prices. However, it is far from being simple to accelerate the computation.

The remaining time, typically not more than 2–5 %, results due to

- Aggregating these prices as specified in the objective function and
- Let the optimization algorithm choose a new set of model parameters.

Still, depending on the quality of the optimization method, more or less product prices have to be calculated. For example there exist various modifications of the downhill simplex algorithm, where some of these use first order derivatives to detect the direction of the steepest descent. This results in additional function evaluations. Also the adaptive simulated annealing algorithm can be combined with a local optimizer. Though this might lead to more function evaluations in the first place, a satisfying local minimum might be found faster. However, it is a priori not clear which effect prevails.

Recommendation: Use a fast optimization algorithm.

In most of the models we considered, the cumulative distribution function of the standard normal distribution $\Phi(\cdot)$ arises. Here, there exist several fast rational approximation algorithms, such as the one given in [8]. Note that it is very important, to have precise values also for small and large arguments, which in general corresponds to the use of deep in-the-money or out-of-the-money options.

Recommendation: Use a fast approximation of the cumulative distribution function.

It is also most useful when speeding up calibration to pre-calculate terms that only depend on the observable market parameters such as $\exp(-rT)$. Further, when evaluating the objective function, terms solely depending on market and model parameters can be pre-calculated, since product parameters are only changed here.

Recommendation: Use pre-calculated terms whenever possible.

The calculation of the swaption price in the Hull-White models requires a root finding algorithm, see Eqs. (2.8) and (2.10). At least in our examples, the functions involved are smooth and monotone in the variable, so that a simple bisection search usually gives satisfying results. Since in this case the derivative can be obtained analytically, the more advanced Newton-Raphson method can also be profitably used. However, it should be combined with a simple bisection algorithm to capture cases of very steep functions.

Recommendation: Use a fast root finding algorithm.

For calculating Eq. (2.3), the evaluation of the cumulative distribution function is replaced by the integral of a characteristic function. Here, the integrand contains a singularity at the lower bound while the upper bound is infinity and it is a priori not clear where to truncate the integration domain. In fact, the integrand is smooth and the integral can be determined by using a suitable quadrature algorithm, for example Gauss-Legendre integration.

There are as well other methods to get the improper integral. For example, adaptive methods could be used, or in [7] a fast Fourier transform method is applied.

The integration problem is also inherent in the pricing formula for the swaption in the two factor Hull-White model, see Eq. (2.9).

Recommendation: Use a suitable integration method.

In order to determine the price of a European exercise feature call in the general model, see Eq. (2.3), the logarithm of a complex number has to be calculated. As distinct from real arguments, this number is not unique. Thus, evaluating the complex logarithm might result in discontinuities, which distort the result, see for instance [10].

Recommendation: Use complex logarithm carefully.

As we already pointed out, it is often worthwhile to sacrifice accuracy for speed. For example, in many situations it is not necessary to stick to double precision. The same principle holds of course in all the numerical approximation steps discussed so far.

Recommendation: Use only as much accuracy as needed.

Finally, we note that ideally a calibration implementation takes into account all of the above points, but still allows the user to be flexible enough to decide which algorithm, accuracy, method, etc. is used in the particular application. Such a methodology is given in [6] for the Heston case.

References

1. Bakshi, G., Cao, C., Chen, Z.: Empirical performance of alternative option pricing models. J. Financ. **52**(5), 2003–2049 (1997)
2. Bates, D.S.: Jumps and stochastic volatility: exchange rate processes implicit in deutsche mark options. Rev. Financ. Stud. **9**(1), 69–107 (1996)
3. Black, F.: The pricing of commodity contracts. J. Financ. Econ. **3**(1), 167–179 (1976)
4. Black, F., Scholes, M.: The pricing of options and corporate liabilities. J. Political Econ. **81**, 637–654 (1973)
5. Brigo, D., Mercurio, F.: Interest Rate Models – Theory and Practice: With Smile, Inflation and Credit. Springer Finance, 2nd edn. Springer, Berlin/New York (2006)
6. Brugger, C., Liu, G., de Schryver, C., Wehn, N.: A systematic methodology for analyzing closed-form Heston pricer regarding their accuracy and runtime. In: Proceedings of the 7th Workshop on High Performance Computational Finance, pp. 9–16. IEEE (2014), New Orleans, Louisiana, USA
7. Carr, P., Madan, D.: Option valuation using the fast Fourier transform. J. Comput. Financ. **2**, 61–73 (1999)
8. Cody, W.J.: Rational Chebyshev approximations for the error function. Math. Comput. **23**(107), 631–637 (1969)
9. Dimitroff, G., Lorenz, S., Szimayer, A.: A parsimonious multi-asset Heston model: calibration and derivative pricing. Int. J. Theor. Appl. Financ. **14**(8), 1299–1333 (2011)
10. Gatheral, J.: The Volatility Surface: A Practitioner's Guide. Wiley, Hoboken (2006)
11. Hadamard, J.: Lectures on Cauchy's Problem in Linear Partial Differential Equations. Yale University Press, New Haven (1923)
12. Hagan, P., Kumar, D., Lesniewski, A., Woodward, D.: Managing smile risk. Wilmott Mag. 84–108 (2002)
13. Heston, S.L.: A closed-form solution for options with stochastic volatility with applications to bond and currency options. Rev. Financ. Stud. **6**(2), 327–343 (1993)
14. Hirsa, A., Neftci, S.N.: An Introduction to the Mathematics of Financial Derivatives. Academic, London (2013)
15. Korn, R., Korn, E.: Option Pricing and Portfolio Optimization: Modern Methods of Financial Mathematics. American Mathematical Society, Providence (2001)
16. Korn, R., Korn, E., Kroisandt, G.: Monte Carlo Methods and Models in Finance and Insurance. Chapman & Hall/CRC Financial Mathematics Series. CRC, London (2010)
17. Lindström, E., Ströjby, J., Brodén, M., Wiktorsson, M., Holst, J.: Sequential calibration of options. Comput. Stat. Data Anal. **52**(6), 2877–2891 (2008)
18. Merton, R.C.: Option Pricing when underlying Stock Returns are discontinuous. J. Financ. Econ. **3**(1–2), 125–144 (1976)
19. Mikhailov, S., Nögel, U.: Heston's stochastic volatility model: implementation, calibration and some extensions. Wilmott Mag. (2003) 74–79
20. Schoutens, W., Simons, E., Tistaert, J.: A perfect calibration! Now what? Wilmott Mag. 66–78 (2004)
21. Tikhonov, A.N., Arsenin, V.Y.: Solutions of Ill-Posed Problems. Wiley, New York (1977)

Chapter 3
Comparative Study of Acceleration Platforms for Heston's Stochastic Volatility Model

Christos Delivorias

Abstract We present a comparative insight of the performance of implementations of the Heston stochastic volatility model on different acceleration platforms. Our implementation of this model uses Quasi-random variates, using the Numerical Algorithms Group (NAG) random number library to reduce the simulation variance, as well as Leif Andersen's Quadratic Exponential discretisation scheme. The implementation of the model was in Matlab, which was then ported for Graphics Processor Units (GPUs), and then Techila platforms. The Field Programmable Gate Array (FPGA) code was based on C++. The model was tested against a 2.3 GHz Intel Core i5 Central Processing Unit (CPU), a Techila grid server hosted on Microsoft's Azure cloud, a GPU node hosted by Boston Ltd, and an FPGA node hosted by Maxeler Technologies Ltd. Temporal data was collected and compared against the CPU baseline, to provide quantifiable acceleration benefits for all the platforms.

3.1 Introduction

The computational complexity requirements of calculating financial derivatives' prices and risk values have increased dramatically following the most recent financial crisis in 2008. The requirements of counterparty risk assessment have also introduced taxing calculations that impose an additional levy in computational time. There is a two-fold need for this rapid calculation; first the ability to price the option value on a given underlying is essential in a fast-paced market environment that relies on dynamic hedging for portfolio immunisation on large books. The second use is in a relatively new sector of the financial market that deals with High-Frequency Trading (HFT). This sector relies on extremely fast computations in order to make algorithmic decisions based on the current market information. This chapter

C. Delivorias (✉)
Aberdeen Asset Management, 40 Princes Street, Edinburgh EH2 2BY, UK
e-mail: Christos.Delivorias@aberdeen-asset.com

C. De Schryver (ed.), *FPGA Based Accelerators for Financial Applications*,
DOI 10.1007/978-3-319-15407-7_3

will focus on the pricing of options rather than the aspects of HFT. The informed reader could care to look into some of the negative aspects of HFT as explained by [3].

This work was a joint project between the University of Edinburgh and Scottish Widows Investment Partnership (SWIP)[1] and aimed at exploring the possibilities in accelerating financial engineering models, with real life applications in the markets. The goal was to evaluate the same model on different platforms, and assess the benefits of acceleration that each platform provided.

This chapter is organised as follows. Section 3.2 introduces the model used in order to evaluate the computational performance of a well known model of the evolutions of equities prices. This model has a known analytical solution which can serve as a cross-check of correctness. This kind of model can have a numerical solution simulated via Markov Chain Monte Carlo (MCMC) simulation, which is explained in Sect. 3.3. The variance reduction using quasi-random numbers, as well as the discretisation scheme are also expanded in this section. Section 3.4 goes into more details on the acceleration platforms of FPGA, GPU, and the Techila Cloud. Section 3.5 details the efficiency and the accuracy of the implementation of the Heston model in Matlab, and finally Sect. 3.6 presents the experimental results and the conclusion.

3.2 Heston's Stochastic Volatility Model

The Heston model extends the Black-Scholes (BS) model by taking into account a stochastic volatility that is mean-reverting and is also correlated with the asset price. It assumes that both the asset price and its volatility are determined by a joint stochastic process.

Definition 3.2.1. The Heston model that defines the price of the underlying S_t and its stochastic volatility v_t at time t is given by

$$\begin{aligned} dS_t &= \mu S_t dt + \sqrt{v_t} S_t dW_t^S, \\ dv_t &= \kappa(\theta - v_t)dt + \sigma\sqrt{v_t} dW_t^v, \\ Cov\langle dW_t^S, dW_t^v \rangle &= \rho dt, \end{aligned} \tag{3.1}$$

where the two standard Brownian Motion (BM)(W^S, W^v) are correlated by a coefficient ρ, κ is the rate of reversion of the volatility to θ – the long-term variance –, σ is the volatility of volatility, and μ is the expected rate of return of the asset.

[1] now Aberdeen Asset Management.

The Heston model extends the BS model by providing a model that has a dynamic stochastic volatility, as described in Eq. (3.1). This model has a semi-analytic formula that can be exploited to derive an integral solution. Additionally if the [5] condition is upheld, this process will produce strictly positive volatility with probability 1; this is described in Lemma 3.2.1.

Lemma 3.2.1. *If the parameters of the Heston model obey the condition*

$$2\kappa\theta \geq \sigma^2,$$

then the stochastic process v_t *will produce volatility such that* $\Pr(v_t > 0) = 1$, *since the upward drift is large enough to strongly reflect away from the origin.*

The stochastic nature of this model provides two advantages for evaluating it with a Monte Carlo simulation. The first is that in its original form, it possesses a closed-form analytical solution, that can be used to evaluate the bias of the numerical solution. The second benefit is that its path-dependent nature can accommodate more complex path behaviors, e.g. barrier options, and Affine Jump Diffusion (AJD).

3.3 Quasi-Monte Carlo Simulation

The reference to Monte Carlo (MC), is due to the homonym city's affiliation with games of chance. The premise of chance is utilised within the simulation in order to provide a random sample from the overall probability space. If the random sample is as truly random as possible, then the random sample is taken as an estimate across the entire probability space. The law of large numbers guarantees [8] that this estimation will converge to the true likelihood as the number of random draws tends to $+\infty$. Given a certain number of random draws, the likely magnitude of the error can be derived by the central limit theorem.

The Feynman-Kač theorem is the connective tissue between the Partial Differential Equation (PDE) form of the stochastic model, and its approximation by a Monte Carlo simulation. By this theorem it is possible to approximate a certain form Stochastic Differential Equation (SDE), by simulating random paths and deriving the expectation of them as the solution of the original PDE.

As an example we can return to the BS model, where the price of the option depends on the expected value of the payoff. In order to calculate the expected value of f it is possible to run a MC simulation with N paths,[2] in order to approximate the actual price of the call option c with the simulated price $\hat{c}_N$,

[2]A path in this context is a discrete time-series of prices for the option, one for each discrete quantum of time.

$$\hat{c}_N = \frac{e^{-rT}}{N} \sum_{i=1}^{N} \left(S_0 e^{(r-\frac{1}{2}\sigma^2)T+\sigma\sqrt{T}N^i_{(0,1)}} - K \right)^+, \tag{3.2}$$

where r is the risk-free interest rate, T is the time to maturity of the option, σ is the volatility, K is the strike price at maturity date T, S_0 is the spot price at $t = 0$, and $N^i_{(0,1)}$ are Gaussian Random Variables (RVs). By the strong law of large numbers we have,

$$\Pr(\hat{c}_N \to c) = 1, \text{ as } \quad N \to \infty. \tag{3.3}$$

3.3.1 Variance Reduction with Quasi-Random Numbers

There are two major avenues to take in order to reduce variance in a MC simulation, one is to take advantage of specific features of the problem domain to adjust or correct simulation outputs, the other by directly reducing the variability of the simulation inputs. In this section we'll introduce the variance reduction process of quasi-random numbers. This is a procedure to reduce the variance of the simulation by sampling from variates of lower variance. Such numbers can be sampled from the so called "low discrepancy" sequences. A sequence's discrepancy is a measure of its uniformity and is defined by Definition 3.3.1 (see [6]).

Definition 3.3.1. Given a set of points $\mathbf{x^1}, \mathbf{x^2}, \cdots, \mathbf{x^N} \in I^S$ s-dimensional unit cube and a subset $G < I^S$, define the *counting function* $S_N(G)$ as the number of points $x^i \in G$. For each $x = (x_1, x_2, \cdots, x_S) \in I^S$, let G_x be the rectangular s-dimensional region,

$$G_x = [0, x_1) \times [0, x_2) \times \cdots \times [0, x_S),$$

with volume $x^1 x^2 \cdots x^N$. Then the *discrepancy* of the points $x^1, x^2, \cdots, x^N$ is given by,

$$D^*_N(x^1, x^2, \cdots, x^N) = \sup_{x \in I^S} |S_N(G_x) - N \cdot (x_1 \cdot x_2 \cdots x_S)|.$$

The discrepancy value of the distribution compares the sample points found in the volume of a multi-dimensional space, against the points that should be in that volume provided it was a uniform distribution.

There are a few sequences that are used to generate quasi-random variates. The NAG libraries provide three sequence generators. The [7, 9], and [4] sequences are implemented in MATLAB with the NAG functions g05yl and g05ym (Fig. 3.1).

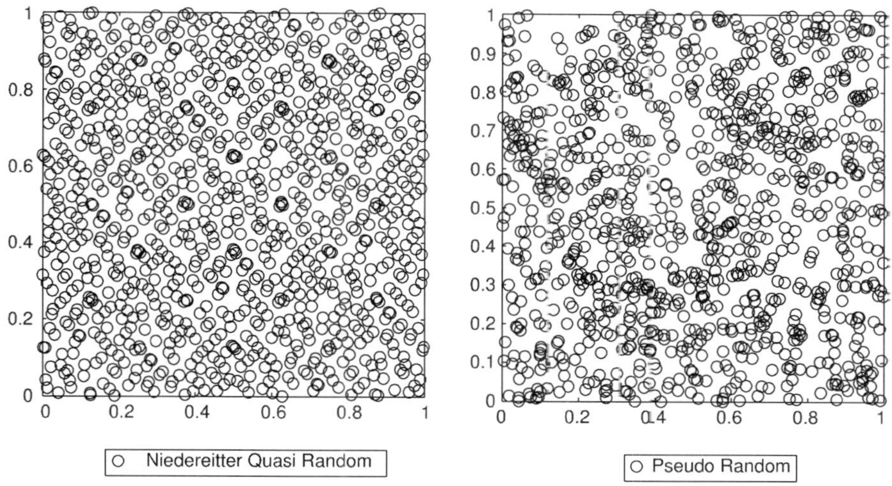

Fig. 3.1 The Niedereitter quasi-random uniform variates on the *left*, versus the pseudo-random unifrom variates on the *right*. The NAG library was used for the quasi-random variates

3.3.2 Descritisation Scheme

In 2005 [1] proposed a new scheme to discretise the stochastic volatility and the price of an underlying asset. This scheme takes advantage of the fact that a non-central χ^2 sampled variate can be approximated by a related distribution, that's moment-matched to the conditional first and second moments of the non-central χ^2 distribution.

As Andersen points out, the cubic transformation of the Normal RV although a more accurate representation of the distribution closer to 0, it introduces negative values of variance. Thus the quadratic representation is adopted as a special case for when we have low values of $V(t)$. Therefore when $V(t)$ is sufficiently large, we get,

$$\hat{V}(t+\Delta t) = a(b+Z_V)^2, \tag{3.4}$$

where Z_V is an $N_{(0,1)}$ Gaussian RV, and a, b scalars that will be determined by moment-matching. Now for the complementary low values of $V(t)$ the distribution can – asymptotically – be approximated by,

$$\mathbb{P}(\hat{V}(t+\Delta t) \in [x, x+\Delta t]) \approx (p\delta(0) + \beta(1-p)e^{-\beta x})dx, \qquad x \geq 0, \tag{3.5}$$

where δ is the, Dirac delta-function, strongly reflective at 0, and p and β are positive scalars to be calculated. The scalars a, b, p, β depend on the parameters of the Heston model and the time granulation Δt, and will be calculated by moment-matching the exact distribution.

To sample from these distributions there are two distributions to take into account:

- Sample from the normal $N_{(0,1)}$ Gaussian RV and calculate $\hat{V}(t+\Delta t)$ from Eq. (3.4).
- To sample for the small values of V the inverse of Eq. (3.5) will be used. The inverse of the distribution function is,

$$\Psi^{-1}(u) = \Psi^{-1}(u;p,\beta) = \begin{cases} 0 & \text{if } 0 \leq u \leq p, \\ \beta^{-1} \ln\left(\frac{1-p}{1-u}\right) & \text{if } p \leq u \leq 1. \end{cases} \tag{3.6}$$

The value of V can then be sampled from

$$\hat{V}(t+\Delta t) = \Psi^{-1}(U_V;p,\beta), \tag{3.7}$$

where U_V is a uniform RV. The rule on deciding which discretisation of V to use depends on the non-centrality of the distribution, and can be triaged based on the value of Ψ. The value of Ψ is,

$$\Psi := \frac{s^2}{m^2} = \frac{\frac{\hat{V}(t)\xi^2 e^{-\kappa\Delta t}}{\kappa}(1-e^{-\kappa\Delta t}) + \frac{\theta\xi^2}{2\kappa}(1-e^{-\kappa\Delta t})^2}{(\theta + (\hat{V}(t)-\theta)e^{-\kappa\Delta t})^2}, \tag{3.8}$$

where m, s^2 are the conditional mean and variance of the exact distribution we are matching. What Andersen showed was that the quadratic scheme of Eq. 3.4 can only be moment-matched for $\Psi \leq 2$ and similarly the exponential scheme of Eq. 3.7 can only be moment-matched for $\Psi \geq 1$. It follows then, that there is an overlap interval for $\Psi \in [1,2]$ where the two schemes overlap. Appropriately Andersen chooses the midpoint of this interval as the cut-off point between the schemes; thus the cut-off $\Psi_c = 1.5$.

Since we've defined the discretisation process for the Quadratic Exponential (QE) scheme, with Eqs. 3.4 and 3.7, and the cut-off discriminator, what is left is to calculate the remaining parameters a, b, p, β for each case. The algorithm for this process is detailed in Algorithm 1.

3.4 Implementations of the Algorithm on Different Platforms

3.4.1 CPU Baseline Model in Matlab

The algorithm described in Algorithm 1 is implemented in MATLAB, and is used for numerical comparisons of acceleration. It will be used hencforth as the baseline calculation cost. All the implementations that are explained below will be compared against this implementation.

Algorithm 1 QE variance reduction

Input: The present value for the variance, $\hat{V}(t)$
Output: The value for the variance in the subsequent time-step, $\hat{V}(t+\Delta t)$

```
Compute m ← θ + (V̂(t) − θ)e^{−κΔt}
Compute s² ← (V̂(t)ξ²e^{−κΔt}/κ)(1 − e^{−κΔt}) + (θξ²/2κ)(1 − e^{−κΔt})²
Compute Ψ ← m²/s²
if Ψ ≤ Ψ_c then
    Compute a ← m/(1+b²)
    Compute b ← 2Ψ^{−1} − 1 + √(2Ψ^{−1}) √(2Ψ^{−1} − 1)
    Generate Normal random variate Z_V
    return V̂(t+Δt) ← a(b+Z_V)²
else
    Compute p ← (Ψ−1)/(Ψ+1) ∈ [0,1)
    Compute β ← (1−p)/m
    Generate Uniform random variate U_V
    if 0 ≤ U_V ≤ p then
        return V̂(t+Δt) ← 0
    else
        return V̂(t+Δt) ← β^{−1} ln((1−p)/(1−U_V))
    end if
end if
```

3.4.2 *Dataflow Programming on FPGAs*

This approach to accelerate code, takes advantage of the fact that most of the time the CPU is busy figuring out the scheduling of the instructions and the branch prediction of the program. The purpose of an FPGA is to provide a customisable "field-programmable" chip that can be optimised to perform calculation for a specific problem domain. This is achieved by allowing the logic blocks on the chip to be re-wirable. This way, even after a board has been purchased, it can still be re-wired and re-purposed.

This re-wiring is achieved via a Hardware Description Language (HDL). This language offers the ability to interconnect the logic blocks into different combinations, cater for complex combinatorial functions, and also manage the on-chip memory.[3]

3.4.2.1 FPGA Versus Intel Multi-core

Thus far, CPU developments have adhered to Moore's law.[4] This assertion has been adhered to up to this point, however limitations of the scale that transistors can

[3]This is mostly in the form of either flip-flops, or more structured blocks of memory.

[4]Moore postulated in 1965 that the transistor density of Intel's semiconductor chips should double roughly every 18 months.

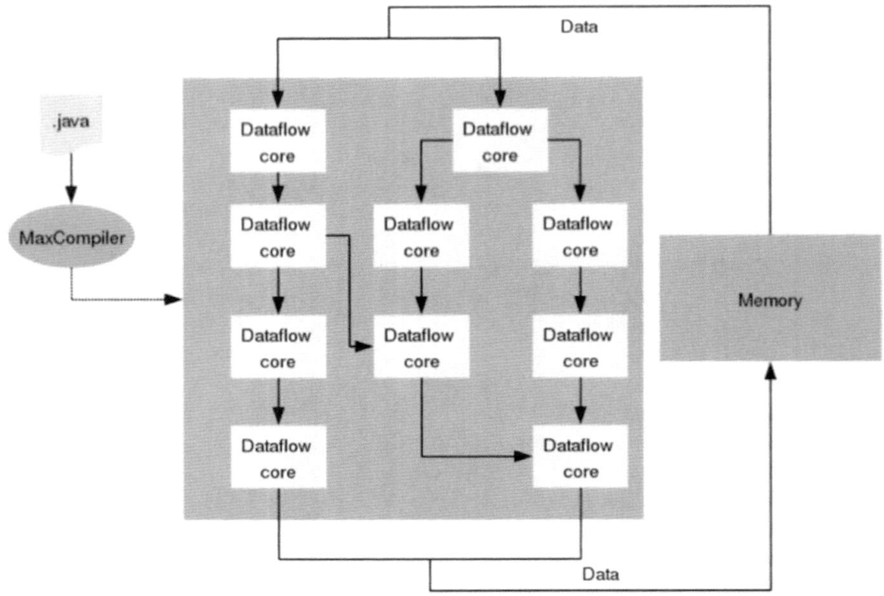

Fig. 3.2 The FPGA architecture as implemented by Maxeler Technologies. The MaxCompiler constructs the DataFlow tree which defines the circuit architecture on the FPGA chip. From then on data from memory gets piped into the different DataFlow cores until it exits the calculation pipe and is committed to memory (Photo used by permission of Maxeler Technologies)

achieve coupled with the issue of power consumption increasing the more transistors are fitted in a chip, casts doubt into its future accuracy. However, what might actually happen instead is that the transistors will double their count every 18 months, mainly because the number of cores in each chip will double. What this means is that the Operating System (OS) will be able to take advantage of multiple cores within a CPU and via efficient scheduling maximise performance while minimising power consumption.

The benefits of the Intel multi-core approach is that the current programming paradigm can abide, and most existing code could be relatively easily – compared to more exotic implementation on GPU and FPGA – ported to the many-core architecture.

The FPGA can leverage two advantages over the CPU approach. First it has more silicon dedicated to calculations compared to the CPU. And second it relies on the DataFlow architecture to do away with the taxing aspects of instruction scheduling and branch-predictions. This way the calculations pipeline is always full and a result is calculated every clock cycle (see Fig. 3.2 for more details). On the other hand the CPU, as shown in Fig. 3.3, needs to handle concurrent threads vying for their turn on the Arithmetic Logic Unit (ALU) in order to progress their calculation status.

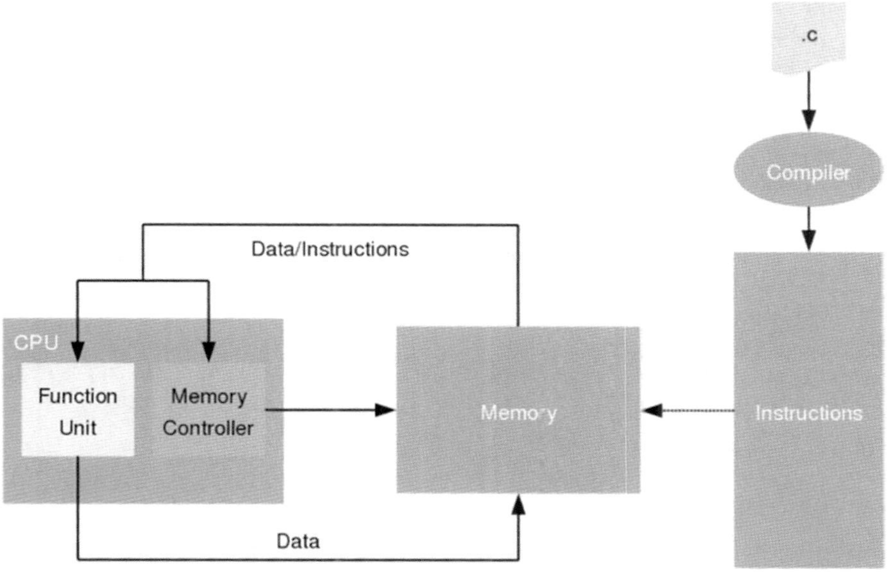

Fig. 3.3 The process architecture on a CPU where the ALU is referred as the Function Unit. Data has to be moved into the Funtion Unit form memory and then moved back into memory for storage (Photo used by permission of Maxeler Technologies)

3.4.2.2 Application to the Heston Model

The implementation of Heston's stochastic volatility model has two parts. First the code that is run on the host, and second the code that defines the circuit architecture of the FPGA and performs the necessary calculations (Fig. 3.4). Since only repetitive calculations can benefit from the DataFlow architecture there are certain elements that need to run on the host and others on the FPGA card. Maxeler Technologies use the nomenclature of a kernel and a manager. The kernel comprises of a set of calculations that produce a distinct result, e.g. a 3-value moving average. The manager's responsibility is to instantiate and administer the life cycle and functions of each kernel that is assigned to it. For this implementation the manager creates numerous pipes within a given MaxCard[5] to handle different operations. The more pipes that can be filled into the available silicon the better the overall performance of the FPGA. The manager is responsible to create and to populate the pipes with kernels to generate random variates from the Gamma distribution, and also kernels that calculate the next values for the variance and the price of the underlying. Once all the prices of the underlying have been generated for every timestep, the results are aggregated back on the host's CPU.

[5]Latest models of Maxeler's FPGA cards provide an ever increasing number of resources on-board the chip.

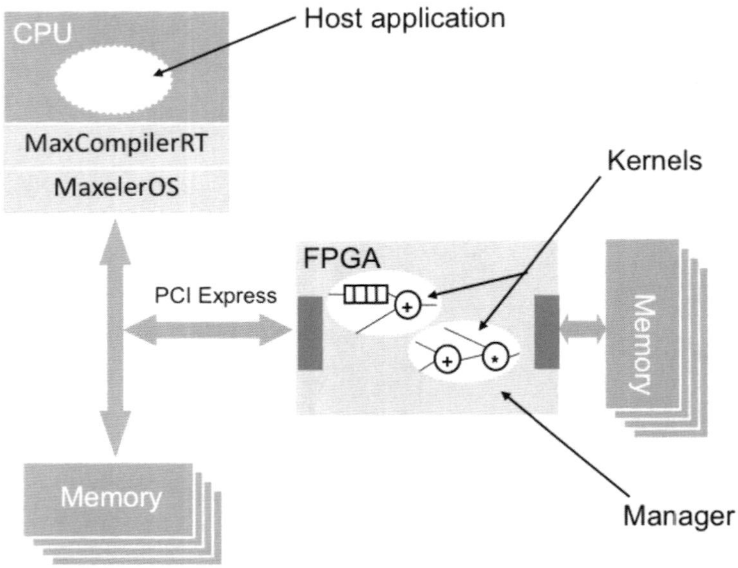

Fig. 3.4 This figure illustrates how code interacts between the host CPU and the FPGA kernels (Photo used by permission of Maxeler Technologies)

3.4.3 Implementation on GPGPUs

The GPU is a highly specialised parallel processor for accelerating graphical computations. The first GPU card that consolidated the entire graphics processing pipeline off-chip for 3D and 2D acceleration was the NVIDIA GeForce256, which was released in 1999 by NVIDIA. The main function of the card was to render a graphical image on the GPU chip to turn a model into an image. The strength of this chip was its specialised computational capacity for floating-point arithmetic.

3.4.3.1 MATLAB Implementation of the Heston Model for the General Purpose Graphics Processor Unit (GPGPU)

Using the existing Heston model, as implemented in Matlab, it is possible to augment and differentiate this model to be able to take advantage of Matlab's Parallel Computing Toolbox to deploy the existing model on multiple GPU. To do this there are several ways to take advantage of a single GPU card, multiple cards per node, and finally multiple nodes in a grid.

The easiest form of access to the GPU is via the overriden functions within the toolboxes acted on a specific class of the *gpuArray*. This method will effectively transfer data over the Peripheral Component Interconnect Express (PCIe) card onto the GPU memory. Methods of transformation, e.g. addition, subtraction, multiplication, etc., would then be applied on the GPU and not on the CPU. The calculated data is then aggregated from the GPU into the CPU memory via the *gather* method. This high-level implementation is useful when quick matrix calculation without significant need for minute control is required. However when the problem dimensionality increases, fine-grained control is required in order to increase the performance of the acceleration. Specifically the problem needs to be properly structured in order to saturate the GPU pipes.

For this reason it is important to distinguish between independent calculations in the model and take advantage of multiple GPU cards in order to accelerate the calculations. The Monte Carlo simulation lends itself quite fittingly to this, because the individual path calculations are independent of each other, hence can be executed on multiple GPU cores and nodes. This process is achieved by executing a *parfor* loop, whose constituent loops are distributed on multiple GPU and even across multiple nodes.

In order to access multiple GPU cards it is necessary to bind them to specific Matlab workers. This is achieved via the *spmd.. end* structure. The following Listing 3.1 shows this in more detail.

```
N=1000;
A = gpuArray(magic(N));

spmd
  gpuDevice(labindex);
end

parfor ix=1:N % Distribute for loop to workers
  % Do the GPU calc
  X = myGPUFunction(ix,A);
  % Gather data
  Xtotal(ix,:) = gather(X);
end
```

Listing 3.1 Example of multiple GPU assignment to distribute a Matlab function to multiple workers

3.4.4 Implementation on the Techila Cloud

Techila is based Finland and deals in the domain of distributed computational problems. The latter are characterised by two kinds of problems. The first are traditional parallel problems where the individual distributed processes are run on different hosts, but require some information to be passed between them. Problems like fluid dynamics or finite element models are examples of parallel problems (Fig. 3.5).

On the other hand are the embarrassingly parallel problems. These kinds of problems do not require communication between the distributed calculations, since

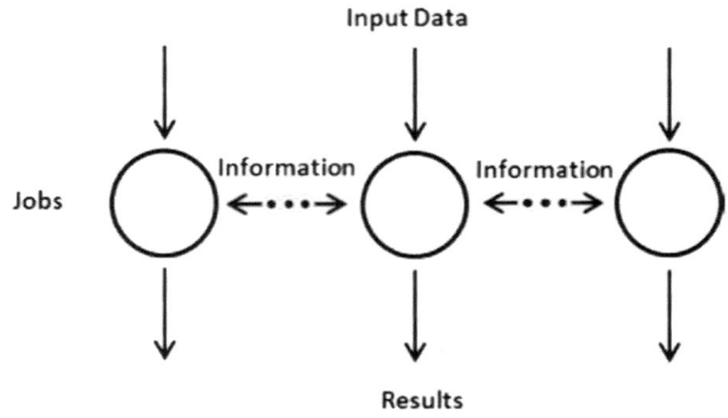

Fig. 3.5 Parallel problem. Each computational job accepts input data and uses this input data to compute results. Communication between jobs is required in order for the individual jobs to be completed (Photo used by permission of Techila)

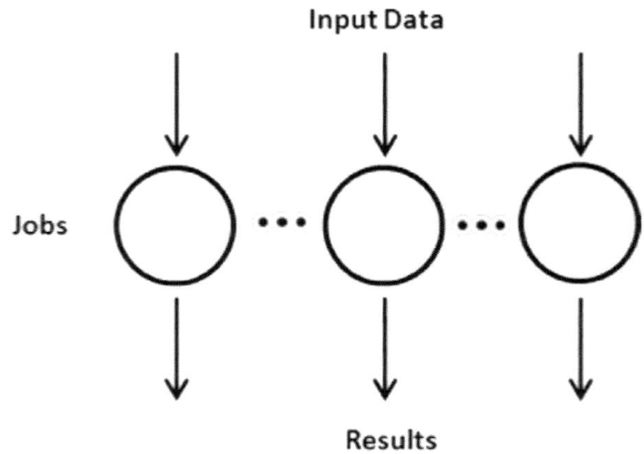

Fig. 3.6 Embarrassingly parallel problem. Each process accepts input data and uses this input data to compute results. Communication is limited to receiving input data and transferring output data; no inter-process communication takes place (Photo used by permission of Techila)

each calculation is independent from all others. Monte Carlo simulations, computer graphics rendering, genetic algorithms, brute force methods in cryptography, Basic Local Alignment Search Tool (BLAST) searches in bioinformatics, and Machine Learning analytics [2], are some examples of these kinds of problems (Fig. 3.6).

Techila has produced a computational platform that deals with heterogeneous distributed computing of embarrassingly parallel problems. The benefit of such an approach is the ability to utilise multiple hardware and operating systems in order to leverage all available computing power for problem solving. The computational resources could be a plethora of instances, from laptops, to unused machines in close proximity, to local servers, and even on-demand instances from cloud providers. In other words Techila ports and manages the code on a hybrid set of computational resources.

Each computational resource is assigned a worker instance that performs the independent computation. All the workers are administered by a central manager which interacts with the end user to collect the problem state. As shown in Fig. 3.7 the End User communicates solely with the server, whose job in turn is to segment the problem in its distributed parts and transfer them to the heterogeneous constituency of computational resources.

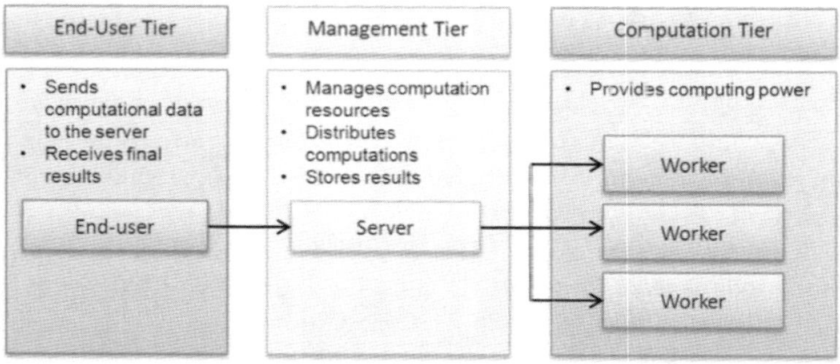

Fig. 3.7 Techila system three-tiered structure. The End-User interacts with the Techila Server, which distributes computations to the Workers. All network communication in the Techila environment are be secured by using SSL (Secure Socket Layer), meaning all transferred data will be encrypted (Photo used by permission of Techila)

3.4.4.1 Description of Platform Implementation

For the present case of the Heston model the cloud implementation was achieved by modification of the existing MATLAB code in order to take account of the Techila parallelisation layer. The computing resources used was an Extra Large instance of a server in the Microsoft Azure cloud. Each instance provices 8 cores @ 1.6 GHz and 14 GB of memory. A total of 354 cores were used for this simulation.

The relative ease with which code can be altered, in order to run on a parallel environment, is one of the platforms greatest strengths. As shown in Listing 3.3 the main difference in the code is to change the *for* command into Techila's *cloudfor* to distribute the for loop to the computational resources.

```
% Main Monte Carlo loop
for pth = 1: paths
  if (NAG==1 || NAG==2)
      ...
  C(pth) = S(pth,end) - K;
end
```

Listing 3.2 Monte Carlo on the local node

```
% Parallel Monte Carlo loop
cloudfor pth = 1: paths
  if (NAG==1 || NAG==2)
      ...
  C(pth) = S(pth,end) - K;
cloudend
```

Listing 3.3 Monte Carlo on the grid nodes

3.5 Efficiency and Accuracy

One could argue that since we have a closed-form analytical solution for these models, why would we opt to evaluate them with a Monte Carlo simulation instead. The main reason we opt to use a Monte Carlo instead of a deterministic analytical solution is twofold. First is the curse of dimensionality[6] of the analytical method. Suppose that we wanted to calculate an integral over ten variables; an integration in 10-dimensional space. By approximating the integral with 20 points in each dimension would yield a total number of integration points of $20^{10} \backsimeq 10^{13}$. The Monte Carlo simulation can approximate with high accuracy using far fewer points, e.g. 10^6. The second reason is the convenience of assessing path-dependent payoffs for more exotic options, e.g. Asian, lookback, and barrier ones.

The Heston model was implemented as a function in Matlab, and a test harness was also developed to create plots and metrics. The test harness performed 21 perturbations of the initial value for the underlying asset and performed the Monte Carlo simulation with certain parameters. Figure 3.8 shows the behaviour and results that the Monte Carlo simulation produced. The upper left figure shows 5,000 paths generated over 1,250 timesteps, for a vanilla European call option price with 5 years maturity (the black lines overlayed on the plot indicate the 99 % and 95 % Confidence Interval (CI)). The upper right figure shows the corresponding variance for each path and is being used to calculate the next price. The bottom left figure shows the prices for all the purturbations of the underlying's price, and finally the bottom right figure shows the standard error for all the prices of the underlying.

3.5.1 *Accuracy of Simulation*

The very very fact that we can't have an infinite number of paths in the Monte Carlo simulation pre-arranges that there will be issues of accuracy of the calculations. Since the discounted stock price process is Martingale, this effectively implies a zero drift on the MCMC process.[7] If we could use a bullseye analogy, then the arrows shot at the target is the initial value of the stochastic process, and the center of the bullseye is the value at maturity, e.g. when the arrow, whichever route it takes, arrives at its target. In this imaginary scenario the bias of the calculation is how far off the target the arrows fall, and the variance shows just how densely concentrated the arrows' spatial distribution is.

---

[6]In numerical analysis the curse of dimensionality refers to various phenomena that arise when analyzing and organizing high-dimensional spaces (often with hundreds or thousands of dimensions) that do not occur in low-dimensional settings such as the physical space commonly modeled with just three dimensions.

[7]A process where the transition probabilities from one timestep to the other are irrelevant from all the previous states, i.e. the process is memoryless.

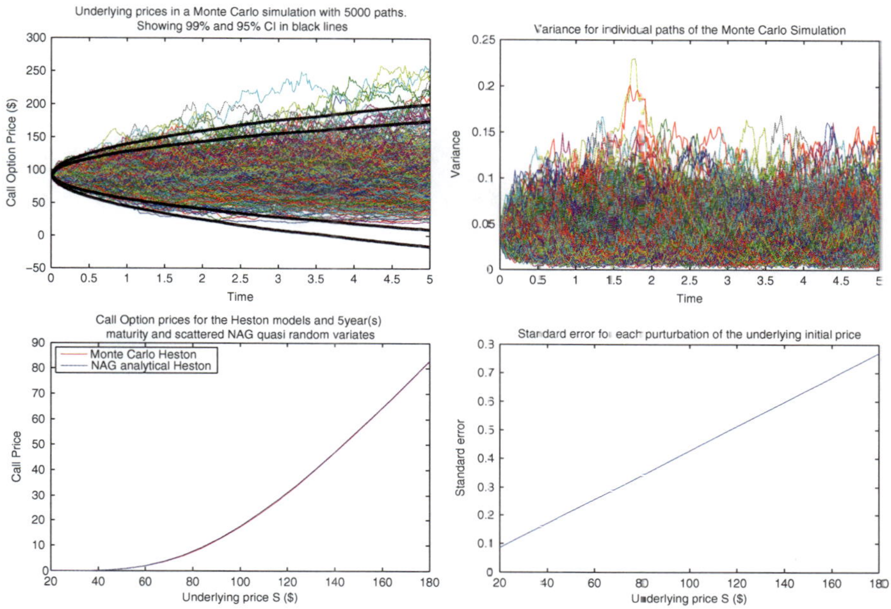

Fig. 3.8 Matlab figures for the Monte Carlo simulation of the Heston model. The parameters for this simulation are: $S_0 = 100, v_0 = 0.04, K = 100, \rho = -0.2, \sigma = 0.2, \theta = 0.04, \kappa = 1.5, T = 5.0, \text{NoSteps} = T * 250, \text{NoPaths} = 1,000, \lambda = 0, r = 0.0,$ and $q = 0.0$. For the bottom two plots the S price is perturbed in the prices $S_0 = [20 \quad 30 \quad \cdots \quad 100 \quad \cdots \quad 170 \quad 180]$

Figure 3.9 plots together different models to calculate the price of a European vanilla call option. The *Heston – NAG* model uses the NAG libraries to calculate the price based on the heston model, as is the *Black Scholes – NAG* model using the same libraries. The *Black Scholes – Analytical* has the analytical solution for what the according BS model would produce. The *Heston – Quasi-MC with QE*, is the present implementation using the NAG libraries to generate Quasi-random variates, and uses the Quadratic Exponential discretisation. Finally the *MaxMC* model shows the calculations on the FPGA using 10^6 paths. There are two sets of plots in Fig. 3.9. One that corresponds to the BS model and another to the Heston stochastic volatility model. Since volatility is constant in the BS model, the call option prices are different near At the Money (ATM) prices for the underlying, between the two models. What is important in this figure is that there is no variation between the different implementation of each respective model.

3.5.2 *Bias and Variance of Results*

Given the Monte Carlo simulation, as was defined in Eq. 3.2, then since $\hat{c}_N$ is a random variable, its bias, equilibrium bias, and variance are defined as follows:

$$Bias_D = \mathbb{E}\left[\hat{c}_N\right] - c_N,$$

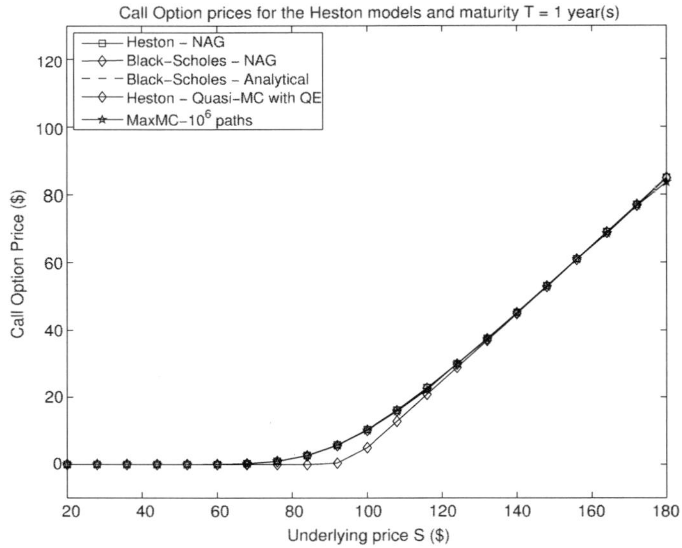

Fig. 3.9 The prices of a call option using different calculation methods, in order to assure the accuracy and the validity of the results

$$
\begin{aligned}
Bias^{eq} &= \lim_{n \to \inf} Bias_D, \\
Var_D &= \mathbb{E}\left[\hat{c}_N - \mathbb{E}\left[\hat{c}_N\right]\right]^2,
\end{aligned}
\tag{3.9}
$$

where $\mathbb{E}$ denotes the expectation with respect to the distribution of $\hat{c}_N$, and c_N is the analytical exact solution of the integral.

However the quality of the estimation of a Monte Carlo simulation is not measured by the bias but an average square estimation error; the standard error.

3.5.3 Standard Error

The central limit states that the bias of the Monte Carlo simulation is normally distributed,

$$
Bias_D = \mathbb{E}\left[\hat{c}_N\right] - c_N \sim \sigma_D Z, \tag{3.10}
$$

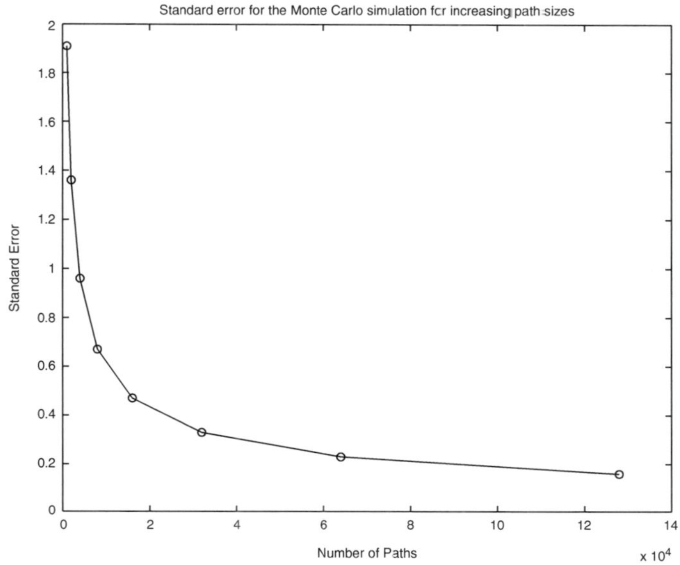

Fig. 3.10 The standard error of the Monte Carlo simulation as it decreases with an increasing amount of paths

where σ_n is the standard deviation of the distribution of $\hat{C}_N$, and $Z \sim N_{(0,1)}$. A simple calculation derives that,

$$\sigma_D = \frac{1}{\sqrt{N}}\sqrt{\sigma^2}, \tag{3.11}$$

where $\sigma^2 = Var_n$ from Eq. 3.9. It is then possible to estimate σ^2 with,

$$\widehat{\sigma_N^2} = \frac{1}{N}\sum_{i=1}^{N}(c_N - \hat{c}_N)^2, \tag{3.12}$$

Thus Eq. 3.12 can be used to calculate

$$\widehat{\sigma_D} = \frac{1}{\sqrt{N}}\sqrt{\widehat{\sigma^2}}. \tag{3.13}$$

Typically the representation of the result of a Monte Carlo simulation is of the form, $c = \hat{c}_N \pm \widehat{\sigma_n}$. For the purposes of the simulation the standard error was measured as a way to test the convergence of the simulation. Figure 3.10 plots the standard error of the simulation from 1,000 up to 128,000 paths. The error is reducing steadily as the number of paths increase.

3.6 Results

In this section we present the data collected from running the Heston model on all available platforms. The CPU∗ that was used was an Intel core i5 at 2.3 GHz clock speed and 4 GB of DDR3 RAM on a MacBook Pro. The GPU server was provided by Boston Limited[8] and contained 2 T M2090 GPUs. The FPGA nodes were located on Maxeler's MaxCloud[9] servers and contained four vectis dataflow engines. Each dataflow engine has up to 24 GB of RAM, giving a total of up to 96 GB of DFE RAM per node. Finally the Techila installation was hosted on Microsoft's Azure cloud on a large server with 4 cores at 1.6 GHz each.

3.6.1 Acceleration Results

Data was collected from running the same model on all of the available platforms. A baseline time measurement was sampled from the CPU in order to compare against the times accrued on the acceleration platforms. The acceleration fold, that is how much faster the Heston model ran under different platforms, is shown in Table 3.1. The same data are plotted against each other in a graphical form in Fig. 3.11.

As shown in Table 3.1 the best results are obtained the closer we get the machine level language. For instance the acceleration of the FPGA, which is coded in C++ and JAVA, severely outperforms the Matlab implementations for a small number of simulation paths.

What is apparent is that the closer we are to the machine level the better the acceleration benefits that can be achieved. The FPGA implementation clearly

Table 3.1 This is a synopsis of the acceleration leverage achieved using FPGA, Matlab's Parallel Computing Toolbox (PCT) for GPU, and Techila Grid platforms

NoPaths	CPU (s)∗	GPU	Techila	FPGA	Std. error
1,000	14.45	× 5.56	× 0.37	× 247.23	1.91
2,000	36.48	× 13.51	× 5.00	× 231.33	1.36
4,000	78.62	× 29.12	× 6.75	× 369.09	0.96
8,000	177.94	× 61.36	× 9.61	× 690.96	0.67
16,000	386.62	× 128.87	× 15.23	× 770.01	0.47
32,000	952.25	× 238.06	× 30.65	× 879.10	0.33
64,000	2009.74	× 401.95	× 45.14	× 975.82	0.23
128,000	3956.29	× 565.18	× 58.23	× 987.35	0.16

[8] www.boston.co.uk

[9] http://www.maxeler.com/products/maxcloud/

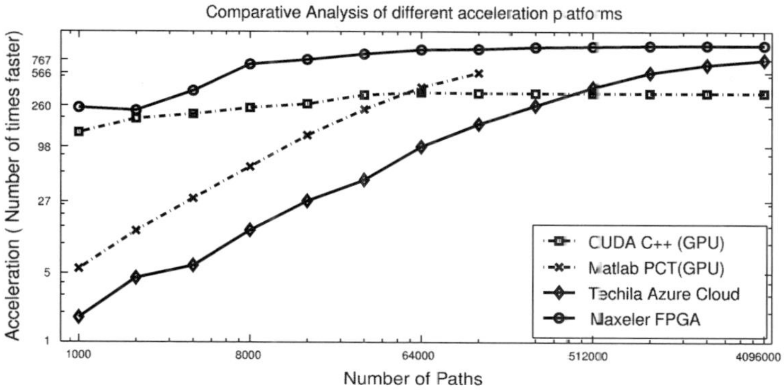

Fig. 3.11 This figure shows the amount faster each platform performed against the CPU implementation

outperforms the others. However the caveat for the GPU implementation is that the model was coded in Matlab using the PCT which could be affecting the performance of the GPU.

3.7 Conclusion

We presented

This chapter describes a novel comparison of different acceleration platform based on the same embarrassingly parallel problem. We have implemented and compared an MCMC simulation of the Heston model, on an FPGA server, a GPGPU server, and a distributed node cluster on Azure. The performance gain was significant on all platforms, making them worthwile options. The lowest cost amongst the platform would be the GPGPU server, who's amortising cost, coupled with a locality factor make it the optimal choice. However, if locality of data is not an option, e.g. when confidentiality is not important and data transfers are small, the Azure cluster would provide a more easily scalable solution. Finally if real-time calculations with complete saturation of the available pipes were required, e.g. in HFT platforms, then the FPGA platform would provide the necessary computational latency.

Future research will include the mapping of qualitative characteristics into a composite metric comprising the Capital Expenses (CAPEX) and Operating Expenses (OPEX) of each platform. The sum of these would provide us with the Total Cost of Ownership (TCO) metric. We should use this metric in order to re-base all performance gains.

Acknowledgements The support of Maxeler Technologies Ltd, Techila Technologies, Numerical Algorithms Group, Nvidia Corp, and Boston Ltd is gratefully acknowledged. I would also like to single out the help and guidance provided by Mr Erik Vynckier then at Scottish Widows Investment Partnership.

References

1. Andersen, L.: Efficient Simulation of the Heston Stochastic Volatility Model (2007)
2. Dejaeger, K., Eerola, T., Wehkamp, R., Goedhuys, L., Riis, M.: Beyond the hype: cloud computing in analytics, pp. 1–7 (2012)
3. Easley, D., Lopez de Prado, M., O'Hara, M.: The Microstructure of the 'Flash Crash': Flow Toxicity, Liquidity Crashes and the Probability of Informed Trading. J. Portf. Manag. **37**(2), 118–128 (2010)
4. Faure, H.: Discrépances de suites associées à un système de numération (en dimension un). Bulletin de la S. M. F **109**, 143–182 (1981)
5. Feller, W.: Two singular diffusion problems. Ann. Math. **54**(1), 173–182 (1951)
6. Levy, G.: An introduction to quasi-random numbers (2002)
7. Nidereitter, H.: Random Number Generation and Monte Carlo Methods. SIAM, Philadelphia (1992)
8. Renze, J., Weisstein, E.W.: Law of Large Numbers – from Wolfram MathWorld
9. Sobol, I.M.: On the distribution of points in a cube and the approximate evaluation of integrals. U.S.S.R. Comput. Math. Math. Phys. **7**(4), 86–112 (1967)

Chapter 4
Towards Automated Benchmarking and Evaluation of Heterogeneous Systems in Finance

Christian De Schryver and Carolina Pereira Nogueira

Abstract Benchmarking and fair evaluation of computing systems is a challenge for High Performance Computing (HPC) in general, and for financial systems in particular. The reason is that there is no *optimal* solution for a specific problem in most cases, but the most appropriate models, algorithms, and their implementations depend on the desired accuracy of the result or the input parameters, for instance. In addition, flexibility and development effort of those systems are important metrics for purchasers from the finance domain and thus need to be well-quantified.

In this section we introduce a precise terminology for separating the *problem*, the employed *model*, and a *solution* that consists of a selected *algorithm* and its *implementation*. We show how the *design space* (the space of all possible solutions to a problem) can be systematically structured and explored. In order to evaluate and characterize systems independent of their underlying execution platforms, we illustrate the concept of application-level benchmarks and summarize the state-of-the-art for financial applications.

In particular for heterogeneous and Field Programmable Gate Array (FPGA)-accelerated systems, we present a framework structure for automatically executing and evaluating such benchmarks. We describe the framework structure in detail and show how this generic concept can be integrated with existing computing systems. A generic implementation of this framework is freely available for download.

4.1 Introduction

The increasing complexity of market models and financial products has led to the lack of closed-form pricing formulas for many products. In general, we have to use numerical approximations to compute product prices with specific market models for many cases. However, not only various algorithms exist for calculating such approximate prices (for example Monte Carlo (MC), finite difference, or tree methods), but all of them come with various flavors and parameters that need to

C. De Schryver (✉) • C.P. Nogueira
Microelectronic Systems Design Research Group, University of Kaiserslautern, Kaiserslautern, Germany
e-mail: schryver@eit.uni-kl.de; nogueira@rhrk.uni-kl.de

C. De Schryver (ed.), *FPGA Based Accelerators for Financial Applications*,
DOI 10.1007/978-3-319-15407-7_4

be carefully set to achieve meaningful results in the end. Another issue is that not all algorithms are suitable for all execution platforms: For example, trees can be very efficient and fast on a Central Processing Unit (CPU), but on reconfigurable architectures such as FPGA MC methods can benefit from the highly available spatial parallelism. In addition, thinking about custom data types or mixed-precision implementations, flexibility, maintainability, or system costs, the *design space*[1] of available solutions for one particular product-model constellation explodes.

A crucial challenge for quants and practitioners in general is to select the most appropriate implementation under given side constraints. Although this holds for nearly all problems we see in computational finance, we focus on pricing tasks in this chapter. Those constraints can for example be expertise or knowledge in a particular algorithmic domain, required flexibility of the implementation, throughput or latency requirements, or simply what is already available in the company. However, in particular tuning the algorithmic parameters to match a specific purpose is a non-trivial tasks in general. For instance, considering a classic MC pricer based on Euler discretization, we already need to specify the step width h of the grid and the number of paths N. Assuming that we compute this on a CPU or Graphics Processor Unit (GPU), we need to select either single- or double-precision floating point arithmetics. In contrast, we can arbitrarily select the precision in each stage of the accelerator data path when implementing on FPGAs. In addition, we need to consider algorithmic flavors that may be available for a particular market model. For example, Marxen et al. have shown that log price simulation with full truncation and continuity correction performs well for European barrier options in the Heston model, but this also depends on the specific market and option parameters [20].

Figure 4.1 illustrates the large design space for constructing an option pricing system. For reasons of lucidity, only a few layers are shown and sub-branches for other algorithms besides MC have not been included. It is obvious that already on application level a very precise selection of the supported products is mandatory to allow a sensible selection of suitable algorithms and execution platforms on the lower levels. A final solution corresponds to one single point in the design space and is specified by a vector of settings for all available parameters.

The question is now: what is the "most appropriate" solution we are looking for in the design space? The answer is in general very difficult to give right away: Is it the fastest or the most energy efficient solution? And how much flexibility is required? And which accuracy of the result do we require under which circumstances? Usually performance and efficiency stand in clear contrast to flexibility, since inflexible solutions can get rid of any overhead not required to solve the specific job they

[1]The design space is the space that depicts all possible solutions to a particular problem. Each adjustable parameter is reflected as one dimension of the design space. One specific implementation (a *solution* for the problem) is exactly one point in the total design space [6].

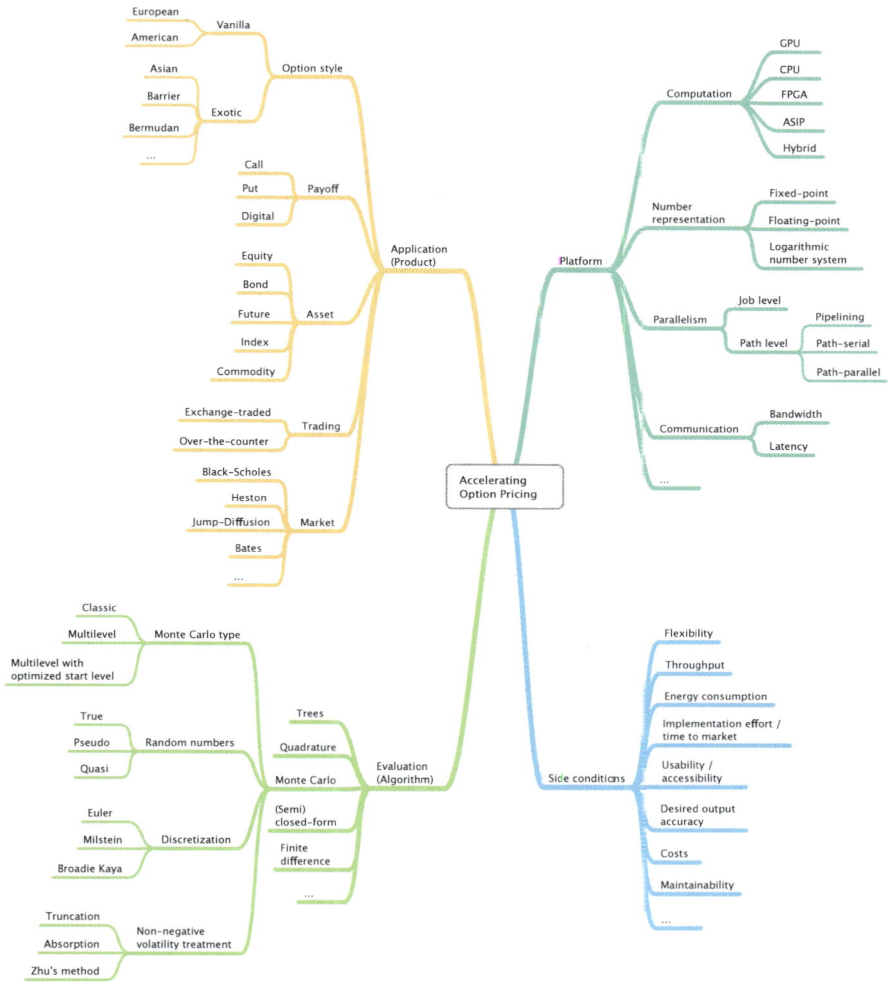

Fig. 4.1 The (cut) design space for constructing an option pricing system. For reasons of clarity only the branch for MC algorithms is depicted in detail

are constructed for. We already see that usually there will be no general answer to this question and that we need to carefully balance between our requirements and the solutions we can construct.

Due to the various different sections of the global design space (such as algorithmic tuning, precision in the data path, flexibility, ...) trying to determine the optimal solution in an analytical way is not feasible. In addition, not for all characteristics there will be formal *hard* metrics that allow us to use an optimizer for this task. Furthermore, in some cases such as pricing with stochastic MC methods it is not even clear what the correct price should be, and of course every solution might for example use different Random Numbers (RNs) and therefore provide different results. And how can we compare systems that do not even use the same technology, for example a tree-based American option pricer on a CPU with a MC pricer on a hybrid CPU/FPGA device? The resolution is to employ problem-specific application-level benchmarks that only rely on settings and metrics based on the *problem*, but not on *algorithms* or *implementations* of those.

4.2 Problem, Model, Solution: A Precise Terminology

The target job of a system should always be: Solve a specific *problem* – it does not matter which algorithm or implementation is used, as long as the result is calculated correctly and the final implementation fulfills all throughput and energy (and further) given constraints. Therefore we should clearly distinguish between the following three terms [6, 8]:

- The *problem* that is tackled (what to solve),
- The employed *model* (an abstract representation of the problem), and
- The final *solution* (an implementation that solves the problem).

Figure 4.2 illustrates how problem, model, and solution depend on each other. A model can be seen as an abstract representation of the actual problem that is used to mathematically represent the problem in accessible dimension. The first challenge is therefore to select an appropriate model for the problem in focus. Usually this task is carried out by application experts such as financial mathematicians or quants for an option pricing example. By (implicitly) navigating through the design space, the application expert in many cases already selects an algorithm that shows a good performance for his or her settings. It is very common that this person uses available high-level implementations for instance in Matlab or Python to explore first performance results of various algorithms in this setting. However, since this exploration is performed on CPU based architectures, it is in general not easy to already estimate the benefits and drawbacks of a specific algorithm for a

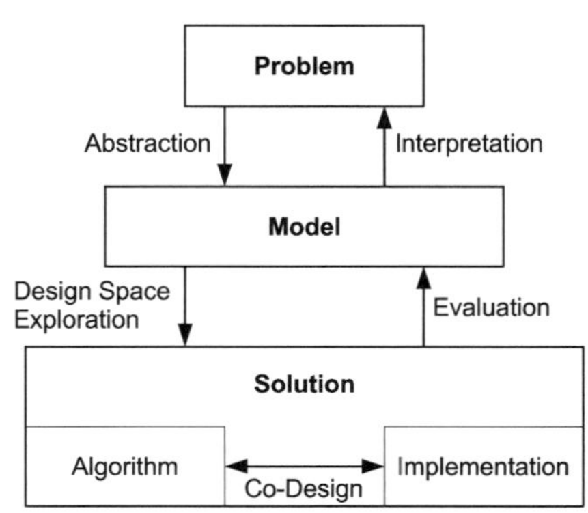

Fig. 4.2 The problem/model/solution layers and their interdependencies in general

high performance implementation later without detailed knowledge about computer architectures and hardware design.

Let us illustrate this categorization with the example *problem* "calculate the price of a European option with two knock-out barriers with a given strike price K and maturity T on a specific market". European knock-out barrier options pay a certain amount of money at a fixed maturity time depending on the value of the underlying asset, in this case only if none of the barriers is hit during the lifetime of the option.

It is obvious that the problem description itself does not yet give any suggestions to its solution. Since the price of an option is tightly coupled to the price of a certain stock at the market, we need a *model* that reflects the underlying stock price behavior of the time. For the selected example, suitable models are for example the Black-Scholes model or the Heston model. The model therefore gives a formal and abstract view of (a certain aspect of) the problem.

The *solution* finally is a dedicated approach for solving a (modeled) problem and producing a result. It is characterized by a specific *algorithm* and its *implementation*. For evaluating the Heston model, e.g. finite difference methods or stochastic MC simulations can be used. Those may be implemented for instance on standard Personal Computer (PC) clusters, GPU, or on FPGA-based platforms. The selected *solution* therefore evaluates the model, and the results of the *model* need to be interpreted and applied to the actual *problem*.

When looking at literature from the financial mathematics community we observe that many authors focus on the development of algorithms for solving specific financial problems, but do not consider their implementation on a particular platform. In contrast, hardware engineers in many cases take algorithms for granted and come up with tailored and optimized architectures for them without questioning if the algorithm is suitable for hardware implementation at all. This illustrates how important it is to always keep the *problem*, *model*, and *solution* domains mentally isolated and to start with the *problem* in focus at the beginning.

4.2.1 Cross-Domain Parameter Dependencies

In order to achieve an optimal solution in the end it is crucial to understand existing dependencies between different parameters. For instance, the selection of the underlying market model fixes the available algorithms that can be used to price a specific option in this model. Another example is the correlation between the available computation units and suitable number representations, since standard CPU and GPU only support a specific set of number types.

Besides parameters characterizing the application and model domain, the algorithm, and the implementation platform, side conditions may exist that influence large parts of the final solution. Such side conditions are for example the desired flexibility and maintainability of the system, throughput and energy requirements, or the Total Cost of Ownership (TCO). The most flexible execution platform is by sure a general-purpose CPU that can never provide the throughput and energy efficiency of a dedicated pricing hardware tailored to one specific product.

4.3 Problem-Specific Application Level Benchmarks

The large design space as illustrated in Fig. 4.1 shows many degrees of freedom to construct one specific option pricing solutions. The high number of possible implementations poses not only the challenge to select beneficial parameter constellations during the design process, but also shows that completely different solutions may exist that solve the same problem in the end. For example, a CPU implementation based on finite difference methods may be able to evaluate European double barrier options in the Heston model, but the same task could be performed by a dedicated MC engine implemented in an FPGA.

Evaluating and comparing technical systems has been (and still is) a hot topic for many years, in particular in the HPC domain [1, 3, 26]. In general, *benchmarking* can be defined as a methodology to reveal the performance of a system (or process, for example in business) using various *metrics*. The metrics strongly depend on the goal and the application of the benchmark. In the HPC domain, mainly Floating-Point Operations per Second (FLOPS) or derived scores are used, focussing on the computational power of a system.

In 2009, van der Steen has identified three main reasons for benchmarking [26]:

- Benchmarking for selling systems,
- Benchmarking for buying systems, and
- Benchmarking for knowledge (for example to understand architectural influences on specific applications).

Those areas not necessarily exclude each other. Depending on the purpose for benchmarking, several characteristics are more important than others. For example, benchmarks can focus on evaluating

- The pure computational power of a system,
- The overall performance for computing one particular application, or
- The costs (in USD or Joule) for solving one or more problems on this machine.

Standard HPC benchmarks like Linpack, STREAM, or matrix transposes are used to evaluate the computational performance of supercomputers, and aim at scoring a system for general purpose use mainly. They are mainly constructed synthetically and try to stress various parts of the systems as much as possible. A commonly employed bundle in this field is the *HPC Challenge benchmark* that incorporates seven standard benchmarks for HPC [25]. Those benchmarks are mainly used to evaluate systems for selling and categorization purposes with standard metrics like FLOPS or scores.

Application-level benchmarks seem to be more interesting for acquisitive buyers of a system since they show the actual performance for one dedicated purpose. Their importance has already been highlighted by Berry et al. in 1989 [3]. In general, those benchmarks use different metrics, for example "How many problems can be solved within one time unit?" or "How does the throughput scale with increasing problem dimensions?". Armstrong et al. have reinforced the importance of application-level benchmarks by stating that "realistic benchmarks cannot be abstracted from real applications" in 2006 [1].

However, in their work Berry et al. also comment on possible pitfalls one might experience by applying application-level benchmarking. One important point that should be noted is that the results of application-level benchmarks are much less generic and applicable to other applications compared to computation-centric benchmarks [3]. This even holds for only slight modifications of the applications that might result in a different problem structure, for example with completely different memory accesses and communication requirements. Furthermore, Armstrong et al. claim that today's application may not necessarily be tomorrow's application, pointing at the sustainability of purchasing or developing an application-tailored system [1].

Application-level benchmarks are established in many fields, for example with standard reference streams for Moving Picture Experts Group (MPEG) decoding, the famous Lenna picture for image processing, or the standardized block- or bit-error curves in communication technology. For the latter domain, Kienle et al. have illustrated the complexity of comparing various implementations of channel decoders on application level in 2011 [17]. All in all, we are convinced that application-level benchmarks are the only way to fairly compare implementations (solutions of a particular problem) over architectural and algorithmic borders.

4.4 Benchmarks for Option Pricing Systems

When looking at the available literature describing option pricing implementations, we see that there are high variations with respect to:

- The covered product range,
- Employed market models and algorithms,
- Underlying execution platforms, and
- Metrics used for evaluation and comparison with other works.

A very common way of characterizing a novel implementation is to compare its speedup to a reference software model. We think that this is a feasible approach, but only if all details of there reference implementation (the source code, all compiler specifications and settings, and the execution platform) are carefully specified as well. In general, at least some of those pieces of information are missing in literature, and therefore do not allow to fairly evaluate the proposed solutions.

In this section we present an application-level benchmark set for European (double) barrier option pricing in the Heston model and show how it is used to fairly compare different solutions of the covered problem-model double. We have introduced this benchmark in 2011 [7, 19], it is freely available for download.[2]

[2]http://www.uni-kl.de/benchmarking

4.4.1 Related Work

Morris and Aubury have already commented on this fuzzy situation and have claimed the need for standard benchmarks in option pricing in 2007 [21].

In parallel to our work Jin et al. have come up with an evaluation methodology for hardware option pricers in 2011 [16]. It is based on speed (derived from the required number of clock cycles) and accuracy of the solution (measured with the Root Mean Squared Error (RMSE) compared to a golden reference solver with a high precision). While their approach is promising for comparing the performance of solvers based on different algorithms, it can only be used for hardware implementations and for example not for software or hybrid architectures. Furthermore, they do not consider energy consumption in any way.

The commercial STAC-A2 benchmark [24] was first presented at the SC12 conference in 2012 [18]. It incorporates a broad range of tasks mandatory for financial institutes, is architecture-independent, and scalable over large clusters. The metrics are speed, efficiency (power and space consumption), quality, and programming difficulty. However, access to the STAC-A2 benchmark is not for free right now.

Implementations of the STAC-A2 benchmark on Intel Xeon E5 and Xeon Phi and their performance have been presented at the WHCPF 2013 by Nikolaev et al. [22] and on WHCPF 2014 by Fiksman and Salahuddin [12].

4.4.2 The Benchmark Settings

Our proposed benchmark set consists of three main components:

- The parameter sets defining the current *market situation* like the current volatility or the correlation between price and volatility,
- The *option parameters* such as the type of option and the strike price, and
- The correct *reference price* or a good approximation thereof, together with a reference precision.

The metrics we suggest for comparing different implementations are:

- The consumed energy for pricing one option in *Joule/option*,
- The number of priced options per real time *options/second*,
- The numerical accuracy that is achieved by the proposed design compared to the presented benchmark results (e.g. the RMSE of the difference), and
- The consumed area on chip for hardware architectures (slices, LUTs or mm^2 on silicon).

Our proposed benchmark is based on the Heston equations as introduced e.g. in Chap. 3 by Delivorias. The starting conditions are $S(0) = S_0$ and $\nu(0) = \nu_0$. For ease of computation we always set $S_0 = 100$ without loss of generality.

Table 4.1 The market parameter settings for the proposed benchmark

	κ	θ	σ	r	v_0	ρ	T
I	2.75	0.035	0.425	0	0.0384	−0.4644	1
II	2	0.09	1	0.05	0.09	−0.3	5
III	0.5	0.04	1	0	0.04	0	1
IV	1	0.09	1	0	0.09	−0.3	5
V	0.5	0.04	1	0.08	0.04	−0.9	10
VI	2.75	0.35	0.425	0	0.384	−0.4644	1

Table 4.2 The 12 options used in the proposed benchmark

Option number	Parameter set from Table 4.1	Detailed description
1	II	ATM European call
2	II	ATM single barrier call with barrier 120
3	IV	ATM single barrier call with barrier 120
4	III	ATM double barrier call with barrier 90 and 110
5	I	ITM double barrier call with barrier 80 and 120, strike 90
6	IV	ATM double barrier call with barrier 66 and 150
7	V	ITM double barrier call with barrier 66 and 150, strike 90
8	VI	ATM double barrier call with barrier 66 and 150
9	I	ATM double barrier put with barrier 80 and 120
10	VI	OTM double barrier call with barrier 66 and 150 and strike 120
11	I	ATM single barrier kick-in call option with barrier 120
12	IV	ATM double barrier digital call with barrier 66 and 150

Table 4.1 shows the six market parameter selections we have chosen for the benchmark (one exception is T that belongs to the option itself). They are taken from literature and describe both standard and corner settings relevant for real business [7, 19]. Derived from the settings in Table 4.1, we have constructed a set of 12 options that makes up our proposed benchmark. It focuses on At the Money (ATM) options since they are most interesting for precise pricing [14, 19], but also includes some In the Money (ITM) and Out of the Money (OTM) constellations. The selected options and their respective payoff functions are given in Table 4.2.

Another important purpose of the benchmark is to validate the functional correctness of an implementation and to check if it computes the right results in the end. Therefore it is important to state that the benchmark contains an executable GNU's Not Unix (GNU) Octave implementation of a standard MC solver that can be run to re-generate the reference results. We have used this model to generate the reference prices that come with the benchmark with their specified reference accuracy (in this case the estimated RMSE). They are given in Table 4.3. Some of the results have also been evaluated with the finite difference method or the (semi-)closed formula for European vanilla options in the Heston model.

We have employed the proposed benchmark to evaluate the performance of different algorithmic tunings in a Multilevel Monte Carlo (MLMC) simulations

Table 4.3 The computed reference results for the benchmark

Number	1	2	3	4
Price	34.9998	0.10280	0.31606	0.74870
Obtained by	Closed form	MC	MC	Closed form
Precision	0.0001	0.0001	0.0003	0.00001
Number	5	6	7	8
Price	5.7576	3.0421	0.017117	0.82286
Obtained by	Finite difference	MC	MC	Finite difference
Precision	0.001	0.005	0.0002	0.001
Number	9	10	11	12
Price	1.5294	0.17167	4.9783	0.16805
Obtained by	MC	MC	MC	MC
Precision	0.0005	0.0005	0.0005	0.0001

[20]. Furthermore, have used the benchmark to validate and characterize our FPGA architectures for European barrier option pricing in the Heston model [6, 9]. Inggs et al. have used our benchmark to evaluate their heterogeneous computing framework for computational finance in 2013 [15].

In the next section we introduce a framework for automated evaluation of various implementation with application-level benchmarks.

4.5 A Framework for Automatic Benchmark Execution on Heterogeneous Platforms

In the previous sections we have highlighted how difficult it is to find an *optimal* solution for a particular problem. To decide which solution is better under given conditions, is necessary to compare the candidates in a standardized way. As we said, a widely spread approach for evaluating implementations are application-level benchmark batteries, i.e. a representative task set that ideally should cover all the main and corner cases in a good real-world balance.

Nevertheless, any modification in either the algorithm or the implementation leads to a new solution that requires to be benchmarked again with the complete battery. This can be a very time consuming process. It is obvious that for a large number of solutions an automated benchmark execution and evaluation system is desirable. Especially if a set of solution candidates shall be compared for a specific task battery, we would like to automatically dispatch those tasks to the implementations, collecting the results to further visualize them in an intuitive way. Ideally should be available an integrated benchmark tool that interconnects and support all those different solutions in order to compare the performance results automatically.

However, besides the *hard* numbers that can be measured (e.g. runtime in seconds or energy per task in Joule), there are *soft* characteristics of implementations like flexibility, maintainability and extensibility, or portability to other platforms. Those aspects need a careful special treatment in order to reflect the overall attributes of the solutions in focus. To analyze the different algorithm implementations, we have used, as comparative base, the hard facts only.

In this section we present a universal approach for integrating a number of available solutions for easy benchmarking with centralized evaluation of all results for further analysis. The main requirements for the design have been: to be as flexible as possible, allowing to integrate different hardware and algorithms with less effort as possible, and easy deployment, resulting in low effort when being integrated into pre-existing infrastructures. Since the soft characteristics are not been measured on the implemented algorithms, our proposed integration tool focus in the soft characteristics. In this sense, the implemented algorithms just have to receive the data to perform the simulations, and the benchmark tool takes care to manipulate the data to be understandable by the implementation. To achieve our goals, we have used *web services*[3] as a software architecture and standard protocols to interconnect and communicate through the infrastructure. In the following section we describe our software architecture in detail as a blueprint for all readers interested in setting up such a framework for themselves.

4.5.1 Software Architecture

The decision of the right software architecture plays a big role during the development of a system and it is the first step after defining the software requirements, representing the earliest design decisions. It is the software architecture which defines how the system elements are going to interact to each other, specifying some general characteristics as how they are interconnected, how resources are allocated and which protocols are used, for example. Keeping this high importance in our minds, we have investigated some options and analysed the benefits, comparing advantages and disadvantages of each approach [23]. We have used as comparative parameters the soft characteristics, since the hard numbers are generated by the algorithms implemented on the working nodes.

During our research, we have checked that web service provides an abstraction layer which allows to interconnect different devices in an homogeneous way. Since the algorithms we are interconnecting are developed aiming different kind of hardware, makes sense to chose this software architecture as base of our implementation. However, usually web services have a big trade off due to this abstraction layer. It means that the data must be processed and formatted in a understandable way for both sender and receiver. Then the ideal solution should have less processing overhead as possible, even with the abstraction layer that is

[3]"Web service is a software system designed to support interoperable machine-to-machine interaction over a network." [13]

really important for easy integration. There are different types of web-services, some of them uses eXtensible Markup Language (XML)-based protocol to exchange information (as SOAP and WSDL, for example). We decided not use them because we wanted to use a standard communication protocol, and XML-based protocols usually are application specific.

Since the elements of a web service by definition communicate to each other over a network, to encapsulate the information that we want to exchange into the network protocol seems to be reasonable. With this idea in our minds, we searched for solutions that use this idea to implement web services. Hypertext Transfer Protocol (HTTP) is the standard application protocol for general propose networks, as the World Wide Web (WWW), for example. It means that every node which belongs to this kind of software architecture is capable to generate and parse HTTP. WWW has another important advantage from our point of view: it has higher degree of flexibility, since at any point of time there are new websites been connected and many other been disconnected to the infrastructure. It is possible due to the stateless nature of its software architecture, where one node must not depend on the other nodes to successfully process a request. The result of this characteristic is a loosely coupled infrastructure. Thus there are many similarities between the WWW model and what we are expecting from our integrate benchmark tool.

An abstraction for the WWW architecture is the Representional State Transfer (ReST) architectural style [11]. This architectural style defines several architectural constrains. It means that to be considered ReST, the system should behave accordantly to those constrains. There are some optional constraints as well, but here we are going to focus in the main ones, explaining how can we make use of them in our software to achieve our goals.

First of all, the software should be client-server. In other words, there is a well-defined separation of concerns: the user interface concern is detached from the data storage concerns. This separation improves the portability of the user interface across multiple platforms, as well as improves *scalability*[4] by simplifying server components. Both improvements are highly valuable for our needs, since they increase overall flexibility.

Besides to be client-server, the interaction between components must be stateless. Stateless in a sense that all the client requests must contain all the information necessary for the server understand it, without taking advantage of any context stored on the server, keeping the communication and data control much easier. On the other hand, the client must keep the session state. This constraint carries along many improvements, but the most important in our context is that the server can quickly free resources, since it do not have to store any context, simplifying its implementation and reducing load processing. It also permits that the infrastructure adds or removes nodes with less impact since the other nodes do not need to be aware of environmental changes all the time. The stateless constrain trade-off is that it may decrease the network performance, by sending repetitive data. Once most

[4] Scalability is the ability of a system to accommodate an increasing number of elements or objects, to process growing volumes of work gracefully, and/or to be susceptible to enlargement [4].

of our server requests are relatively small and are by nature somehow stateless (perform new simulation, get a simulation result), the advantages of a stateless approach are still greater than the disadvantages.

A constraint added to reduce the number of redundant data passing through the network is to allow data to be cacheable. The response of a request must be labelled, implicitly or explicitly, as cacheable. This gives the right to the client decide whether it wants to keep this information for reuse it in equivalent requests. As a consequence, some interactions can be partially or completely removed, improving the efficiency and the user perceived response time.

In addition to those constraints, there is another requirement: the uniform interface. All the system components must communicate one to another using a uniform interface, decoupling the implementation from the provided service. Each component can evolve without the need to worry about compatibility, since the interface is uniform and the way to exchange data never changes. The cost for having such a flexible and independent interfaces is a efficiency degradation, since the information is transferred in a standardized way and not in a application specific method. As we stated at the beginning, we were aiming the use of standard protocols even with this trade-off.

The last but not less important constraint is that the system should be layered. In a sense that each system layer is agnostic to the overall interactions. The layer can only see its own direct interactions, but not beyond them. This restrict knowledge to a single layer promotes extract independence, given the possibility to change layers without have to worry about how it will impact to the others. A great example to see the advantages is the idea that the user interface does not know whether it is direct connected to a unique final server, or to an intermediate server, or to a cache, etc. The main advantage is to improve scalability, allowing to implement load balance mechanisms, for example. However, the layers add an overhead and latency to process the data [5], due to this abstraction, but this can be overcomed by the use of intermediate shared caches.

In our unified benchmark platform, we have integrated different option pricers solvers to compare their performance as study case, but it is not restrict to only this kind of solver, since it has flexibility as main requirement. This was possible because we have been using ReSTful Application Programming Interfaces (APIs). Taking in account the ReST main constraints, we can check that it provides a loosely coupled approach to client-server model. So, all the components of the infrastructure have none or little knowledge about the definitions of other separated system's components. Thus when a component is changed, it provides a lower overall impact. This proposal aims to maximize the independence and scalability of the architecture components, and also to minimize latency and network communication. The communication between all the components is done over the standard network protocol HTTP to interconnect the available resources. Each resource [2] has its own identification on the system, called Uniform Resource Identifier (URI), which allows its use and access. All interactions of a resource are done by URIs and no other way is allowed, ensuring the uniform access.

Each transaction of our unified benchmark tool contains all data necessary to complete all needed requests, keeping the communication and data control much easier. It also permits that the infrastructure adds or removes nodes with less impact since the other nodes do not need to be aware of environmental changes all the time. For a better understanding of how it impacts, it is important to take a look on the infrastructure and how the elements are related.

4.5.2 Infrastructure

The proposed infrastructure is composed by four distinct elements as illustrated in Fig. 4.3:

1. Front End: is the part of the system from where the user is able to access the framework in order to, for example, check results, compare them and execute simulations;
2. Back End: is responsible for receive data from the front end, process them, communicate with the database and dispatch the simulations to the working nodes;
3. Database: all important informations of the framework are stored on the database, as simulation result, for example;
4. Working Node: is the node (FPGA, CPU, GPU, etc.) which simulates an implementation of an algorithm with hardware acceleration and generate results for further analysis.

The elements are interconnected by HTTP and all the information that a node need to complete a request is encapsulate inside the payload of the protocol's header. Using this technique instead of an XML file, for example, we have less communication overhead, sending only relevant data. In addition, there are less

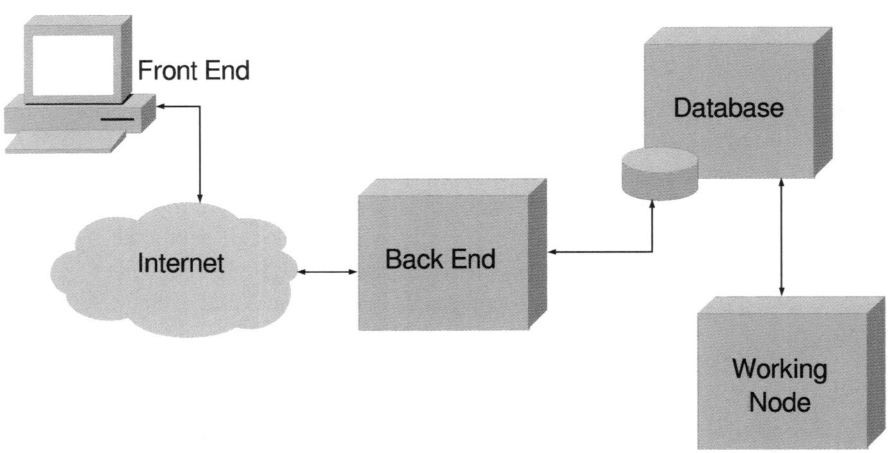

Fig. 4.3 Proposed Infrastructure

processing overhead on the nodes, since it does not have to create and parse a file to extract the data from its content.

Not only the communication protocol and the way that elements exchange data is standard, also the operations allowed over each resource are pre-defined. Since our framework uses HTTP as base to transfer data, the available operations to interact with an resource are the same as the most frequent used ones available for this protocol: POST, GET, PUT and DELETE. These methods correspond to Create, Read, Update and Delete (CRUD) operations, respectively, and it is enough to perform all the needed system actions.

Whenever the back end receives an request, it parses the header to check which CRUD operation is being requested. It is necessary to be authenticated to access any resource, for security reasons. GET operations have pre-defined patterns, avoiding to expose unnecessary information. Those patterns include regular expressions and permit to execute simple requests, for example returning a register of a table, as well as complex request, as returning only certain fields from a join operation of many tables with some constraints. If it is an UPDATE or PUT request, the back end stores all the relevant information on the database, so it will be available for all the nodes belonging to the current infrastructure.

As we previously said, all relevant data is stored on the database. The database model is flexible, since each job is composed by many simulation associated to market parameters, option parameters and the user name of the person who started it. Market and option parameters are independent from platform and they must be the same if we want to compare different implementations and does not make sense to compare the results of an FPGA and a cluster with different input parameters, because it can cause biased results. For this reason, a benchmark set is the combination between those two parameters. Each simulation has its own particularity, so it is associate to a job, an algorithm parameter entry, a result and a working node. The result is empty until the simulation finishes to execute. Working node is the place where the simulation was performed. Based on the results of a simulation, we can numerically and graphically compare the implementations, using energy consumption, runtime, the price and precision as parameter. When all simulations of a job is finished, an e-mail is sent to the user with the id of the result. There are background tasks periodically pooling to check if a result is available.

To develop this framework, we have used web2py [10] which has a ReST API and a simple task scheduler. For a new working node join the infrastructure and start to perform simulations, it is necessary to register on a group and start a scheduler worker. A task will be on the ready queue whenever the user start a new job, after all the simulations have been added to the database. Each task is assigned to a working group, which can have one or more working nodes. The scheduler defines a working node to perform a certain task associated with its working group. If a working node receives a task to perform, we say that the task was assigned and when it starts to execute, we say that the task is running. The result of a task is stored on the database, so it is possible to check why does a task failed to run, or either if everything run according the expected. It means that, at the end of a job execution, we have not only the results, but also the complete run log of all related tasks for debug purpose.

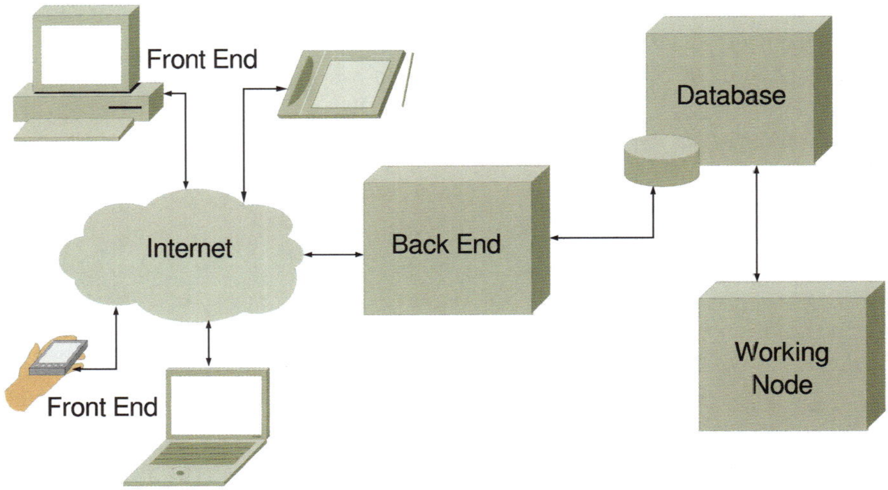

Fig. 4.4 Infrastructure with many front ends

4.5.3 Deployment Scenarios and Requirements

In order to deploy the framework, there are no big changes required on an already defined infrastructure. This is the main advantage of this tool, since it can be deployed with less effort and less impact on the pre-existing infrastructure. No specific database is required, since we use Database Abstraction Layer (DAL) to access the database and it supports most of the current used ones. There exist a string connection where explicitly says which database is going to be used, but this is the only place where it really matters. After this connection, all transactions and operations are performed through the DAL. Despite this, the only requirement are the ones related to specific benchmark performing and the working nodes.

The simplest deployment scenario is the one presented on Fig. 4.3. What is important to notice is that, since our front end is a web interface, we represent as only one front end, but it allows multiple client connections, so many users can use the system at the same time.

Another possibility from the front end point of view is to develop a different front end, that can or cannot be web based, which access the back end to perform the simulation. Since the communication is standardized and the execution is stateless, there is no need to implement different concurrence control from the ones natively implemented on database. The operations performed by one front end, does not directly affect the other one. Both are going to access the back end through URI and the shared resource in this case will be the database. Figure 4.4 shows how does it looks like.

The database is really important since it keeps all the relevant data and, as it is presented until now, is a single point of failure. To avoid data lost, we strongly

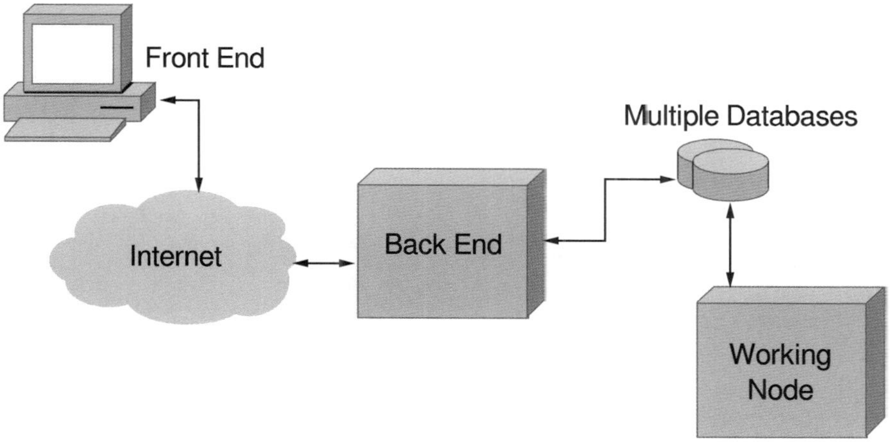

Fig. 4.5 Infrastructure with multiple databases

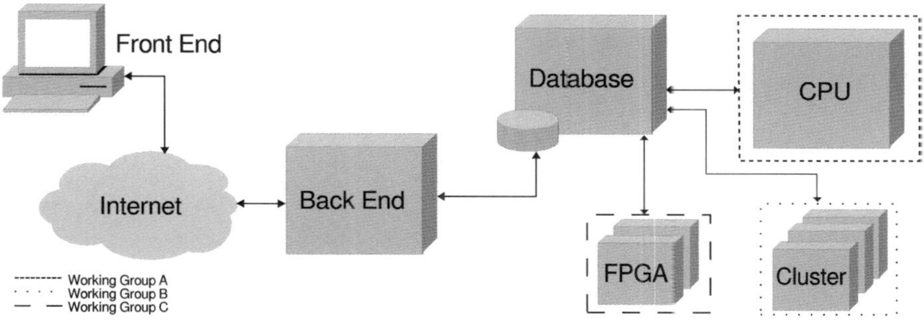

Fig. 4.6 Infrastructure with multiple working groups

recommend to have a multiple databases (Fig. 4.5), achieving data redundancy. Also, it permits to implement load balance and distribute the workload among the available database servers in a master-slave configuration. We have mentioned the working group which contains one or many working nodes along this chapter. In our prototype, each group represent a different implementation of a MC algorithm with hardware acceleration. This means that we can have many different working groups connected to our dispatcher, centralizing the information, becoming easier to either start a simulation or to compare their results. The working nodes which belong to the same group do not have to be physically on the same location, giving more freedom to the network topology. Figure 4.6 shows how those working groups are located in the infrastructure.

Due to the high modularity of the benchmark tool, it is possible to add new elements to improve the perceived performance, for example caches. Caches can be included between the front end and the back end, storing static data and reducing the number of requests to the back end. The stateless constraint of ReST is not

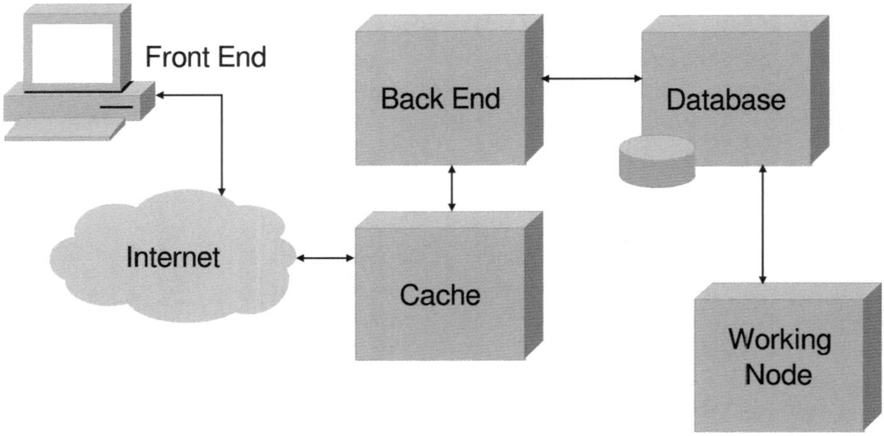

Fig. 4.7 Infrastructure with cache

violated since the requests state is still not responsibility of the receiver (back end in this case) and all requests contain all the needed information to complete. Adding a cache (Fig. 4.7) on the system also reduces the load of requests on the server, since static data could be directly retrieve from the cache.

Combining one or more of the presented scenarios together is also possible, which leads to a wider range of possibilities, and thereby system can be adapted to the needs of different deployment sites with a lower effort, since it has been designed with flexibility as the main goal. In case the number of performed benchmarks increase, or either different kinds of benchmarks have to be available from a certain period of time, the loosely coupled infrastructure provides the scalability capability, allowing infrastructure changes with low or none impact for all the other components within the system.

4.5.4 Improvements Aggregated for Current State of Benchmarking

Nowadays in order to compare different implementations and algorithms for market simulation, application-level benchmark batteries are been used. Integrating those benchmark batteries and the implementations in a unified platform reduces the time to compare the obtained results. Thus, our proposed unified approach aims to fill this gap on the comparative process, making the comparing process easier and less time demanding. In addition, the way this software has been developed allows cooperation between different institutes to work together, without be in the same physical location. Those advantages permit the research to focus on developing better and more efficient algorithms and implementations.

4.6 Summary

Application-level benchmarks are the only feasible way to characterize, evaluate, and compare computing systems used to solve a specific problem independent of their underlying technology. In particular, FPGA-accelerated systems that can be reconfigured depending on the currently active task prohibit executable standard benchmarks written in programming languages for CPUs and GPUs.

Upcoming commercial approaches like the STAC-A2 benchmark show that this topic becomes more and more important in the finance business at the moment. In this section we illustrate for the example of European barrier option pricing how such benchmarks can look like and how they can be designed platform- and algorithm-independent. One remaining challenge is that *soft* metrics like flexibility or the effort to adapt of enhance such systems during their lifetime cannot be quantified easily in numbers. For those characteristics we could for example apply batteries of tasks covering a broad range of applications and assign penalty points for non-supported features.

Finally, executing application-level benchmarks for a large number of solutions with the goal of comparing their performance is a time-consuming task, especially if the result accuracy is pre-given and various algorithms and their implementations are candidates and need to be investigated. For this task we introduce a framework for automatically evaluation an arbitrary number of implementations, independent of the underlying technology or internal structure. The framework is based on standard protocols, can integrate any kind of compute units, and communicate with any kind of sources or result collection/visualization tools. A core implementation that implements all the basic features is freely available for download.

Acknowledgements We gratefully acknowledge the partial financial support from the Center of Mathematical and Computational Modelling (CM)^2 of the University of Kaiserslautern, from the German Federal Ministry of Education and Research under grant number 01LY1202D, and from the Deutsche Forschungsgemeinschaft (DFG) within the RTG GRK 1932 "Stochastic Models for Innovations in the Engineering Sciences", project area P2. The authors alone are responsible for the content of this work. Furthermore, we thank Gordon Inggs from Imperial College London for his helpful inputs related to the current benchmark state and scheduler expertise.

References

1. Armstrong, B., Bae, H., Eigenmann, R., Saied, F., Sayeed, M., Zheng, Y.: HPC benchmarking and performance evaluation with realistic applications. In: SPEC Benchmarking Workshop, Austin (2006)
2. Berners-Lee, T., Fielding, R., Masinter, L.: Uniform Resource Identifiers (URI): Generic Syntax (1998)
3. Berry, M., Chen, D., Koss, P., Kuck, D., Lo, S., Pang, Y., Pointer, L., Roloff, R., Sameh, A., Clementi, E., Chin, S., Schneider, D., Fox, G., Messina, P., Walker, D., Hsiung, C., Schwarzmeier, J., Lue, K., Orszag, S., Seidl, F., Johnson, O., Goodrum, R., Martin, J.: The perfect club benchmarks: effective performance evaluation of supercomputers. Int. J. High Perform. Comput. Appl. **3**(3), 5–40 (1989)

4. Bondi, A.B.: Characteristics of scalability and their impact on performance. In: Proceedings of the 2nd International Workshop on Software and Performance (WOSP '00), Ottawa, pp. 195–203. ACM, New York (2000)
5. Clark, D.D., Tennenhouse, D.L.: Architectural considerations for a new generation of protocols. In: Proceedings of the ACM Symposium on Communications Architectures & Protocols (SIGCOMM '90), Philadelphia, pp. 200–208. ACM, New York (1990)
6. de Schryver, C.: Design methodologies for hardware accelerated heterogeneous computing systems. PhD thesis, University of Kaiserslautern (2014)
7. de Schryver, C., Jung, M., Wehn, N., Marxen, H., Kostiuk, A., Korn, R.: Energy efficient acceleration and evaluation of financial computations towards real-time pricing. In: König, A., Dengel, A., Hinkelmann, K., Kise, K., Howlett, R.J., Jain, L.C. (eds.) Knowledge-Based and Intelligent Information and Engineering Systems. Volume 6884 of Lecture Notes in Computer Science, pp. 177–186. Springer, Berlin/Heidelberg (2011). Proceedings of 15th International Conference on Knowledge-Based and Intelligent Information & Engineering Systems (KES)
8. de Schryver, C., Marxen, H., Schmidt, D.: Hardware accelerators for financial mathematics – methodology, results and benchmarking. In: Proceedings of the 1st Young Researcher Symposium (YRS) 2011, pp. 55–60. Center for Mathematical and Computational Modelling $(CM)^2$, $(CM)^2$ Nachwuchsring (2011). ISSN 1613-0073, urn:nbn:de:0074-750-0
9. de Schryver, C., Shcherbakov, I., Kienle, F., Wehn, N., Marxen, H., Kostiuk, A., Korn, R.: An energy efficient FPGA accelerator for Monte Carlo option pricing with the Heston model. In: Proceedings of the 2011 International Conference on Reconfigurable Computing and FPGAs (ReConFig), Cancun, Dec 2011, pp. 468–474 (2011)
10. Di Pierro, M.: web2py, 5th edn. Experts4Solutions (2013)
11. Fielding, R.T.: Architectural styles and the design of network-based software architectures. PhD thesis, University of California, Irvine (2000)
12. Fiksman, E., Salahuddin, S.: STAC-A2 on Intel architecture: from scalar code to heterogeneous application. In: Proceedings of the 7th Workshop on High Performance Computational Finance (WHPCF '14), New Orleans, pp. 53–60 (2014). IEEE, Piscataway
13. Haas, H., Brown, A.: Web services glossary. W3C note, W3C, February (2004) http://www.w3.org/TR/2004/NOTE-ws-gloss-20040211/.
14. Hull, J.C.: Options, Futures, And Other Derivatives, 8th edn. Pearson, Harlow (2012)
15. Inggs, G., Thomas, D., Luk, W.: A heterogeneous computing framework for computational finance. In: 2013 42nd International Conference on Parallel Processing (ICPP), Lyon, Oct 2013, pp. 688–697 (2013)
16. Jin, Q., Luk, W., Thomas, D.B.: On comparing financial option price solvers on FPGA. In: 2011 IEEE 19th Annual International Symposium on Field-Programmable Custom Computing Machines (FCCM), Salt Lake City, May 2011, pp. 89–92 (2011)
17. Kienle, F., Wehn, N., Meyr, H.: On complexity, energy- and implementation-efficiency of channel decoders. IEEE Trans. Commun. **59**(12), 3301–3310 (2011)
18. Lankford, P., Ericson, L., Nikolaev, A.: End-user driven technology benchmarks based on market-risk workloads. In: Proceedings of the 2012 SC Companion: High Performance Computing, Networking, Storage and Analysis (SCC), Salt Lake City, Nov 2012, pp. 1171–1175 (2012)
19. Marxen, H.: Aspects of the application of multilevel Monte Carlo methods in the Heston model and in a Lévy process framework. PhD thesis, University of Kaiserslautern (2012)
20. Marxen, H., Kostiuk, A., Korn, R., de Schryver, C., Wurm, S., Shcherbakov, I., Wehn, N.: Algorithmic complexity in the Heston model: an implementation view. In: Proceedings of the Fourth Workshop on High Performance Computational Finance (WHPCF '11), Seattle, Nov 2011, pp. 5–12. ACM, New York (2011). ISBN 978-1-4503-1108-3
21. Morris, G.W., Aubury, M.: Design space exploration of the European option benchmark using hyperstreams. In: International Conference on Field Programmable Logic and Applications (FPL 2007), Aug 2007, Amsterdam, pp. 5–10 (2007)

22. Nikolaev, A., Burylov, I., Salahuddin, S.: Intel® version of STAC-A2 benchmark: toward better performance with less effort. In: Proceedings of the 6th Workshop on High Performance Computational Finance (WHPCF '13), Portland, pp. 7:1–7:7. ACM, New York (2013)
23. Nogueira, C.P.: Benchmark management of option pricer implementations with hardware accelaration based on the heston model. Bachelor's thesis, Universidade Federal do Rio Grande do Sul, (2014)
24. Securities Technology Analysis Center LLC: STAC-A2 Central. https://stacresearch.com/a2 (2012). Last access 14 Feb 2015
25. The University of Tennessee Knoxville: HPC challenge benchmark. http://icl.cs.utk.edu/hpcc. Last access 14 Feb 2015
26. van der Steen, A.: The art of benchmarking HPC systems. http://www.hpcresearch.nl/talks/benchm_tech.pdf (2009). Last access 14 Feb 2015

Chapter 5
Is High Level Synthesis Ready for Business? An Option Pricing Case Study

Gordon Inggs, Shane Fleming, David B. Thomas, and Wayne Luk

Abstract High-Level Synthesis (HLS) tools for Field Programmable Gate Arrays (FPGAs) have made considerable progress in recent years, and are now ready for deployment in an industrial setting. This claim is supported by a case study of the pricing of a benchmark of Black-Scholes (BS) and Heston model-based options using a Monte Carlo Simulations approach. Using a high-level synthesis (HLS) tool such as Xilinx's Vivado HLS, Altera's OpenCL SDK or Maxeler's MaxCompiler, a functionally correct FPGA implementation can be developed from a high level description based upon the MapReduce programming model in a short time. This direct source code implementation is however unlikely to meet performance expectations, and so a series of optimisations can be applied to use the target FPGA's resource more efficiently. When a combination of task and pipeline parallelism as well as C-slowing optimisations are considered for the problem in this case study, the Vivado HLS implementation is 9.5 times faster than a sequential CPU implementation, the Altera OpenCL 221 times faster and Maxeler 204 times, the sort of acceleration expected of custom architectures. Compared to the 31 times improvement shown by an optimised Multicore CPU implementation, the 60 times improvement by a GPU and 207 times by a Xeon Phi, these results suggest that HLS is indeed ready for business.

5.1 Introduction

In the last few years HLS has made great strides, with the consensus being that FPGA designs can be created from a high level of abstraction. By abstracting onto commonly used programming languages such as C, custom hardware design and use by non-FPGA specialists is now a real possibility in industrial settings.

G. Inggs (✉) • S. Fleming • D.B. Thomas
Department of Electrical and Electronic Engineering, Imperial College London, London, UK
e-mail: g.inggs11@imperial.ac.uk; s.fleming06@imperial.ac.uk; d.thomas1@imperial.ac.uk

W. Luk
Department of Computing, Imperial College London, London, UK
e-mail: w.luk@imperial.ac.uk

C. De Schryver (ed.), *FPGA Based Accelerators for Financial Applications*,
DOI 10.1007/978-3-319-15407-7_5

In this chapter, our aim is to convince financial engineers that HLS is a viable means to use FPGAs in a production environment. To make our argument, we provide an introduction to the use of HLS in computational financial applications. This introduction is illustrated using a case study of an option contract pricing application using a Monte Carlo Simulations-based algorithm.

In our case study, we describe how the Monte Carlo Simulations option pricing can be implemented in three leading HLS tools from Altera, Xilinx and Maxeler. To demonstrate the versatility of HLS, we apply this approach to a benchmark of 13 exotic option pricing tasks that are based upon both the BS and Heston asset price evolution models.

Furthermore, we describe three common optimisations in a form accepted by the HLS tools. These optimisations increase the task, data and pipeline parallelism of the designs created, and so improve the throughput of the resulting designs considerably.

We characterise the development and resulting performance of our Monte Carlo Simulations case study using the three HLS tools both qualitative and quantitatively. We also compare the performance of the HLS implementations to other FPGA implementations created by hardware experts using a Hardware Description Language (HDL) as well as competing acceleration technologies, namely Central Processing Unit (CPU), Graphics Processor Unit (GPU) and hybrid architectures such as Intel's Xeon Phi.

With the evaluation of our case study, as illustrated in Fig. 5.1, we show that HLS is indeed ready for business, providing FPGA implementations for computational finance that are comparable to competing technologies both in terms of performance and development effort. This recommendation is not without caveats however, as care needs to be taken to ensure that the programming paradigm of HLS tool used is aligned with the algorithm being implemented.

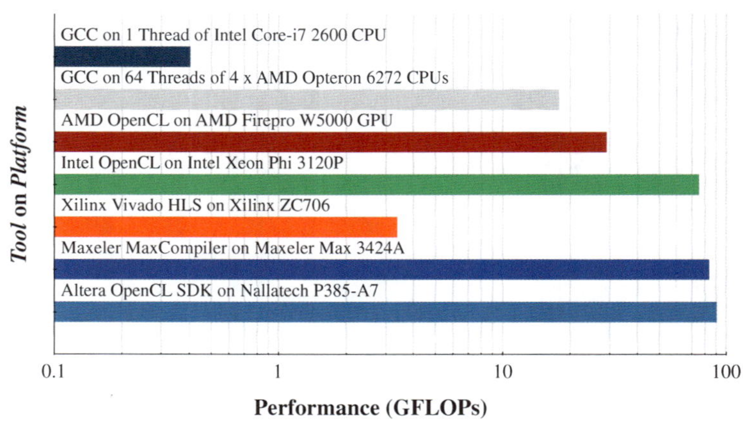

Fig. 5.1 Performance of option pricing benchmark implementations

A limitation of our study is that we have sought to cover the width of the current state of the art and so the three HLS tools surveyed target three FPGAs with different architectures and transistor technologies. As a result, the performance reported does not just reflect the capability of the particular HLS tools, but are also influenced by the devices used.

Our focus is on FPGA-based accelerators created with HLS as a whole, and not on comparing between implementations achieved using different HLS tools. Hence, we have not normalised the results across different FPGA devices; we do however try to explain differences in results in the relevant sections.

In this chapter, we will:

- Use a case study of forward looking option problems to demonstrate how HLS tools may be used in a computational finance context.
- Describe task, data and pipeline parallelism optimisations for FPGAs and how these may be expressed within HLS.
- Evaluate three leading HLS tools both quantitatively and qualitatively. This evaluation includes a comparison to FPGA expert implementations as well as competing accelerator technologies.

In the following section, we provide the background to the Monte Carlo Simulations approach as well as the option types, BS and Heston models that we use in our case study. The computational implementation of this approach is discussed, particularly its rendering as a programming pattern as well as previous work on FPGAs.

We then describe HLS implementations using our case study as an example. First we comment on an initial, direct source code entry into HLS tools, and then how optimisations that extend the task, pipeline and data parallelism of a design can be introduced.

In the Case Study Setup section, we detail the HLS tools and the experimental setup utilised in our study. This setup includes the software framework utilised as well as the platforms, both FPGA and competing accelerator technologies.

We then report and discuss the results of the Case Study Evaluation of the HLS tools, including a comparison with competing accelerator platforms. Finally, we present our conclusions on the maturity of the HLS tools surveyed.

5.2 Case Study Background

5.2.1 Option Pricing

For our case study, we have considered forwards looking option contracts. These are agreements where a holder pays a premium to the contract writer in order to obtain rights with regards to an underlying asset, such as 100 shares of a stock. This right either allows the holder to buy or sell the underlying asset at a defined strike price (K) at a defined exercise time (T). The holder has bought the right to exercise the option contract if they so choose, and is in no way obligated. Hence, in option pricing, the value of the option (c in the case of a call option, p for put options), is the difference between the strike price and spot price of the asset at exercises (S_T), or zero, whichever is higher [5].

5.2.2 Monte Carlo Simulations Option Pricing

5.2.2.1 The Monte Carlo Pricing Algorithm

The popular Monte Carlo Simulations approach for option pricing uses random numbers to create potential scenarios or paths for the underlying asset based upon a model of its spot price (S) evolution. The average outcome of these scenarios is then used to approximate the most probable option value [5], i.e.

$$c = e^{-r(T-t)} \int_W c(W) d\mathbf{P}(W) \approx e^{-r(T-t)} \frac{1}{N} \sum_i^N c_i$$

Where $e^{-r(T-t)}$ is the time discount factor and $P(W)$ is the probability space defined by the underlying asset. Although computational expensive, this approach is robust, and capable of tolerating asset models with many more variables than competing methods [5, 11].

5.2.2.2 Computational Implementation

Another advantage of the Monte Carlo Simulations approach is that it is extremely amenable to parallel execution. It is an "Embarrassingly Parallel" algorithm [1] that can easily be expressed as the MapReduce programming pattern [3]. Each underlying asset's price evolution and its interaction with the derivative product comprises the map operation, while the reduction is the average of the payoffs. We have described this algorithm in the MapReduce pattern in C code in Listings 5.1 and 5.2.

```
void monte_carlo_map(int seed,float *value){
    state_t state;
    for(i=0;i<N;++i){
        state = path_init(seed+i);
        for(j=0;j<D;++j) state = path(state);
        value[i] = payoff(state);
}
```

Listing 5.1 Map behaviour of Monte Carlo option pricing as MapReduce

```
float monte_carlo_reduce(float *value){
    float value = 0.0;
    for(i=0;i<N;++i) result += value[i]/N;
    return value;
}
```

Listing 5.2 Reduce behaviour of Monte Carlo option pricing as MapReduce

The map function is the more computationally intensive of the two. The map function body is comprised of two loops: an outer loop with no dependencies between iterations, that is bound by the *N* variable as per Listing 5.1. Each outer loop iteration contains the simulation initiation, an inner simulation lifetime loop and the product payoff behaviour at the simulation termination. The inner lifetime loop, which is bound by *D*, evolves the underlying price and any accompanying option product behaviour over the time period specified, and so is data-dependent.

5.2.3 Monte Carlo Simulations upon FPGAs

Imperial College London has published work on the FPGA acceleration of the pricing of Exotic Options using the BS model [12]. The BS model is computationally complex with its price evolution given by [5]

$$\hat{S}_t = S_0 e^{\gamma_t}$$

$$\gamma_t = \sum_{i=1}^{t} [(\mu - \frac{\sigma^2}{2})h + \sigma X_i^1 \sqrt{h}]$$

We have captured the BS model in code in Listing 5.3. The Imperial College work has shown that through the use of FPGAs, both the latency and energy utilisation of option pricing computations can be minimised.

```
void path(state_t *state){
    float con = (RFIR-pow(SIGMA,2)/2)*H;
    float x = gaussian_rng()
    float vol = SIGMA*x*sqrt(H);

    state->gamma += con + vol;
    state->time += H;
}
```

Listing 5.3 BS model path description

Table 5.1 Overview of option pricing problems

Option type	Path	Payoff
European	None	$c_E = max(K - \hat{S}_T, 0)$
Barrier	Check barrier	If knock-in and barrier crossed, or knock-out and not crossed, $c_B = c_E$, else $c_B = 0$
Double barrier	Check both upper and lower barrier	Same as barrier
Double digital barrier	Same as double barrier	If $c_B > 0.0$, $c_{DDB} = 1.0$, else $c_{DDB} = 0$
Asian	$\hat{S}_{sum,t} = \hat{S}_{sum,t-1} + \hat{S}_t$	$c_A = max(K - \frac{\hat{S}_{sum,T}}{D}, 0)$

The University of Kaiserslautern has published a benchmark of forward looking option pricing problems[1] based upon the Heston model [4] for use in evaluating the performance of accelerators of option pricing problem. Along with the benchmark, they have published work on the FPGA acceleration of these problems, also demonstrating latency and energy optimisations [9].

The Heston model used in the Kaiserslautern work is even more computational complex than the BS model, as the volatility, v, varies stochastically as well as the asset price. Hence,

$$\hat{v}_k = \sum_{i=1}^{k} [\kappa(\theta - \sigma_k)h + \sqrt{\sigma} X_i^2]$$

Where X_i^2 correlates with the asset's X_i^1 with factor ρ.

A summary of the option product behaviours which interact with the underlying simulations covered in both papers is provided in Table 5.1. Both of these bodies of work refer to HDL implementations, where the computational advantages of FPGAs are only realised through the specialist experience and knowledge of the researchers undertaking these implementations.

5.3 HLS Implementation Approach

5.3.1 Initial HLS Implementation

A reasonable first HLS implementation would be to identify the computationally intensive component of an application and then implement it in the FPGA using the HLS tool. For the Monte Carlo Simulations approach this would be the map

[1] http://www.uni-kl.de/en/benchmarking/option-pricing/

function as described in Listing 5.1, with the reduce behaviour in Listing 5.2 being performed upon a CPU that interfaces with the map function implemented upon the FPGA.

5.3.2 *Optimisating the HLS Implementation*

While the implementation outlined above for Monte Carlo Simulations approach would achieve functionally correct designs, its unlikely that much of the FPGA resources would be utilised, or that those resources utilised would be put to the most effective use. This is not dissimilar to how a software compiler that would not automatically achieve a highly optimal, parallel implementation.

Below we outline three optimisations that we illustrate using the Monte Carlo Simulations example that allow for the target FPGA's resources to be utilised more efficiently.

5.3.2.1 Task Parallelism

```
void monte_carlo_map_tp(int seed,float *value){
    for(p=0;p<P;++p){
        for(i=0;i<N;i+=P){
            state_t state = path_init(seed+i+p);
            for(j=0;j<D;++j) state = path(state);
            value[i+p] = payoff(state);
        }
    }
}
```

Listing 5.4 Making the potential task parallelism explicit

The first optimisation we describe, Task Parallelism, will be familiar to most programmers, thanks to the popularity of software libraries such as OpenMP, Pthreads and Threaded Building Blocks.

In a similar manner to how additional, identical software worker threads may be spawned, multiple instances of the design can be implemented alongside one another in the FPGA fabric, with additional control and communication logic. We have illustrated task parallelism for our case in Listing 5.4 by introducing a third parallel loop bound by *P*.

5.3.2.2 Pipeline Parallelism

```
void monte_carlo_map_pp(int seed,float *value){
    for(i=0;i<N;++i){
        state_t state = path_init(seed+i), state2;
        for(j=0;j<D;j+=2){
            state2 = path(state);
            state = path(state2);
        }
    value[i] = payoff(state);
    }
}
```

Listing 5.5 Doubling the potential pipeline parallelism

Computationally intensive FPGA designs, such as our use case, take the form of pipelines of operations that are performed by discrete units of logic. As these discrete logic units can operate simultaneously, by filling the target design's pipeline each cycle, a high calculation throughput can be achieved.

Our second optimisation is to 'unroll' the data dependent inner loop of our code so as to extend the length of the operation pipeline, and further exploit pipeline parallelism as demonstrated in Listing 5.5.

5.3.2.3 C-slowing

```
void monte_carlo_map_cs(int seed,float *value){
    state_t state[C];
    for(k=0;k<N;k+=C){
        for(i=0;i<C;++i)state[i]=path_init(seed+k+i);
        for(j=0;j<D;++j)
            for(i=0;i<C;++i)state[i]=path(state[i]);
        for(i=0;i<C;++i)value[k+i]=payoff(state[i]);
    }
}
```

Listing 5.6 C-slow transformation

The final optimisation that we consider, C-slowing, is reconfigurable computing-centric. This technique uses memory resources to 'hide' the latency of operations in the design pipeline. Ordinarily a pipeline is constrained by the slowest operation, however by inserting a memory array between each operational unit, each operational unit can operate on all of the values in the array before requiring outputs from the previous unit in the pipeline.

In our Monte Carlo Simulations case this is achieved by adding a new outer loop which chunks the set of simulations by factor *C*, which is the degree of C-slowing. The simulation initiation, lifetime and payoffs behaviour are then all performed within independent loops bound by *C*. An appropriate memory structure must also be added to maintain the loop state. We have illustrated this optimisation in Listing 5.6.

5.4 Case Study Setup

5.4.1 Tasks

For our case study, we consider both the BS model-based Asian Option from the Imperial College work as well as the 12 Heston model-based Barrier options from the Kaiserslautern benchmark, as discussed in Sect. 5.2.

By implementing all of these heterogeneous problems, we demonstrate the versatility and flexibility enabled by HLS tools.

5.4.2 HLS Tools Surveyed

For our case study, we utilised the three tools detailed below. Two are from well-known FPGA vendors, Xilinx and Altera while the third is from Maxeler, which targets FPGA devices from both of vendors.

Xilinx's Vivado HLS – Designs are inputted using a subset of the C programing language, with user-added "directives" that can influence how the design is implemented within the FPGA architecture. Vivado HLS prioritises functional equivalence with the inputted source code, and relies upon the hardware developer to transform their code in order to enhance the efficiency of their designs. The large number of diverse directives allow for this to be done effectively [13].

Altera OpenCL SDK – The SDK is the first offering from a FPGA manufacturer to support the OpenCL standard [10] which is supported by many CPU and GPU manufacturers. Similar to Vivado HLS a reasonably large subset of C is supported. In addition to source code pragmas, there are compiler options which influence how the design is mapped into hardware. The SDK is only supported by a set of compatible boards from preferred Altera partners such as Nallatech. The programming paradigm of the Altera OpenCL SDK is to map the task parallelism made explicit in the OpenCL model into pipeline parallelism within the FPGA.

Maxeler's Tools – this is a third party vendor that makes use of both Xilinx and Altera's FPGAs to provide both the hardware platform and the accompanying software tools for high performance, reconfigurable computing. Designs or "dataflow engines" are described using a purpose-built Java library in a dataflow paradigm, allowing for both behavioural and architectural features to be described by the developer. Similar to the Altera OpenCL SDK, the dataflow paradigm ensures that the designer makes the degree to which pipeline parallelism can be exploited in the design as explicit as possible.

5.4.3 Supporting Software Framework

For our case study we extended the Forward Financial Framework(F^3), an open source, Domain Specific application framework for computational finance that targets heterogeneous computing platforms [6, 7].

We performed our evaluation within an exiting framework as we believe this reflects a similar development process to what would be undertaken in an industrial setting. Once suitably extended, F^3 provides the ability to generate, compile and execute computational finance problems from a single description upon a variety of platforms including CPUs, GPUs and FPGAs.

5.4.4 Competing Accelerator Technologies

In our case study we have compared the optimised HLS implementations of Monte Carlo Simulations option pricing problems against other classes of computing platforms popular in financial engineering including CPUs, GPUs and hybrid accelerators.

The Monte Carlo Simulations pricing algorithm is well suited to parallel architectures, with our expectation being that the latency of each platform (L_p) relative to the sequential processor (L_s) would be given by the degree of floating point computational parallelism of each platform (P_p) and ratio of the clock rate ($\frac{C_s}{C_p}$), as per Amdahl's Law i.e.

$$L_p \approx \frac{L_s C_s}{P_p C_p}$$

5.4.5 Metrics

5.4.5.1 Development Metrics

We consider four development metrics in our case study. We captured the average code length for the computational kernel for each HLS implementation, the development time that we spent on the applications in question, the time spent integrating the implementations into F^3 and finally the compile time range that we observed for that particular HLS tool.

5.4.5.2 Performance Metrics

Latency, the primary performance metric used in our case study, is measured using the wall time from when the requisite execution on the host system is first executed until it returns the price result to the user. Thus, the latency metric captures all of the communication time between the host and the FPGA, the overhead incurred in initialising the computation, as well as the computation itself.

5.4.6 *Experimental Platforms*

5.4.6.1 Host Systems

We utilised three comparable reconfigurable computing platforms to evaluate the HLS implementations as detailed in Table 5.2. The Nallatech P387-A7 and Maxeler Max3425A are standalone Peripheral Component Interconnect Express (PCIe) cards hosted within commodity computing platforms. In the case of the Maxeler tools, the host system is a modified desktop grade system, with a Intel Core-i7 CPU running CentOS Linux. The Nallatech board was hosted in a server grade system with an Intel Xeon CPU also running CentOS Linux. The Xilinx ZC706 Zynq development board is more self-contained with both the host CPU and FPGA fabric contained within a single integrated circuit with an Advanced eXtensible Interface (AXI) bus between the two. The host CPU is an ARM dualcore CPU, which is running Ubuntu Linux.

5.4.6.2 Experimental FPGAs

The resources of the FPGAs used are detailed in Table 5.3 The Stratix V has the most logic resources, however the least dedicated Digital Signal Processor (DSP) units; the Xilinx Virtex 6 FPGA targeted by the Maxeler tools has the most DSP

Table 5.2 Experimental platforms

Vendor	Name	FPGA	Communication Technology	HLS tool
Xilinx	ZC706 1.1	Xilinx Zynq 7Z045	AXI	Xilinx Vivado HLS 2013.4
Nallatech	P385-A7	Altera Stratix V GXA7	PCIe	Altera OpenCL SDK 13.0
Maxeler	Max 3424A	Xilinx Virtex 6 XC6VSX475T	PCIe	Maxeler MaxCompiler 13.2.2

Table 5.3 Comparison of experimental FPGA resources

FPGA	CMOS size (nm)	Targeted clockrate (MHz)	Lookup Table (LUT)s (k)	Flip-Flop (FF)s (k)	Block RAM (BRAM)s	Digital Signal Processor (DSP)s
Xilinx Zynq 7Z045	28	100	218.6	437.2	545	900
Altera Stratix V GXA7	28	250	622	939	2,304	768
Xilinx Virtex 6 XC6VSX475T	40	200	297.6	595.2	1,064	2,016

Table 5.4 Comparison of reference platforms

Platforms	CMOS size (nm)	Clockrate (GHz)	Memory (GBs)	Threads	Tool
Intel Core i7-2600S	32	2.8	16	1	GCC 4.8
AMD Opteron 6272	32	2.1	128	32	GCC 4.8
AMD Firepro W5000	28	0.825	2	768	AMD OpenCL SDK 2.9
Intel Xeon Phi 3120P	22	1.1	6	256	Intel OpenCL SDK 2014

units, however is a larger process technology; the Zynq platform has the least resources of the three however is the most integrated, with the reconfigurable logic and host CPU sharing the same silicon. When measuring resource utilisation, we used a percentage of the most constrained resource: DSP units in the case of ZC706 and P385-A7 platforms and Lookup Table (LUT)s in the case of the Max 3424A.

5.4.6.3 Competing Accelerator Technology Platforms

The competing accelerator technology platforms are detailed in Table 5.4. The CPU comparisons include highly optimised sequential and multicore C code implementations compiled with GNU's Not Unix (GNU)'s Cross Compiler targeted at an Intel Core-i7 and at AMD Opteron CPU respectively. For the GPU and Xeon Phi implementations the same OpenCL code for the Altera OpenCL SDK was used, however the code was optimised to the platforms, an AMD Firepro W5000 GPU and an Intel Xeon Phi 3120P Co-processor. Finally, we also compared our HLS efforts against the HDL implementations reported in the studies described in Sect. 5.2.3.

5.4.7 *Case Study HLS Implementations*

5.4.7.1 Initial Implementations

Across all implementations we used the Combined-Tausworthe random number generator with the Box-Muller transformation to produce the Gaussian random numbers required [2, 8]. We seeded the random number generator for each path using random numbers generated on the host CPU. As suggested in the previous section, the reduction operation to calculate the average option value was also done upon the host CPU, and was performed in parallel with the map operation upon the FPGA.

In all cases arithmetic was done using single precision floating point operations, which is common practise in this application domain. All results were verified by ensuring that there was significant overlap between the price distributions across the platforms surveyed.

5.4.7.2 HLS Optimisations

All three HLS tools allow for **task parallelism** to be expressed explicitly by the programmer. Source code pragmas were used in the Altera OpenCL SDK and an architectural loop description in the Maxeler tools. In Vivado HLS, multiple function instantiations in the source code were required to achieve task parallelism.

Pipeline parallelism was introduced using source code pragmas in the Altera OpenCL SDK and a combination of directives and source code transformation in Vivado HLS. There is no provided technique in the Maxeler tools, so the source code generation capabilities of F^3 were utilised to create longer operation pipelines.

In the Maxeler tools **C-Slowing** is implemented using the dataflow stream addressing functionality while inverting the loops, while in the Vivado HLS and Altera OpenCL SDK this was implemented by also inverting the loops while explicitly creating the memory resource for storing the loop state in the source code.

5.4.7.3 Competing Accelerators Implementations' Optimisation

For both the sequential and multithreaded CPU C code implementations we used algorithmic optimisations, such as exiting simulations early if a barrier event occurs, as well as compiler optimisations, such as relaxed mathematical operations.

For the GPU and hybrid accelerator implementations the main optimisation was the agglomeration of multiple simulation paths within a single OpenCL work item so as to reduce the communication overhead. Furthermore we used platform-specific mathematical functions.

5.5 Case Study Evaluation

5.5.1 Development

Table 5.5 provides the developmental results for the three HLS tools considered in our Monte Carlo Simulations case study.

Generally, the development metrics reported support the claim that the HLS tools are ready for industrial deployment, with a relatively short design time. We recognise the relatively long compile times will be of concern to developers new to FPGAs, and hence highlight the need for the use of effective debugging tools throughout the design process.

A notable exception is however the Xilinx Vivado HLS implementation. We found Vivado provides the most faithful transformation of the naive source code into resource efficient hardware implementations – the Vivado HLS code was the closest to Listing 5.1. However, we often found this fidelity of transformation results in under-use of the resources available, and as a result, poor throughput. It is left up to the hardware developer to identify and exploit optimisations through the use of source code transformation and directives.

5.5.2 Performance

Figure 5.2a,b provide the performance results for the case study's Maxeler Max-Compiler results according to the optimisations implemented. Figure 5.3a, b provide the performance results for the Altera OpenCL SDK as implemented upon the Nallatech P385-A7 board. Finally, Fig. 5.4a, b provide the same for the Xilinx Vivado HLS implementations upon the ZC706.

For the Monte Carlo Simulations case study, we found a combination of task parallelism and C-slowing provided the best performance for the Maxeler and Xilinx HLS tools and platforms, while loop unrolling and C-slowing provided the

Table 5.5 Development metrics of Monte Carlo Simulations case study

HLS tool	Code length (LoC)	Development time (weeks)	Integration time (weeks)	Compile time range (h)
Xilinx Vivado HLS	63	≈ 12	≈ 1	[2;6]
Altera OpenCL SDK	72	≈ 1	≈ 0.5	[2;8]
Maxeler MaxCompiler	130	≈ 4	≈ 2	[2;48]

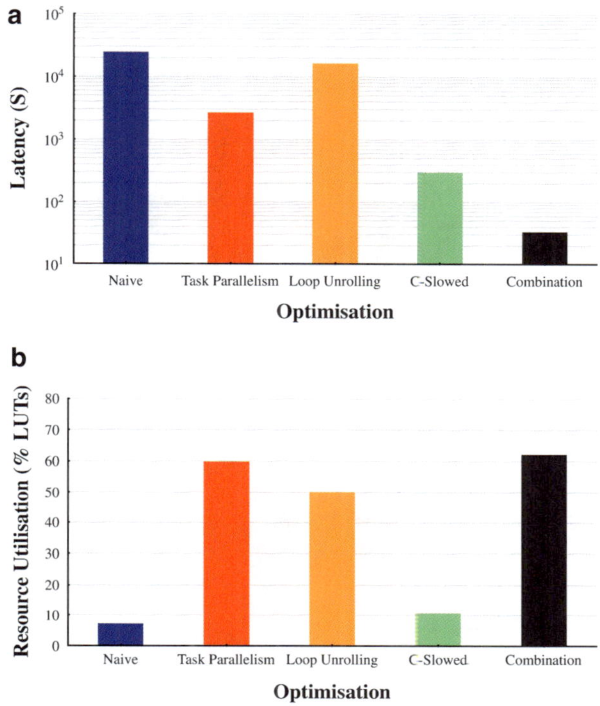

Fig. 5.2 Maxeler MaxCompiler results. (**a**) Latency of MaxCompiler implementations. (**b**) Resource utilisation of MaxCompiler implementations

best performance for the Altera tool and platform. The increased task or pipeline parallelism allows for FPGA resources to be traded for better throughput, while C-slowing improves the efficiency of the resources that are utilised.

In the Maxeler case, as the Virtex 6 FPGA we used is significantly older than the other platforms utilised, hence the logic resources proved to be the limiting factor.

We found that the Xilinx tool and platform performed poorly, which we attribute to a mismatch between the algorithm and the direct translation programming paradigm of Vivado HLS. A further contributing factor to this platform's relative under-performance was the inability to implement a design with a clock rate higher than 100 MHz. This was due to the nature of the AXI communications infrastructure we were utilising.

5.5.3 Comparing HLS to Competing Accelerators

The absolute latency performance results reported in Table 5.6 suggest that the HLS implementations generally provide superior or competitive performance results to

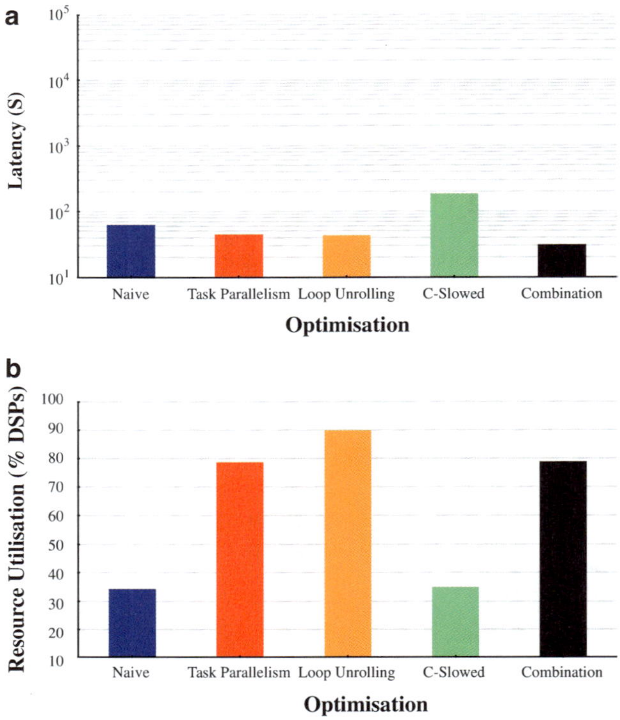

Fig. 5.3 Altera OpenCL results. (**a**) Latency of Altera OpenCL SDK implementations. (**b**) Resource utilisation of Altera OpenCL SDK implementations

other platforms. The relative latency results in Table 5.7 provide further insight into how the HLS tools compare to other implementations as part of the same software framework.

We found the acceleration relationship of the multicore CPU to the sequential is close to what is predicted by the equation we developed in Sect. 5.4.4 as it uses the identical implementation albeit multithreaded. The GPU implementation underperforms by a factor of approximately 3, while the Xeon Phi performs better by approximately 2 times, suggesting the extra distributed control structures in the latter are a key architectural feature in performance.

The HLS implementations of Altera, Maxeler and even Xilinx similarly show acceleration beyond that which is explained by the clock rate and degree of task parallelism. We attribute this performance to both the customisation of the architecture to the problem under consideration, as well as the exploitation of fine-grained parallelism within the algorithm which is being captured by the tools utilised.

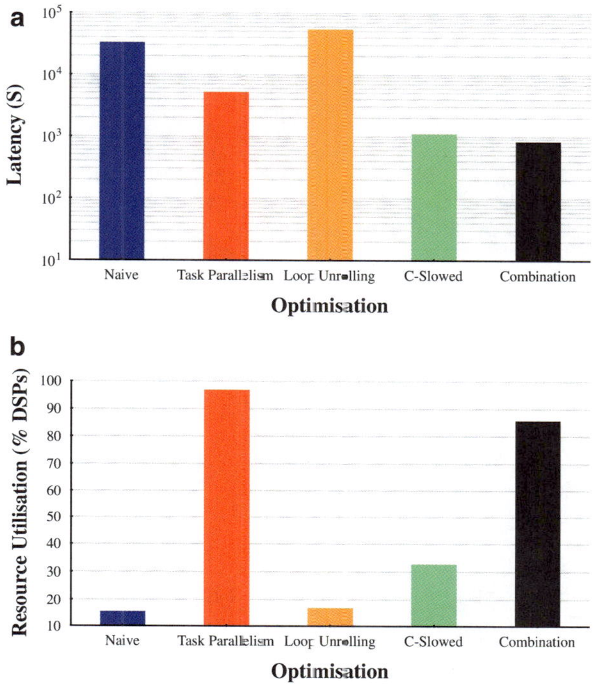

Fig. 5.4 Xilinx Vivado HLS results. (**a**) Latency of Xilinx Vivado HLS implementations. (**b**) Resource utilisation of Xilinx Vivado HLS implementations

Table 5.6 Absolute latency performance of HLS implementations, competing accelerator technologies and references (S)

	Target platform	Heston European	Heston barrier	Heston double barrier	Heston double digital barrier	Black-Scholes Asian
HLS	Max 3424A	38.41	38.62	38.37	38.36	19.21
	P385-A7	32.49	29.70	32.67	30.50	29.50
	ZC706	753.53	958.94	959.23	959.70	588.33
Competing accelerator technologies	Sequential CPU	11,560.68	6,763.17	5,957.36	6,111.69	5,308.41
	Multicore CPU	316.81	203.56	175.16	185.73	234.26
	GPU	123.54	129.99	131.23	131.19	67.09
	Xeon Phi	18.08	47.94	57.09	58.55	13.63
Expert FPGA Reference	Various[a]	287.00	287.00	287.00	287.00	31.68

[a]Please see [9, 12]

Table 5.7 Relative latency of experimental implementations

	Sequential CPU	Max 3424A	P385-A7	ZC706	Multicore CPU	GPU	Xeon Phi
Sequential CPU	–	0.005	0.005	0.121	0.032	0.017	0.005
Max 3424A	204.81	–	0.925	24.775	6.521	3.382	0.988
P385-A7	221.46	1.081	–	26.789	7.051	3.657	1.068
ZC706	8.27	0.040	0.037	–	0.263	0.137	0.040
Multicore CPU	31.41	0.153	0.142	3.799	–	0.519	0.152
GPU	60.561	0.296	0.273	7.326	1.928	–	0.292
Xeon Phi	207.292	1.012	0.936	25.075	6.600	3.423	–

5.6 Conclusion

In this chapter we have introduced the use of HLS tools for FPGAs in a financial engineering industrial setting. Our argument for doing so rested upon a case study of a benchmark of option pricing problems using the Monte Carlo Simulations approach upon a variety of HLS tools and competing accelerator technologies.

Unsurprisingly we found that naively entering source code into the surveyed tools generally results in inefficient designs. However, through a combination of the exploitation of task and pipeline parallelism, as well as the use of C-slowing, we could make better use of the FPGA's available resources. We also found that the resulting optimised HLS implementations could compete with alternative accelerator technologies.

A further insight is that the interaction between the programming paradigm of the HLS tool and the algorithm being implemented disproportionally impacts the development effort required to realise optimal performance. We found the tools offered by Maxeler and Altera are well-suited to accelerating parallel-friendly algorithms such as the Monte Carlo Simulations pricing algorithm. Xilinx's offering is however currently better suited to small, functional unit prototyping.

Acknowledgements We would like to acknowledge the support of the Maxeler, Altera and Xilinx University Programs, as well as Nallatech for use of their test server. We would also like to thank the University of Cape Town's Radar and Remote Sensing Group for the use of their Xeon Phi server.

Funding support was generously provided by the South African National Research Foundation as well as the Oppenheimer Memorial Trust.

References

1. Asanovic, K., Catanzaro, B.C., Patterson, D.A., Yelick, K.A.: The landscape of parallel computing research: a view from Berkeley. EECS Department University of California Berkeley, Technical report, UCBEECS2006183 (UCB/EECS-2006-183), p. 19 (2006)
2. Box, G., Muller, M.E.: A note on the generation of random normal deviates. Ann. Math. Stat. **29**(2), 610–611 (1958)
3. Dean, J., Ghemawat, S.: Mapreduce: simplified data processing on large clusters. In: Proceedings of the 6th Conference on Symposium on Opearting Systems Design & Implementation (OSDI'04), San Francisco, vol. 6, pp. 10–10. USENIX Association, Berkeley (2004)
4. Heston, S.L.: A closed-form solution for options with stochastic volatility with applications to bond and currency options. Rev. Financ. Stud. **6**(2), 327–343 (1993). http://www.jstor.org/stable/2962057
5. Hull, J.C.: Options, Futures and Other Derivatives, 8th edn. Pearson, London (2011)
6. Inggs, G., Thomas, D.B., Luk, W.: A heterogeneous computing framework for computational finance. In: Proceedings of the 42nd International Conference on Parallel Processing (ICPP), Lyon, pp. 688–697 (2013). doi:10.1109/ICPP.2013.82
7. Inggs, G., Thomas, D.B., Luk, W.: A domain specific approach to heterogeneous computing: from availability to accessibility. In: Proceedings of the First International Workshop on FPGAs for Software Programmers (FSP 2014), Munich (2014). http://arxiv.org/abs/1408.4965
8. L'Ecuyer, P.: Maximally equidistributed combined tausworthe generators. Math. Comput. Am. Math. Soc. **65**(213), 203–213 (1996)
9. de Schryver, C., Shcherbakov, I., Kienle, F., Wehn, N., Marxen, H., Kostiuk, A., Korn, R.: An energy efficient FPGA accelerator for Monte Carlo option pricing with the Heston model. In: 2011 International Conference on Reconfigurable Computing and FPGAs (ReConFig), Cancun, pp. 468–474 (2011). doi:10.1109/ReConFig.2011.11
10. Stone, J., Gohara, D., Shi, G.: OpenCL: a parallel programming standard for heterogeneous computing systems. Comput. Sci. Eng. **12**(3), 66–73 (2010). doi:10.1109/MCSE.2010.69
11. Thomas, D.B., Luk, W.: A domain specific language for reconfigurable path-based Monte Carlo simulations. In: Proceedings of the International Conference on Field-Programmable Technology, Kitakyushu, pp. 97–104 (2007)
12. Tse, A.H., Thomas, D.B., Tsoi, K.H., Luk, W.: Efficient reconfigurable design for pricing Asian options. SIGARCH Comput. Archit. News **38**(4), 14–20 (2011). doi:10.1145/1926367.1926371. http://doi.acm.org/10.1145/1926367.1926371
13. Winterstein, F., Bayliss, S., Constantinides, G.A.: High-level synthesis of dynamic data structures: a case study using vivado hls. In: 2013 International Conference on Field-Programmable Technology (FPT), Kyoto, pp. 362–365. IEEE (2013)

Chapter 6
High-Bandwidth Low-Latency Interfacing with FPGA Accelerators Using PCI Express

Mohammadsadegh Sadri, Christian De Schryver, and Norbert Wehn

Abstract The need for high performance computing dictates constraints on the acceptable bandwidth of data transfer between processing units and the memory. Consequently it is crucial to build high performance, scalable, and energy efficient architectures capable of completing data transfer requests at satisfactory rates. Thanks to increased transfer rates obtained by exploiting high-speed serial data transfer links instead of traditional parallel ones, PCI Express provides a promising solution to the problem of connectivity for todays complex heterogeneous architectures. In this chapter, we first cover the principals of interfacing using PCI Express. To illustrate a practical situation, we select the Xilinx Zynq device and develop an example architecture which allows the x86 CPU cores of the host system, the ARM cores of the Zynq device, and the hardware accelerators directly realized on the FPGA fabric of the Zynq to share the available DRAM memory for efficient data sharing. We provide estimates on possible data transfer bandwidths in our architecture.

6.1 Introduction

As the energy efficiency requirements (e.g. GOPS/W) of silicon chips are growing exponentially, computer architects are seeking solutions to continue application performance scaling. One emerging solution is to use specialized functional units (accelerators) at different levels of a heterogeneous architecture. These specialized units cannot be used as general-purpose compute engines. However, they provide enhanced execution speed and power efficiency for their specific computational workloads [3]. There exist numerous applications for accelerators in both of the embedded and high performance computing markets. However, to make them accessible, a fast and flexible interconnect mechanism to the host system is crucial.

Efficient sharing of data in a heterogeneous architecture which contains different types of integrated elements is a challenging task. A versatile method should

M. Sadri (✉) • C. De Schryver • N. Wehn
Microelectronic Systems Design Research Group, University of Kaiserslautern, Kaiserslautern, Germany
e-mail: sadri@eit.uni-kl.de; schryver@eit.uni-kl.de; wehn@eit.uni-kl.de

C. De Schryver (ed.), *FPGA Based Accelerators for Financial Applications*, DOI 10.1007/978-3-319-15407-7_6

be devised to act as the interconnect between different components in such a system. The method should provide enough data transfer bandwidth meeting the requirements for high performance Central Processing Unit (CPU) cores and hardware accelerators. It should be scalable, allowing addition of new components to the system easily and without degrading overall performance. Furthermore, it is required to guarantee certain levels of robustness and reliability in data transfers. When private caches of CPU cores and dedicated memory of accelerators are used to store local copies of data, it is crucial to ensure that every processing element has a consistent view of the shared memory space. Consequently, the communication method should provide suitable means of ensuring data consistency among different components of the system.

When the components residing in a single chip should communicate with each other, considering the short distances between them and vast on-die routing resources, wide parallel data buses clocked at high frequencies can be used. Examples of such communication schema are Intellectual Property (IP) cores interfacing based on the Advanced eXtensible Interface (AXI) specification developed by ARM.

For inter-chip communications in which the signals should travel long distances, wide parallel data buses are not a suitable choice since the clock frequency of data transfer will be confined to few hundred MHz to ensure correct capturing of all data bits at the receiver side. When transferring data bits over a single link in a continuous serial stream however, the aforementioned problem does no more exist and the clock frequency of data transmission can be increased to multiple GHz. In this case it is crucial to ensure that the receiver can recover each single bit of transmitted data at a suitable time thus it is required to embed the information related to clock signal into the data stream itself. This is usually the source of an additional overhead in transmission of data e.g. in front of each 8 data bits which should be transmitted, practically 10 bits will be sent to ensure correct recovery of data at the receiver. In spite of this, the increase in bandwidth by serial methods is still far beyond parallel methods. Peripheral Component Interconnect Express (PCIe) or shortly PCI Express is currently the leading choice for connecting different hardware units in computational platforms.

6.1.1 History: From ISA to PCI Express

From the early days of appearance of IBM PC computers, different standards were introduced to allow extension cards and peripherals to be added to a computing platform and to communicate to its main CPU through an I/O bus. Examples of such buses include Industry Standard Architecture (ISA), Extended Industry Standard Architecture (EISA), and Micro-Channel. These interfaces had at most a data bus width of 32 bits and were running at frequencies below 10 MHz.

With the appearance of Peripheral Component Interconnect (PCI) the clock frequency of data transfers increased to 33 MHz and above. For applications with higher demand on bandwidth, 64 bits wide versions of PCI were also introduced. The PCI based graphics card soon replaced the old ISA cards. However, in a short duration the bandwidth provided by these cards was also not enough for high definition 3D graphics, as a result a superset of conventional PCI called Accelerated Graphics Port (AGP) replaced PCI graphics cards. A performance improved version of PCI called Peripheral Component Interconnect Extended (PCI-X) dedicated to servers has a data bus width of 64 bits and operates at frequencies up to 533 MHz. This results in a total bandwidth of 4.26 GB/s over the bus. This is the highest bandwidth achievable by the PCI standard.

PCIe first introduced in 2004 is practically a replacement for all of the previous standards. It uses point-to-point high speed serial data links instead of a shared parallel bus. PCIe uses lower number of pins and smaller physical footprint in comparison to PCI while providing higher level of bandwidth. Thanks to the point-to-point connection topology, it provides better performance scaling as well as improved error detection and reporting mechanisms.

The glsPCIe connection between two nodes in the system can contain from 1 to 32 *lanes*. Each lane contains a separate high-speed serial link for receiving and another link for transmitting the data. Each high-speed serial link consists of two wires that transfer the data using differential signaling. As of the first version of PCIe (which was called PCIe Gen1) each link was capable of transferring 250 MB/s of data in each direction. The latest publicly available version (PCIe Gen3) increased the per-lane per-direction rate to 985 MB/s. For example a Gen3 PCIe card with 16 lanes (a 16X PCIe card) is cable of transferring up to 15.75 GB/s in each of the read and write directions at the same time.

6.2 Essential Basics of PCI Express

PCIe uses Transaction Layer Packets (TLPs) to transfer data between two nodes in the system. Each read or write transaction involves a series of one or more packet transmissions. These packets are responsible for transferring data, configuration parameters, messages, and event information between a PCIe device and the host. This also includes the interrupts generated by the PCIe device which should be delivered to the main CPU. Each TLP contains a header of around 16 bytes and payload of up to 4,096 bytes. The header contains information related to the type of the packet, its length, the ID of requester, its destination address, and so on. Two types of TLPs exits:

1. *Request TLPs*, which contain a request for an operation to a PCIe node in the system and
2. *Completion TLPs* that are generated by the completer and contain the response to the request.

For example, when a PCIe device decides to write to a specific I/O address, it generates a write request TLP which contains the destination address in its header

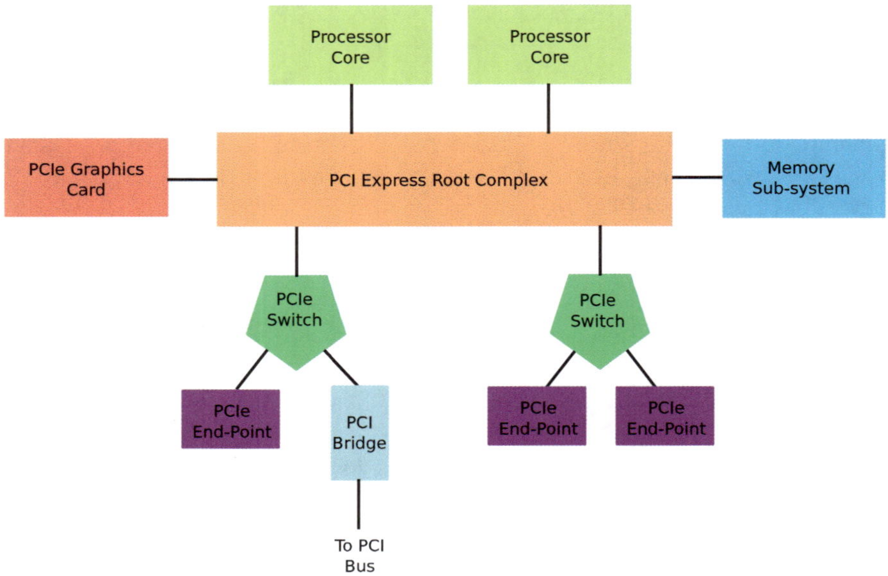

Fig. 6.1 Example architecture of a PCI express based platform

and the write data in its payload. The PCIe sub-system routes the TLP to its destination by looking up the address value in the header. On conclusion of write operation at the I/O device, it returns a completion TLP as the response to the requester to confirm the successful data transfer.

Figure 6.1 shows an example structure of a PCIe sub-system. The basic building elements of a PCIe system are in detail:

1. The *PCIe Root Complex* is usually responsible for connecting the processor and memory sub-system to the PCIe switches. The root complex generates PCIe transactions on behalf of the processor. It usually contain more than one PCIe Express ports.
2. *Switches* are responsible for routing incoming PCIe TLPs towards their suitable destination. The destination will be defined either by the address in the header of the TLP, by the ID of the destination peripheral, or based on the type of the packet (e.g. broadcasts from the Root Complex).
3. *End-Points* are practically the peripherals, boards, and devices installed on the hardware platform.
4. *Bridges* are used to allow hardware components not implementing PCIe directly to be added to the system. The bridge is responsible for performing translation between the other protocol and the PCIe.

PCIe devices are not available only as hardware boards which should be installed on a main board with PCIe backplane. A PCIe peripheral can also be a separate hardware unit in its own box and get connected to another platform through PCIe

Fig. 6.2 One MaxExpansion Gen3 X16 PCIe expansion kit containing one PCIe external cable and its adapter boards

external cables. Figure 6.2 shows a Gen3 X16 PCIe expansion kit which contains 2 adapter boards and a 3 m long PCIe external cable. This setup is capable of transferring data at rates near to 15.75 GB/s. It should be noted that both ends of the cable do not necessarily need to end up an adapter card. For example, it is possible to have one end of the cable connected to the adapter installed in the server machine and the other end directly enter a chip containing an integrated PCIe interface.

With the aid of PCIe fiber optic cables it is possible to extend the physical range of PCIe peripherals for one single platform up to 100 m easily. As an example, the hardware accelerator blocks for a high-end server can be located in another building while they are present to the rest of the system as Gen3 X16 capable PCIe peripherals.

6.2.1 Address Spaces and Base Address Registers

Every hardware component in a computing platform occupies a range of the available physical addresses in the system. Access to that hardware component is done through its base address and according to its address range. For example, each of the Dynamic Random-Access Memory (DRAM) memory, storage devices, and PCIe peripherals have their own specific base address and address range. The Basic Input/Output System (BIOS) is responsible for assigning addresses to the present hardware components at boot time or – for hot-plugging – when the hardware component is plugged into the system. For PCIe, recognition of available PCIe devices, identifying the capabilities and properties of each one, and assigning one or a set of addresses to the device is done through a process called *enumeration*.

Todays computing platforms running operating systems such as Windows or Linux use virtual addresses to manage system memory. Indeed, every process running on the system is given a range of virtual addresses by the Operating System (OS) that it uses for its execution tasks. For every process, accesses to the memory or any of the hardware components in the system will be done by accessing specific locations in the process virtual address range. The OS is then responsible for converting the virtual address to the real physical one and initiating the transaction to the target. To perform the address translation fast and efficiently the OS uses a hardware unit called *Memory Management Unit (MMU)*.

The advantages of using virtual addresses are numerous, for example:

- Memory protection mechanisms can be implemented by the OS to disallow accesses to memory regions of other processes.
- Libraries that contain widely used routines by all processes can be loaded only once and easily be shared among all processes.
- Access to hardware components being used by several processes at the same time can be better governed by the OS.

However, this at the same time makes the task of software development for communicating to the PCIe hardware more challenging. At the first step, the driver which is responsible for talking to the PCIe component obtains the physical address of the device and its address range. These values are calculated at boot time by the OS. It then requests the OS for a region in the virtual address space to use for communicating with the device. Then the driver remaps the physical address of the device to the obtained virtual address. This way, by performing read and write transactions to virtual address locations, the driver can practically access the physical address locations of the PCIe peripheral. We further describe the basic architecture of a PCIe peripheral Linux Kernel driver in Sect. 6.7.

Now consider the fact that a PCIe peripheral has usually integrated CPU cores that are running an operating system themselves. They also have their own MMU. Moreover, each hardware component within the PCIe peripheral has its own internal physical address. Similar to the main system, the MMU is responsible for converting the virtual addresses generated by processes running on CPUs to equivalent physical ones. However, the difference is that this time every thing is happening within the PCIe peripheral.

Consider a simplified architecture like the one shown in Fig. 6.3. Suppose that the host CPU of the system wants to share an array of data with CPU cores within a PCIe peripheral. In order to do that, the host CPU can copy the data to the memory located inside the PCIe peripheral. Several address translation steps are required to accomplish this task:

1. The virtual address of the memory location which holds the array on the host system should be converted into its equivalent physical address.
2. The virtual address through which the driver running on the host system talks to the PCIe peripheral should be converted into its physical equivalent as well. At this stage a transaction can be initiated to transfer the data from the memory to the PCIe peripheral. This transaction can be initiated by a Direct Memory Access (DMA) engine which we describe later in more detail.

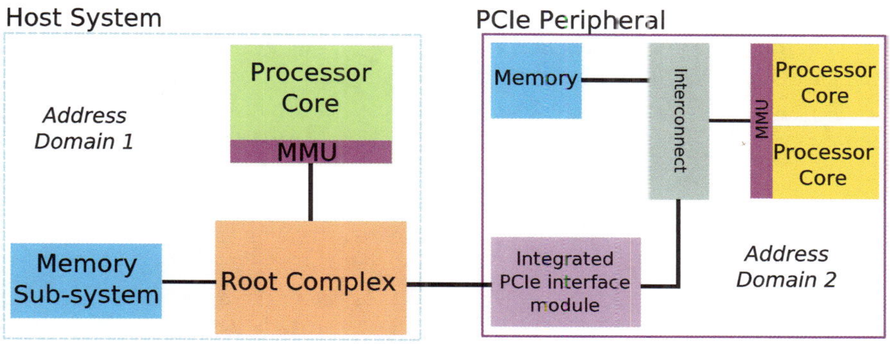

Fig. 6.3 A simplified block diagram of a host system and a PCIe peripheral which contains a set of computing elements inside

3. When the transaction passes the integrated PCIe interface module, its address should be substituted with the correct physical address within the PCIe peripheral hardware subsystem. This physical address usually resides some where in the range of memory address.
4. Finally, for the CPU cores within the PCIe peripheral to access the shared data, a conversion between the virtual address of the shared data array and its corresponding physical address should be done. This will happen using the MMU within the PCIe peripheral.

As we see the address translation can be a tedious task. As a result it is crucial to make sure that it is happening only when it is required and then it is performed in an efficient manner.

The PCIe Base Address Registers (BARs) have a special meaning: they are the base address values assigned to the PCIe peripheral by the host at boot time. However, our example integrated PCIe interface module from Fig. 6.3 has two sets of base address registers: One set representing the physical address of the peripheral for the host system and another set representing its base address as it appears to the local CPU cores within the card. When performing data transactions initiated by the host and targeting the PCIe peripheral or vice versa, it is crucial to have fast translation between these address domains. To improve the performance, the address translation task is usually directly implemented in the hardware of integrated PCIe module. There exist configuration registers within the module where the required translations between two address domains can be defined.

6.3 Interfacing to PCI Express Using FPGAs

With parallel I/O data transfer schemes reaching their bandwidth limits for chip-to-chip communications, high-speed serial interfaces are replacing them wherever possible. A high-speed serial interface transmits the data over differential signal lines in self-synchronous mode. This means the information related to the clock signal of the transmitter is integrated into the data stream itself. At the receiver side the clock signal will be extracted from the data stream using a Clock Data Recovery (CDR) circuit, which mainly contains a Phase Lock Loop (PLL). The principal component used in high-speed serial links is a Serializer/Deserializer (SerDes). A SerDes is a hardware unit responsible for converting the received parallel data into serial, adding clock information and putting it on the transmission line, and on the other hand, receiving the serial data from outside, extracting the clock information and converting the data into parallel.

Field Programmable Gate Array (FPGA) devices that are widely used by engineers in numerous products should also provide the possibility of talking to outside world through high-speed serial links. The shift from parallel data transfer to high-speed serial interfaces, however, does not come without challenges. In fact, designing a high-speed serial interface from scratch can be so complicated and time consuming that the engineers may prefer to continue using traditional parallel transfer solutions. This made FPGA manufacturers to integrate the rapid serial I/O interfaces as ready-to-use hard-core IP blocks into their products. Thus, all the designers need to do is to configure the IP core according to their serial data transfer specifications and implement suitable hardware modules on the FPGA for interacting with the rapid serial I/O block.

For example, Xilinx has integrated high-speed serial interfaces into its FPGAs from the early Virtex-II devices [2]. Primarily they were called *Rocket-I/Os*. This name was later mostly replaced by Multi-Gigabit Transceiver (MGT). Table 6.1 shows available MGTs in Series-7 and UltraScale Xilinx products [8]. The MGTs are the basis for realization of all of the famous high-speed communication protocols using Xilinx FPGAs. Examples include PCIe, Serial AT Attachment (ATA), 10 Gb Ethernet, Infiniband, and so on.

Table 6.1 Multi-gigabit transceivers available in UltraScale and Series-7 devices. For each family only the highest speed MGT is shown

Device	Transceiver type	Max performance (Gbits/s)	Max transceivers	Total bandwidth (TBits/s)
Kintex UltraScale	GTH	16.3	64	2
Virtex UltraScale	GTY	32.75	120	5.8
ZYNQ 7000	GTX	12.5	16	0.4
Artix-7	GTP	6.6	16	0.21
Kintex-7	GTX	12.5	32	0.8
Virtex-7	GTZ	28.05	96	2.8

6.3.1 Integrated Block for PCI Express

To have a fully operational PCIe end-point or root complex, in addition to MGTs, several other blocks are required as well. These additional modules are responsible for realization of transaction and data-link layers of thePCIe protocol. This includes tasks like configuration management, generation and processing of TLPs, flow-control, power management, data protection, error checking, and status tracking. One possibility is that the developer implement all of these additional functionality in Register-Transfer Level (RTL) and therefore using FPGA resources. But also ready-to-use soft-IP cores [10] can be used. However, this results in valuable FPGA resources being consumed by the core, limiting the freedom of designer to implement his own custom logic. Therefore, in some FPGA devices the PCIe block is available as a hard-IP core, meaning that the required logic is already integrated into the silicon and is ready to be used.

Regardless of being a hard- or soft-IP core, the interfaces to the PCIe block are always the same. This allows a developed design to be adapted to different FPGA devices in a short time. Basically, the main interfaces of a PCIe block for transferring data to other modules within the FPGA fabric are based on AXI stream protocol Figure 6.4 shows the symbolic representation of a Gen3 8X PCIe block for a Virtex-7 device and its main connections to other modules on the FPGA and also to the

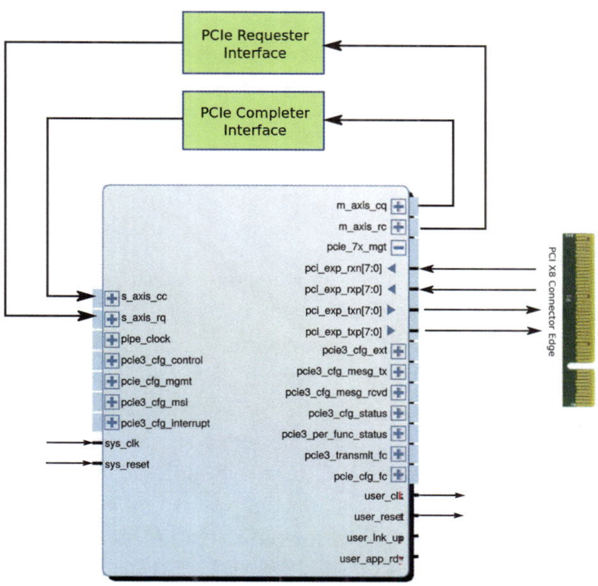

Fig. 6.4 A simplified block diagram of a PCIe block and its main connections to other modules on the FPGA and the outside world. The PCIe block symbol is taken from a Xilinx Vivado project consisting a Gen3 X8 PCIe interface

outside world. In this figure, the PCIe block is shown in blue and then rectangles in green are the logic implemented by the user for talking to the PCIe block.

As described in previous sections, in every PCIe transaction one node in the system is the requester which produces a read or write request and another node is the completer which responds to the requester by performing the required action. The PCIe block shown in Fig. 6.4 is capable of playing both roles. When the user decides to act as the bus master and to initiate a transaction to another PCIe node in the system, it does so through its *PCIe Requester Interface* unit by sending a suitable request packet to the *Requester reQuest (RQ)* interface of PCIe block. The PCIe block forwards the request to the PCIe backplane. When the PCIe block received the response from the completer it sends the response over the *Requester Completion (RC)* interface to the user's Requester Interface module.

The PCIe block can also act as a completer. When another node in the system decides to initiate a transaction to our PCIe peripheral it sends the request with the physical address of our PCIe peripheral. Upon receiving the request TLP, the PCIe block forwards the request to the user's *Completer Interface* through its *Completer reQuest (CQ)* port. The user developed block performs required actions indicated by the request and sends back the response to the *Completer Completion (CC)* interface of the PCIe block. For a Gen3 X8 PCIe block, each of the mentioned ports have a data bus width of 256 bits and are running at 250 MHz. The `sys_reset` and `sys_clk` signals are fundamental to the operation of the PCIe block. These signals can be obtained from the PCIe backplane. There is also the possibility of generating these signals locally on the PCIe peripheral if needed. The `user_reset` and `user_clk` signals are generated by the core and can be used for the logic that user develops on the FPGA fabric for communicating to the PCIe block.

6.4 Introduction to AXI

Looking at a typical System on Chip (SoC) design we see a large number of different modules instantiated and connected together to deliver the required functionality of the product. In order to realize those connections easily and fast with the goal of building a new architecture it is crucial that all of the modules interface to outside world based on a same language. This means that all of the modules should obey a same set of rules and should use the same set of signals as their interface to the rest of the logic. Having a library of different modules meeting this requirement in hand, a designer can build new systems very easily by connecting the required components together. This is the basic motivation behind creation of SoC buses like WishBone [6], IBM CoreConnect [5], and ARM AXI [1].

A SoC bus is not necessarily a data transfer medium which is shared by several modules. It can also be an architecture which provides point-to-point connection between different units of the system, allowing them to initiate transactions concurrently. This is the case for the AXI interconnect, for example.

The basic elements of an AXI based architecture are as follows:

1. *AXI Masters* are units that initiate read or write transactions.
2. *AXI Slaves* are units that receive read or write transactions and produce the suitable response to them.
3. *AXI Interconnects* play the role of switching elements between AXI Masters and AXI Slaves. They are responsible for routing the transactions and data from an AXI Master to the specified destination.

There exist two types of AXI interfaces:

1. *AXI stream* and
2. *AXI memory mapped*.

An *AXI stream* interface is suitable for modules which receive a stream of data, perform some processing on it, and generate another stream of data as the output. In AXI stream interfaces, the source from which the data is coming and the destination to which the data is going are defined from the beginning. Thus the transactions do not need to carry the address for the destination. This simplifies the logic required for interfacing to AXI stream extensively. Figure 6.5 shows the main signals between an AXI stream master and an AXI stream slave plug. In AXI stream, the direction of data is always from the master to the slave.

The `TVALID` signal indicates to the slave unit that a new data is available. `TREADY` is generated by the slave and indicates that the AXI stream slave is ready to receive the data. The transmission of data happens when both of the signals are active. `TDATA` is the actual data being transferred and `TLAST` indicates if the current data which is being transferred is the last data of the packet. In such a configuration, both modules work at the same clock domain. To transfer data between one AXI stream master and one AXI stream slave where both are in different clock domains, an AXI stream asynchronous First in, First Out (FIFO) can be used.

The simple AXI stream interface does not answer all of the connectivity needs in a complex SoC. This is because in many situations the computational task is more complicated than just performing a set of calculations on an incoming stream of data. Indeed, the task usually involves performing read and write accesses to different variables and array elements that are stored in a memory. Consequently, the read and write transactions initiated by the module need to contain an address to indicate to the target location in the memory for the read or write operation.

Furthermore, the simple case in which one module is always receiving its input data from another fixed module is not always true. It happens that a module needs to talk to a different module in the architecture each time. Again, this highlights then need for an address in the initiated transaction for indicating the target module.

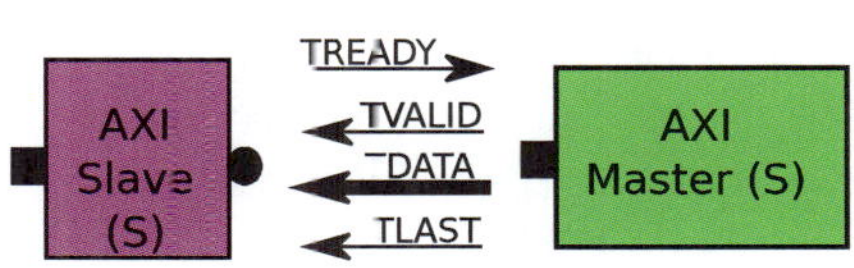

Fig. 6.5 Main signals between an AXI stream master and an AXI stream slave module

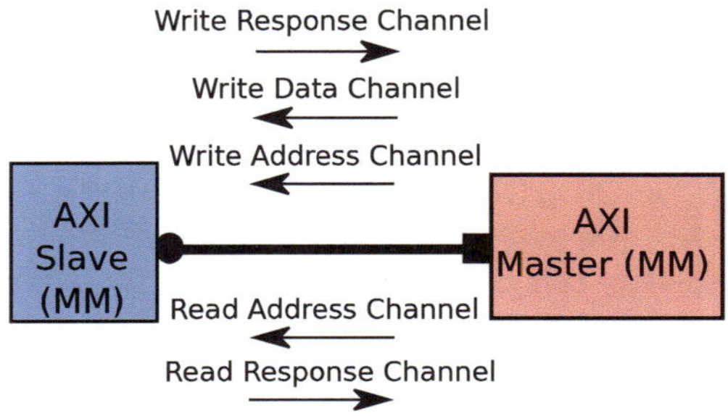

Fig. 6.6 Main channels of an AXI memory mapped interface

The *AXI memory mapped* interfaces solve these issues by adding the address channels to the AXI interface. In contrast to an AXI stream interface that contains only one channel for data transmission and in a fixed direction, an AXI memory mapped interface consist of five channels. The AXI memory mapped interface can perform either reads or writes and for each of these operations since it has a separate read address and write address channels. To improve the performance of data transfers, AXI interfaces support burst transfers of up to 256: For a read or a write transaction, only one address is indicated over the corresponding address channel and the rest of the addresses are automatically calculated at the slave side.

Figure 6.6 shows the five channels of an AXI memory mapped interface. Each channel contains a similar set of signals as one AXI stream interface. In each channel, there are additional signals to carry the information related to transactions bursts, quality of service, memory protection, caching of data, and so on.

As we see in Fig. 6.6, the *address channels* are always outputs from the AXI memory mapped master and inputs to the slave. The *read data channel* that carries the data read of the specified address by the master is an input to the master. Also the *write response channel*, which informs the master if the write transaction has been successful or not, is an input to the master module.

A typical read or write transaction begins first by putting the address on the read or write address channels. After the address is accepted by the slave, the actual data transfer will be performed. For each of the read or write data transfers there are signals which indicate if the transfer has been successful or not. To improve the performance of data transfers it is possible to use all of the five channels in parallel. For example, while an AXI master is receiving data of the previously initiated read transaction, it initiates a write transaction over the write address channel.

Furthermore, there exist the possibility that an AXI slave responds to incoming requests in a out-of-order fashion, i.e. that it receives a set of requests from different masters and it generates the suitable responses to each request as soon as it had the data ready for that specific request. To enable this functionality, a mechanism is required to identifying each transaction. In fact when the slave produces a response to the request of an AXI master, it should be able to indicate that the current response belongs to which of the received requests. The AXI protocol provides an ID signal over each channel. Basically each incoming request over either read or write address channels to an AXI slave has its own specific ID. When the AXI slave produces the

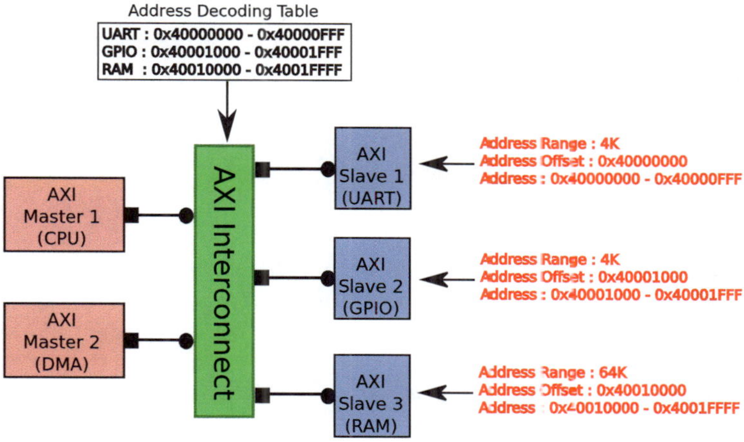

Fig. 6.7 An AXI interconnect operates as the switching element in an architecture consisting modules with AXI memory mapped interfaces

response to the request over the read data channel or the write response channel, it also reflects related ID value of that transaction over the response channel.

In an architecture consisting of modules with AXI memory mapped interfaces, switching mechanisms are required to route transactions to suitable destination based on the transaction address. *AXI interconnects* are responsible for this task. Figure 6.7 shows an example design featuring 2 AXI masters and 3 AXI slaves. Each slave in the architecture has a base address and an address range. The AXI interconnect contains a decoding table which is basis for deciding to which of the slaves an incoming transaction should be routed. The AXI masters can practically initiate transaction to every location they want in the address space. Obviously, only those transactions which fall in the address range of one of the AXI slaves will be responded.

6.4.1 AXI PCIe Bridge

Consider an FPGA based PCIe card which acts as a hardware accelerator and its main operation is to receive an incoming stream of data, to perform a defined set of calculations on the data stream, and then to return back the results to the host. For such an accelerator, the integrated block for PCIe described in Sect. 6.3.1 is a suitable choice. In fact, many of the applications deal only with streams of data. For example, they need to process a stream of data, or to transfer the incoming stream of data to another location e.g. to a fiber optics channel. In these cases, being able to randomly access data, or to keep a history of data in a memory and to retrieve it later if needed, is not required.

However, there exist also many applications in which memories and random accesses to memory are heavily required. For example, imagine a PCIe card which is responsible for accelerating data base operations. The card stores the tables in its internal DRAM to which the FPGA has a very low-latency and high-bandwidth access. Then it receives different query requests over the PCIe interface, performs the required operations using the tables stored on its DRAM and sends back the results. In such a scenario, the stream based architecture is no more efficient. Indeed the logic for the hardware accelerator implemented on the FPGA will mostly be based on memory mapped interfaces. As a result, the output of the integrated block for PCIe should be connected to a unit which is responsible for receiving PCIe packets over AXI stream, looking up their properties and destination address, and converting them into equivalent AXI memory mapped transactions. For this case e.g. Xilinx provides an integrated PCIe IP core with AXI memory mapped interfaces, eliminating the need for using the AXI stream based blocks.

Figure 6.8 shows the symbolic representation of a PCIe AXI unit and also a simplified example architecture representing how this block gets connected to the rest of the system. This figure also contains a processing core and some modules with AXI slave and master plugs which are implemented on the FPGA fabric. In addition, there exist modules that may have more than one AXI plug. AXI interconnects are responsible for connecting the AXI interfaces to each other.

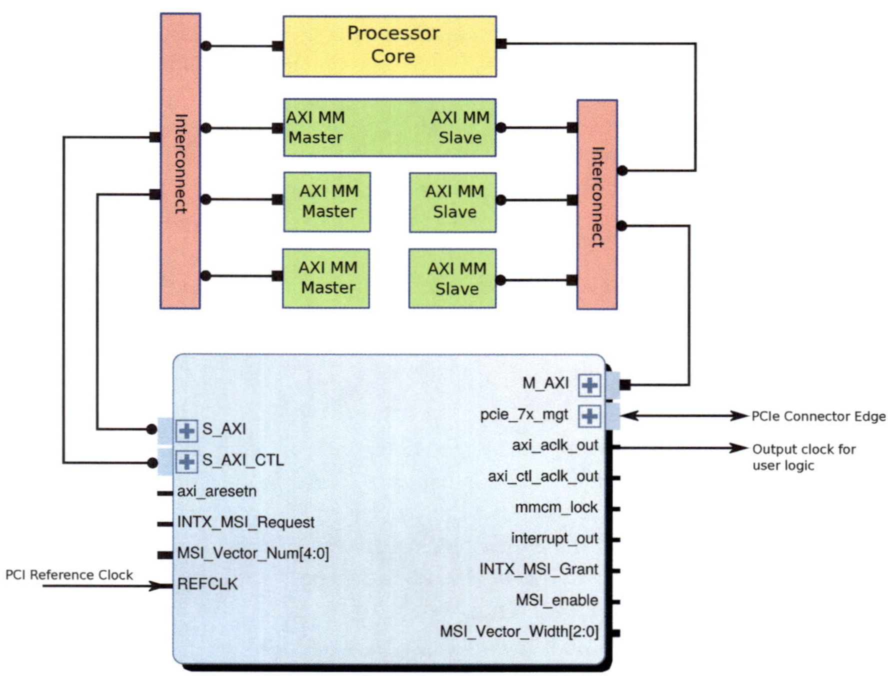

Fig. 6.8 Simplified block diagram showing how a PCIe AXI bridge gets connected to the other modules on the FPGA fabric and also to the outside world

In Fig. 6.8, the `pcie_7x_mgt` port is in fact the interface port of MGTs and should go to the PCIe edge connector. `M_AXI` is an AXI memory mapped master port which will initiate read and write transactions based on the received TLPs at PCIe side. The AXI master port will go to an AXI interconnect from which it gets connected to AXI slaves in the architecture. The `S_AXI` port of PCIe AXI bridge is used when the modules on the FPGA decide to initiate transactions to different address locations over the PCIe bus. For example, the processor core or one of the AXI masters may perform a read or write operation to a specific location in the host's DRAM memory through the `S_AXI` port. In this case the PCIe card acts as a bus master on the PCIe backplane.

Finally, the `S_AXI_CTL` port is an AXI slave port of PCIe AXI bridge which is used for configuring the bridge. Usually this is a task performed by a processing core. For example, as described in Sect. 6.2.1 the PCIe AXI bridge converts the address of incoming transaction from the address space of the host computer to the address space of the embedded architecture on the PCIe card. This address translation can be configured through the `S_AXI_CTL` port using the processor core. Furthermore, the AXI slave port can be used for reading the status and configuration information of the bridge. If the local processing unit is running an operating system like Linux, then a kernel level driver should also be developed to allow the software to access the resources of PCIe AXI bridge.

6.5 Xilinx Zynq Architecture

This section introduces the basic features of the Xilinx Zynq All Programmable SoC required to understand the PCIe example presented in Sect. 6.6.

The Xilinx Zynq All Programmable SoC is a hybrid device with a fast interconnect [12]. It consists of two parts:

1. The *Programmable Logic (PL)* that is roughly a full-featured FPGA and
2. The *Programmable Systems (PS)*, a complete sub-system with ARM CPU cores and different peripherals.

The PS contains the following items:

- An ARM Cortex MPCore-A9 dual core processing engine which also contains NEON Single Instruction Multiple Data (SIMD) units. Each ARM core has its own *L*1 data and instruction caches. Each cache block has a size of 32 KB.
- One *L*2 cache with the size of 512 KB which is shared between two CPU cores. The ARM PL310 cache controller is used for implementation of this unit.
- A *Snoop Control Unit (SCU)*, which ensures coherency between the contents of the caches.
- An On-Chip Memory (OCM), a multi-port memory block of 256 KB that can be accessed by the CPU or other ports and units in the system.

- A DMA controller that can be used for transferring data between peripheral and DRAM memories.
- A multi-port memory controller, which is responsible for connecting to DRAM memories and receiving read/write requests from different sections of the hardware and passing them to DRAM.
- A large ensemble of different peripheral such as Universal Asynchronous Receiver/Transmitter (UART), Gigabit ethernet, Universal Serial Bus (USB) peripheral and host, Controller Area Network (CAN) bus, Inter-Integrated Circuit (I^2C) Bus, Secure Digital (SD) Card interface, General-Purpose Input/Output (GPIO), and so on…, which can be configured and used by the ARM CPU cores very easily.
- An interconnect based on ARM NIC-301 design that connects different blocks of hardware inside PS together.
- A set of AXI interfaces (as shown in Fig. 6.9) are implemented to make the communication between PS and the PL logic possible.

Basically, these AXI interfaces divide into two groups:

- AXI memory mapped master interfaces (GP), connect to AXI slaves residing on the PL. The CPU is able to initiate read/write transactions over these AXI masters to transfer data to PL modules. There are two 32 bits AXI master ports available in the Zynq device: GP0 and GP1.
- AXI memory mapped slave interfaces (High Performance (HP), Accelerator Coherency Port (ACP) and SGP), connect to the implemented AXI masters on the PL. There exist four HP ports and one ACP. Each of these interfaces implements

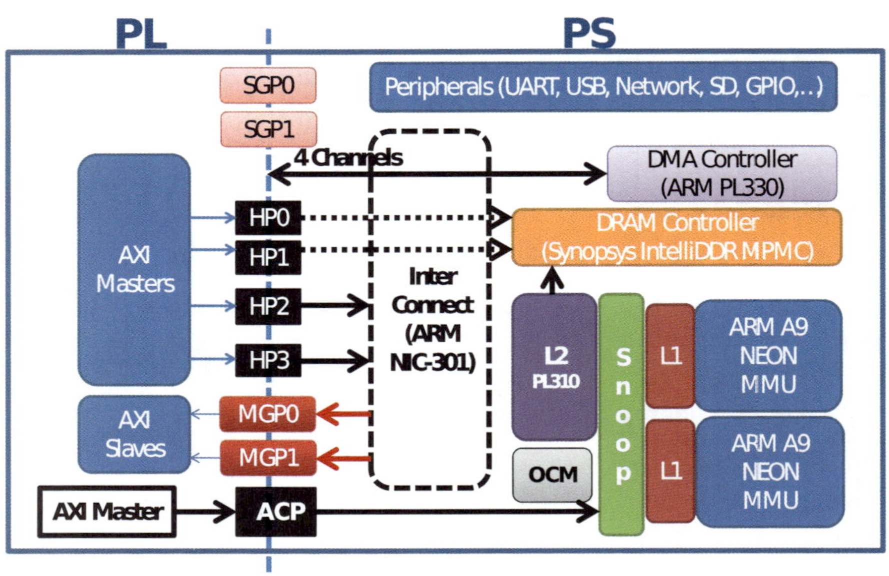

Fig. 6.9 A block diagram representing important elements of the Xilinx Zynq device

a full-duplex 64 bits connection, meaning that at every clock cycle, total 16 bytes of data can be transferred on AXI read and AXI write channels concurrently. The two SGP0 and SGP1 interfaces implement 32 bits connections.

There exists a defined memory map for the Zynq device [12] that indicates the address range of each logic block. Every AXI slave unit implemented on the PL will also occupy a part of this address range. It should be noted that except the CPU cores and their $L1$ instruction caches the rest of the system is using physical address values.

The ACP is connected to the ARM SCU. Thus, it provides the possibility of initiating cache coherent accesses to the ARM sub-system. Careful use of the ACP can improve overall system performance and energy efficiency. However, inappropriate usage of this port can adversely affect execution speed of other running applications because the accelerator can pollute precious cache area [7].

6.6 Example PCI Express Design Based on Zynq

We develop an example architecture on the Xilinx Zynq to allow the host system and the Zynq ARM subsystem to share their DRAM memory spaces over PCIe. When the data sharing between the host system and the processing elements on the PCIe card is complete, additional hardware accelerator modules can be added to the Zynq PL allowing for accelerating computational tasks. Our target hardware for this project is the Xilinx ZC706 board that contains one Zynq 7045 device. The card provides a PCIe Gen2 X4 interface over a PCIe connector. The developed architecture tries to keep every thing as simple as possible while providing the possibility of reaching acceptable bandwidth on the PCIe interface.

Figure 6.10 represents the developed architecture. This architecture consists of three main elements:

1. The Zynq PS,
2. The PCIe AXI bridge, and
3. A DMA engine.

For configuring the PCIe AXI bridge and reading its status information through the `S_AXI_CTL` port, we use the `GP1` AXI master port of the Zynq PS. Through this port, the ARM CPU cores can have access to configuration and control registers of the PCIe AXI bridge. For this part of the circuit, `axi_ctl_aclk_out` generated by the core is used as the clock signal. For the rest of the logic `axi_aclk_out` is used as the clock.

Transactions initiated by other PCIe nodes in the system and received by our card will arrive at `M_AXI` port of PCIe AXI bridge. From there these transactions will be routed to the DRAM memory space through the `HP0` AXI memory mapped slave port. Obviously, suitable address translation should be done before the transaction enters the AXI architecture. Practically for the incoming transaction address, the

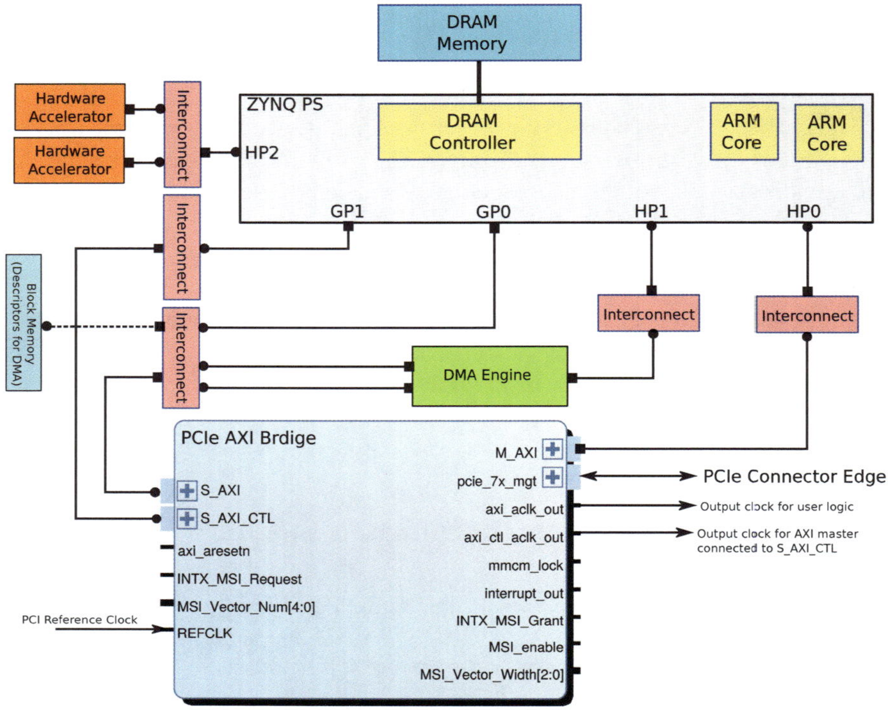

Fig. 6.10 A block diagram representing the architecture of our PCIe based hardware to share the DRAM memory of the host system with the ARM CPU cores on the Zynq and to share the DRAM memory connected to the Zynq PS with the host system. *Squares* are AXI master plugs and *circles* are AXI slaves

base address of the PCIe device should get dropped and instead a suitable base address according to address arrangements in the AXI architecture should be added.

The `S_AXI` port of the PCIe AXI bridge has basically two masters: The ARM CPU cores and the DMA engine. The ARM cores can directly initiate read and write transactions to the address space of the host through the `GP0` port. Again, the destination address for the transaction here will be the address assigned to `S_AXI` port of PCIe AXI bridge, and the translation to the address space of the host will be done within the bridge. Here, the kernel level driver of our PCIe peripheral which is running on the host system, will allocate an amount of memory on the host's DRAM. It then obtains and passes the physical address of the allocated memory to the ARM host on our PCIe peripheral. The ARM host configures the address translation mechanism within the PCIe AXI bridge so that the generated transactions by this block over the PCIe backplane end up the memory range allocated by our driver on the host's DRAM memory.

A similar procedure as what we described holds true when the host CPU wants to access the DRAM memory of the ARM host. This time the kernel level driver running at the ARM side is responsible for memory allocation. Again based on the obtained address values the address translation mechanism within the PCIe AXI bridge will be configured.

When it is required to transfer a large amount of data with a high bandwidth, using the CPU cores for handling data transfers is not efficient. This is true for both CPU cores of the host as well as the ARM CPUs of the PCIe peripheral device. Generating read or write transactions using the CPU core, even when it is done without any interruption, does not eventually result in a satisfactory performance. Consequently, there should exist a dedicated hardware responsible for performing data transfers to and from the PCIe AXI bridge. The DMA engine instantiated in our example architecture is responsible for initiating data transfers to the host without disturbing the CPUs.

In our architecture, the ARM CPU cores are responsible to define the set of data transfer tasks that the DMA engine should do. Each transfer task for the DMA is defined in a structure called descriptor. Through `GP0` port, the ARM CPU core copies the descriptors to a dual-port block memory. The other port of the block memory is practically connected to the DMA engine, which has a dedicated master port to read the descriptors. Each descriptor contains a source address for the transfer which should be done, a destination address, and the length of the transfer. The block memory can keep a large number of descriptors written to it by the CPU. The DMA engine can operate in a cyclic manner, meaning that by reading and executing the transfer for the last descriptor in the block memory, it can jump back to the first descriptor and continue its operation.

Our DMA engine has two AXI memory mapped master ports: One port is practically connected to the `S_AXI` port of the PCIe AXI bridge initiating transactions over PCIe. Another port is connected to the `HP1`. For a read operation, the DMA engine initiates the read transaction over PCIe and when it received the read data it writes the incoming data to the DRAM memory through the HP1 port. For a write operation the DMA engine reads the write data from the DRAM memory through the HP1 port, then it initiates a write transaction over the PCIe bus by writing the data to a suitable address location of the `S_AXI` port.

6.7 Linux Kernel Level Driver

As described in Sect. 6.4.1, our architecture consists of two sets of CPU cores: The host CPU cores that are usually x86 and the PCIe device CPU cores, which in our architecture are ARM v7 cores. We assume that the Linux OS is running on the host CPU cores. For the PCIe device, in most common situations, Linux is also running on the ARM CPUs. In order for user-level applications running on the Linux to be able to interact with the PCIe device, a Linux kernel level driver is needed. As a result, for our architecture two Linux kernel level drivers should be developed separately.

The driver running on the host system should be able to find the PCIe device, allocate I/O and memory regions for it, register required interrupt lines and their corresponding interrupt handlers, and to enable user level applications to communicate with the device. For this purpose, the driver uses structures and functions integrated into the Linux kernel that facilitate interacting with PCI based peripherals. In fact, from the view point of the host driver, our hardware is a PCI peripheral. However, from the view point of the driver running on the ARM host, the PCIe AXI bridge is just an AXI peripheral similar to the other components in the architecture.

For the driver running at the host side, it first announces to the Linux kernel the vendor and device ID of the PCIe device it is seeking for. In addition to the ID values, the driver also gives the Linux kernel pointers to two implemented subroutines inside the driver. The first subroutine is executed when a PCIe device is found in the system with the same ID values as the ones indicated by the driver and the second subroutine is executed when this PCIe device gets removed from the system. The ID values and pointers to subroutines is given to the Kernel through a structure called `pci_driver`. Code Listing 6.1 shows an example definition of this structure.

```
static struct pci_driver pci_drv_template =
{
  .name = "pci_drv_template",
  .id_table = pci_drv_ids,
  .probe = device_probe,
  .remove = device_deprobe,
};
```

Listing 6.1 Example code showing how pci_driver struct is defined.

During the `device_probe` subroutine, the driver obtains the physical address at which the PCIe device is located. This is done through `pci_resource_start` Linux kernel call. The driver then asks the Kernel for permission to access the physical address range of the PCIe device. If granted, a remapping of the physical address of the PCIe device to virtual address space will be done by the driver. This is done with the aid of `ioremap` Linux kernel call. Having the virtual addresses, the driver is able to perform read and write transactions to the address space of the PCIe device using `writel` and `readl` system calls. Code Listing 6.2 shows an example implementation of the `device_probe` routine.

```
static int device_probe
( struct pci_dev *dev, const struct pci_device_id *id )
{
  // get the first i/o region
  ioport = pci_resource_start( dev, 0 );
  iolen = pci_resource_len( dev, 0 );

  // request memory region
  request_mem_region( ioport, iolen, "ZC706");

  // map hardware memory space to virtual space
  pci_bar_vir_addr = ioremap(ioport, iolen);

  // the rest of the code
  // ...
```

Listing 6.2 Example implementation of the device_probe subroutine.

Initial memory allocation can also be done within the `device_probe` routine. For this purpose, the `pci_alloc_consistent` Kernel call can be used. The physical address for the allocated memory can then be transferred to the ARM cores in the PCIe device. The address will be used for configuring the address translation logic within the PCIe AXI bridge.

Both of the drivers represent the device to their Linux system as a character device which supports basic file operations. These file operations include open, close, read, write, i/o control and mmap. In the simplest data transfer scenario, the user level application opens the character device and performs reads and writes to the device using usual file I/O operations. From the host side, the driver can redirect the incoming read and write transactions toward pre-known address offsets in the PCIe device. The same holds true for the driver running at the ARM side. The driver receives the data from the user level application running on the ARM host and transfers it to pre-known offset on the memory allocated by the host driver.

When it is required to transfer and process data at a high rate, the simple file I/O operations can no more be efficient. In fact, during file I/O data transfers the data always gets copied from the user space memory to the kernel memory and from there it gets transferred to the hardware. However, it is possible to share the memory which is obtained by the driver with the user level application directly, eliminating the need for copying data from user space to kernel space and vice versa. This is done through mmap file operation. Usage of mmap for sharing the memory space with user level application is again similar for the drivers at both sides. Code Listing 6.3 shows an example implementation of mmap routine for the host driver. The `io_remap_pfn_range` function receives the user virtual address range and also the physical address range of the device. It then maps the provided range of physical addresses to the provided virtual address range by creating a new set of page tables which essentially perform the address translation between these two ranges. For further information regarding character devices and mmap please refer to [4].

```
static int chr_device_map
(struct file * filep, struct vm_area_struct * vma)
{
  io_remap_pfn_range (vma,
    vma->vm_start,
    ioport >> PAGE_SHIFT,
    vma->vm_end - vma->vm_start,
    vma->vm_page_prot);
  vma->vm_flags |= (VM_DONTEXPAND | VM_DONTDUMP);
  vma->vm_page_prot = pgprot_noncached(vma->vm_page_prot);
  return 0;
}
```

Listing 6.3 Example implementation of the mmap file operation routine (Linux kernel 3.7).

6.8 Experimental Results and Discussion

The performance of a PCIe hardware accelerator card is directly affected by the bandwidth of data transfers between the card and the host system. Considering the PCIe card, it is also crucial to know how fast this architecture can transfer data between its own main memory and the integrated PCIe block. These parameters directly affect the processing performance of the hardware accelerator.

Through a set of stress tests it is possible to obtain estimates on the performance of PCIe interface. For the ZC706 board, which features a Gen2 X4 PCIe interface, Table 6.2 shows obtained estimates on each of the read and write data paths [11]. In this test the performance is measured while sweeping the size of the DMA packets being transferred between the PCIe card and the host system from 512 bytes to 32 KB. Obtained values are representing the net data transfer rates after omitting all of the overheads related to packet headers, acknowledge packets, flow control, and other similar items.

As we see, the performance of the PCIe link changes significantly with the packet size. In fact, the efficiency of data transfers to a hardware accelerator using PCIe is heavily dependent on the size of data chunks being transferred. If the acceleration task involves processing of big continuous blocks of data, the PCIe link can be used with acceptable bandwidth efficiency. Instead, if the acceleration task requires a large number of small data chunks being transferred between the host system and the PCIe card, the efficiency of the link can be degraded seriously. In these situations, increasing the PCIe link width (e.g. from X4

Table 6.2 Approximate read and write bandwidth to the host for ZC706 Gen2 X4 PCIe interface

Packet size (KB)	0.5	1	2	4	32
PCIe to host (MB/s)	250	612	762	937	975
Host to PCIe (MB/s)	250	525	750	930	970

Table 6.3 Approximate read plus write bandwidth to the DRAM memory residing on the ZYNQ PS from ZYNQ PL through each of HP1 and ACP ports

Packet size (KB)	4	16	64	128	256
HP0	600	1,250	1,600	1,700	1,730
ACP	600	1,250	1,600	1,650	1,600

to X8), improving caching mechanisms, and exploiting several concurrent threads for initiating required transactions can enhance overall performance.

In the second test, we stress the local DRAM memory available on the accelerator card. We obtain estimates on how fast the data can be transferred between the PCIe block and the local DRAM memory. This is important since in our architecture the host system shares the data with the hardware accelerator on this DRAM memory.

Table 6.3 shows the obtained read plus write bandwidth values between the DMA engine residing on the Zynq PL and the DRAM memory connected to the Zynq PS [7]. Similar to the previous test, we sweep over the size of data chunks being transferred between two nodes to report the bandwidth. We perform accesses to the DRAM through each of the HP1 and ACP ports. As we see, with the increase in the size of packets being transferred the bandwidth improves. However, the increase in bandwidth gets saturated at around 1,700 MB/s. This total bandwidth is almost evenly divided between read and writes.

For the ACP, accesses practically end up the caches of the CPU for smaller packet sizes. However, as the packet size grows and more memory is needed, the extra accesses to DRAM reduce the ACP performance. The real advantage of ACP over HP1 can be seen in scenarios where the accelerators on the Zynq PL want to share data with the ARM CPU cores of the Zynq PS. Considering the obtained estimations we conclude that a careful design is needed to be done so that the Zynq PS and DRAM memory subsystem sustain the traffic generated by the Gen2 X4 PCIe block completely. This involves definition of descriptors for DMA transfers as well as utilization of suitable ports on the Zynq PS to access the DRAM.

6.9 Conclusion

In this chapter, we summarize the basics of PCIe as one of the most popular interfaces for communicating with accelerators in heterogeneous systems. One reason is that – in contrast to other interfaces like Ethernet – relatively low latencies can be achieved with PCIe. We introduce the basic architectural setup and protocol features of a PCIe communication infrastructure. We describe that PCIe peripherals do not necessarily need to be located in one single box together with the host CPU, but can also be physically apart and be connected through PCIe cables. This is in particular beneficial for external accelerator units that need to be attached to a host device.

We also introduce the fundamentals of using FPGAs for interfacing to PCIe. We briefly look at the PCIe integrated block and the PCIe AXI bridge IP cores provided by Xilinx that are available through the Xilinx Vivado design suite. We briefly discuss how these components can be utilized to create a PCIe accelerator card based on FPGAs. One important aspect for achieving good performance is the use of DMA engines besides the CPU.

From the software point of view – both on the host and the accelerator device part – we briefly describe the mechanisms for accessing the PCIe infrastructure and highlight different sections of an appropriate Linux kernel level driver created for talking to a PCIe peripheral. As a practical example, we have used a Xilinx ZC706 for demonstrating how a PCIe design can be realized for the Xilinx Zynq All Programmable SoC device on this board. We show estimated bandwidth values for key parts of this architecture.

Significant improvements in the IP cores and software drivers created to ease designing with PCIe have made building PCIe peripherals easier than ever before. As an example, the Xilinx SDAccel [9] environment provides the designers with a complete CPU/Graphics Processor Unit (GPU) like development experience on FPGAs. If a financial hardware accelerator card comes with relaxed transfer bandwidth and latency requirements, the entire design can be realized by hardware or software engineers with a common level of design skills.

Nevertheless, for some applications it is required to push the transfer bandwidths to their possible maximum. In these cases designing for PCIe is a challenging task. It includes the electronics design of the board, design and implementation of the hardware architecture, and also software aspects of the design. In these situations it is usually a wise decision to leave the project in the hand of professionals who have constant experience of designing PCIe peripherals.

References

1. ARM.: AMBA AXI and ACE Protocol Specification, d edition (2011) http://ee427plblabs.groups.et.byu.net/wiki/lib/exe/fetch.php?media=ihi0022d_amba_axi_protocol_spec.pdf, last access: 2015-05-07
2. Athavale, A., Christensen, C.: High-Speed Serial I/O Made Simple, A Designer's Guide, with FPGA Applications. Xilinx, San Jose (2005)
3. Cascaval, C., Chatterjee, S., Franke, H., Gildea, K.J., Pattnaik, P.: A taxonomy of accelerator architectures and their programming models. IBM J. Res. Dev. **54**(5), 5:1–5:10 (2010)
4. Corbet, J., Rubini, A., Kroah-Hartman, G.: Linux Device Drivers, 3rd edn. O'Reilly Media, Beijing/Sebastopol (2005)
5. IBM. The CoreConnect^{TM} Bus Architecture, (1999) https://www-01.ibm.com/chips/techlib/techlib.nsf/techdocs/852569B20050FF77852569910050C0FB/$file/crcon_wp.pdf, last access: 2015-05-07
6. OpenCores. WISHBONE System-on-Chip (SoC) Interconnection Architecture for Portable IP Cores, b.3 edn. (2002) http://cdn.opencores.org/downloads/wbspec_b3.pdf, last access: 2015-05-07
7. Sadri, M., Weis, C., Wehn, N., Benini, L.: Energy and performance exploration of accelerator coherency port using xilinx zynq. In: Proceedings of the 10th FPGAworld Conference (FPGAworld '13), New York, pp. 5:1–5:8. ACM (2013)

8. Xilinx. High speed serial. http://www.xilinx.com/products/technology/high-speed-serial.html
9. Xilinx. Sdaccel development environment. http://www.xilinx.com/products/design-tools/sdx/sdaccel.html
10. Xilinx. LogiCORE IP Endpoint for PCI Express, User Guide (2010) http://www.xilinx.com/support/documentation/ip_documentation/pci_exp_ep_ug185.pdf, last access: 2015-05-07
11. Xilinx. ZC706 PCIe Targeted Reference Design (UG963) (2013) http://www.xilinx.com/support/documentation/boards_and_kits/zc706/14_7/ug963-zc706-pcie-trd-ug.pdf, last access: 2015-05-07
12. Xilinx. Zynq-7000 All Programmable SoC Technical Reference Manual (UG585) (2013) http://www.xilinx.com/support/documentation/user_guides/ug585-Zynq-7000-TRM.pdf, last access: 2015-05-07

Chapter 7
Pricing High-Dimensional American Options on Hybrid CPU/FPGA Systems

Javier Alejandro Varela, Christian Brugger, Songyin Tang, Norbert Wehn, and Ralf Korn

Abstract In today's markets, high-speed and energy-efficient computations are mandatory in the financial and insurance industry. As American options are amongst the most frequently traded products in the derivatives market, it becomes essential to place the focus on their pricing process. Calculating the price of an American option in particular is a challenging task due to the freedom the holder is given in terms of exercise date and the involved trading strategy. A well known algorithm that solves this task is the Longstaff-Schwartz (LS) algorithm, which applies least-squares linear regression on simulated Monte Carlo (MC) paths. This work presents a novel way to price high-dimensional American options, coined Reverse LS, using techniques of the embedded community. The proposed architecture targets hybrid Central Processing Unit (CPU)/Field Programmable Gate Array (FPGA) systems, and it exploits the FPGA reconfiguration to deliver high-throughput. With a bit-true algorithmic transformation based on recomputation, it is possible to eliminate the memory bottleneck and access costs present in a straightforward implementation. The result is a pricing system that is 16× faster and 268× more energy-efficient than an optimized Intel CPU implementation.

7.1 Introduction

In the financial world, Over-the-Counter (OTC) derivatives markets trade an average annual volume of approximately USD 700 trillion [12], which increases every year. Increasing competition and stringent regulations lead to a steady growth of computing requirements. Today, financial institutions operate huge clusters to

J.A. Varela (✉) • C. Brugger • N. Wehn
Microelectronic Systems Design Research Group, University of Kaiserslautern, Kaiserslautern, Germany
e-mail: varela@eit.uni-kl.de; brugger@eit.uni-kl.de; wehn@eit.uni-kl.de

S. Tang • R. Korn
Stochastic Control and Financial Mathematics Group, University of Kaiserslautern, Kaiserslautern, Germany
e-mail: tangs@mathematik.uni-kl.de; korn@mathematik.uni-kl.de

C. De Schryver (ed.), *FPGA Based Accelerators for Financial Applications*,
DOI 10.1007/978-3-319-15407-7_7

satisfy these computing needs. Due to their high costs, the financial industry has a high incentive to investigate efficient ways of performing the required computations, both in terms of speed and power consumption.

Not surprisingly, it has become a particular field of research among the engineering community in recent times, due to the challenges involved. In this regard, one approach is to build specialized computing architectures. While more effort is required to design them, they are able to perform computations much more efficiently compared to general-purpose architectures. In this regard, FPGAs have been demonstrated high performance and energy-efficiency when used to speed up financial simulations [5, 15].

While many numeric algorithms map nicely to FPGAs, there often remain parts that are best executed on CPUs. Hybrid devices combine CPUs and FPGA fabrics on a single device, delivering the best of both worlds. One recent example is the Xilinx Zynq All Programmable System on Chip (SoC) based on ARM cores. These devices are able to host fully featured operating systems like Linux and allow programs to reconfigure the FPGA fabric during runtime. A key challenge of such heterogeneous computing systems is to carefully balance all aspects of the hardware, including communication, reconfiguration times, memory bandwidth, FPGA area and CPU loads.

Among the products that are currently offered in the derivatives markets, options are particularly attractive to investors. In general terms, an option is a contract that gives the right, but not the obligation, to buy or sell the underlying asset at a fixed price and date. What makes it attractive is the potential gain associated with the contract, while presenting a limited risk to the buyer, which is equivalent to the premium paid at the moment of purchase. And it is precisely the computation of this premium (the option price) what concerns financial institutions.

American options present the additional challenge that the holder is allowed to exercise the option at any time from purchase until the expiry date, in contrast to the European option style, which can only be exercised at a fixed date. This freedom makes its pricing much more challenging, since now the estimation of an optimal exercise strategy comes into play.

The LS algorithm, which is implemented in this work, has been designed to address the problem of finding such a strategy and deriving from it the option price [9]. This is accomplished by working backwards, from maturity to the initial day, on simulated MC paths by means of the least-squares regression method. For multi-dimensional options, which derive their price from multiple underlying assets, MC is currently the only known method that can be efficiently used to price them.

The quality of the simulated paths also depends on the mathematical model used to describe the evolution of an underlying asset in the market. For high-dimensional options, the Black-Scholes (BS) model has been used extensively due to its relatively light-weight computation (only one parameter cannot be observed: volatility). Besides, its results are close enough to the observed market values, and it can be easily extended to make the model flexible enough for practical cases.

When it comes to implementation, several issues need to be addressed. At first sight, the LS algorithm does not present a clear way to perform hardware-software partitioning. It is also a computationally intensive algorithm. The choice of certain basis functions that work on the simulated paths influences the final price, and has to be matched to the option being priced. Besides, the method used to solve the least-squares process has an impact on the overall runtime. The chosen number of simulated MC paths, and the number of days in which the option can be exercised, define the amount of generated data. Storing this data temporarily in an external memory chip is a straightforward approach, but faces a certain bandwidth limitation and a considerable power consumption.

This work investigates custom computing solutions for the above mentioned LS method [9]. The proposed solution targets hybrid computing systems, like Xilinx Zynq, and is able to perform high-precision and energy-efficient computations. Besides the classical approach, a novel algorithmic improvement called Reverse LS is presented. This new approach does not require the storage of all intermediate steps for all paths, but recomputes them on the fly. Recomputation is a well known technique in embedded system to avoid energy-costly memory accesses [6, 7]. This allows us to reduce the energy consumption by trading-off memory bandwidth with FPGA resources, effectively moving less data across the board.

7.2 Background

This section covers the theoretical background and related work specifically relevant to the content of this chapter. For the general background of financial computations refer to Chap. 1 by Desmettre and Korn.

7.2.1 American Options

In simple terms, a financial derivative is a type of contract which derives its value from the performance of an underlying entity (e.g. an asset) There are many types of derivatives, being one of them the so called *options*. An option contract gives the buyer the right, but not the obligation, to buy or sell an underlying asset at a specified strike price and a specified date. In this regards, there are several exercising styles, being two of them:

- European options, which can only be exercised at the expiry date (also called maturity)
- American options, which could be exercised at any time before or at the expiry date

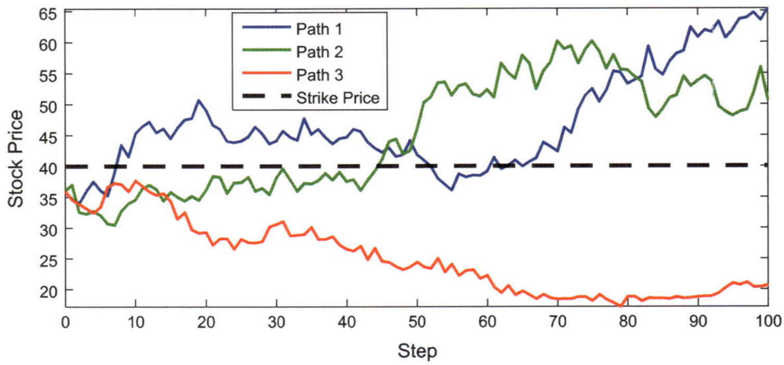

Fig. 7.1 Simulated paths using BS model, with initial price 36 [ad], strike price 40 [ad], American call option

The option gives the holder the right to either:

- Sell the underlying assets: *put option*
- Buy the underlying assets: *call option*

Consider the example presented in Fig. 7.1, where different simulated scenarios are presented for a given American call option, strike price and maturity. The holder of the option needs to decide at each time step on whether to exercise the option or hold it until a future date.

With $S(\tau)$ the value of the underlying asset at time τ, and K the strike price, the intrinsic value at the current time step τ is calculated as Eq. (7.1) for call options, and Eq. (7.2) for put options:

$$payoff(\tau) = max(S(\tau) - K, 0) \qquad \text{call} \tag{7.1}$$

$$payoff(\tau) = max(K - S(\tau), 0) \qquad \text{put} \tag{7.2}$$

The option is then said to be *In the Money (ITM)* if:

- $(S(\tau) > K)$ for a call option
- $(S(\tau) < K)$ for a put option

Following Fig. 7.1, whenever the option is ITM the holder has the choice of executing an early exercise of the option or holding it until further steps in an attempt to maximize its profit.

This right (to sell or buy) given by the option comes at a price, a premium that the buyer pays the seller at the moment of the purchase. The price of an American call/put option is given by Eqs. (7.3) and (7.4) respectively:

$$P = \sup_{\tau \in \mathcal{T}\{t_1,\ldots,t_m\}} \mathbb{E}(e^{-r\tau}(S(\tau) - K)^+) \qquad \text{call} \tag{7.3}$$

$$P = \sup_{\tau \in \mathcal{T}\{t_1,\ldots,t_m\}} \mathbb{E}(e^{-r\tau}(K - S(\tau))^+) \qquad \text{put} \tag{7.4}$$

where:

- $(x)^+$ means $\max(x,0)$
- K is the strike price
- T is the maturity of the option
- $\{t_1,\ldots,t_m\} = \{\frac{T}{m} \times 1,\ldots,\frac{T}{m} \times m\}$ are potential exercise dates of the option
- $\mathcal{T}\{t_1,\ldots,t_m\}$ is the set of stopping times with values in $\{t_1,\ldots,t_m\}$
- r is the risk-free interest rate
- $S(\tau)$ can be simulated with an appropriate mathematical model, for example using BS, as it will be covered in later sections
- In the case of multi-dimensional options, their value is derived from several underlying assets (therefore dimensions)

A note is made on the fact that when the time interval is discretized as in Eqs. (7.3) and (7.4), the option is then called Bermudan options.

The main complexity associated to American options resides in their pricing. As mentioned before, this style of options can be executed not only at maturity (expire date), like in the case of the European style, but also at intermediate steps. This freedom that the option holder is given makes the estimation more complex. The seller of the option (normally a bank or financial institution) has to estimate its price expecting the worst case scenario where the holder would follow a sound strategy at each step that maximizes its return. And this is exactly where the LS algorithm comes into play [9].

7.2.2 Black-Scholes Model

The BS model assumes, among other considerations, that the stock price follows a random walk, which implies that the stock price at any future time has a log-normal distribution (meaning its logarithm has a normal distribution) [4]. It describes the stock price $S(t)$ by means of the Stochastic Differential Equation (SDE):

$$dS(t) = S(t)(r-q)dt + S(t)\sigma dW(t), \tag{7.5}$$

where: r = risk-free interest rate, q = dividend yield, σ = constant volatility of stock's returns, and $W(t)$ is the associated Brownian motion

The BS model is based on certain assumptions [4]. In particular, it assumes constant volatility, which might not be the case in the real market. However, it is still used nowadays due to its simplicity, ease of extension, and because it is a good approximation of how much profit the holder could expect.

7.2.3 Monte Carlo (MC) Methods

Simulating the BS model in Eq. (7.5) requires the application of an appropriate discretization scheme. In this work we have applied the *Euler discretization*. Discretizing into m steps with equal step sizes $\Delta t = \frac{T}{m}$ leads to:

$$\hat{S}_{t_{i+1}} = \hat{S}_{t_i} \exp\left(\left((r-q) - \frac{\sigma^2}{2}\right)\Delta t + \sigma\sqrt{\Delta t}\,\Delta W_i\right), \tag{7.6}$$

with ΔW_i being independent standard normal random variables.

The classic MC algorithm estimates the price P as the sample mean of simulated instances of the discounted payoff values $g(\hat{S})$. The complexity of MC methods depends only linearly on the number of dimensions, which makes them an excellent candidate for high-dimensional problems or a method of last resort for options with no other numerical scheme.

MC results depend heavily on the number of simulated paths, due to its slow convergence. This is based on the fact that the standard deviation of the error only decreases as the square root of the number of simulations [8]. Therefore, the higher the number of paths, the more accurate the result it yields. As an example, a showcase is designed to price an American maximum call option on two correlated stocks (correlation parameter $\rho \neq 0$) under the BS model Eq. (7.6) by means of the LS algorithm. The optimal expected discounted payoff is given by:

$$P = \sup_{\tau \in \mathcal{T}} \mathbb{E}\left[e^{-r\tau}(\max\{S_1(\tau), S_2(\tau)\} - K)^+\right], \tag{7.7}$$

with input parameters: $S_1(0) = S_2(0) = 100$, $K = 100$, $r = 0.05$, $q_1 = q_2 = 0.10$, $\sigma_1 = \sigma_2 = 0.2$, $\rho = 0.1$, $T = 1$, $m = 365$, $\mathcal{T} = \{\frac{T}{m} \times 1, \frac{T}{m} \times 2, \ldots, \frac{T}{m} \times m\}$, $N =$ 10,000.

The influence of the number of simulated paths N on the accuracy of the option price for the given example is displayed in Fig. 7.2, where the benchmark option value is found at 10.12 (unspecified currency) using the binomial tree method [4]. The boxplots show the distribution of the option values obtained for 100 runs of the LS algorithm. A comparison to the benchmark value of 10.12 clearly suggests a minimum number of paths at around 10K.

7.2.4 Paths Generation

The BS model requires a sequence of normally distributed random numbers to generate the paths. Furthermore, because the underlying assets (dimensions of the option) coexist in the same market, these random numbers need to be correlated to each other. In this work the following processes are executed in the given order:

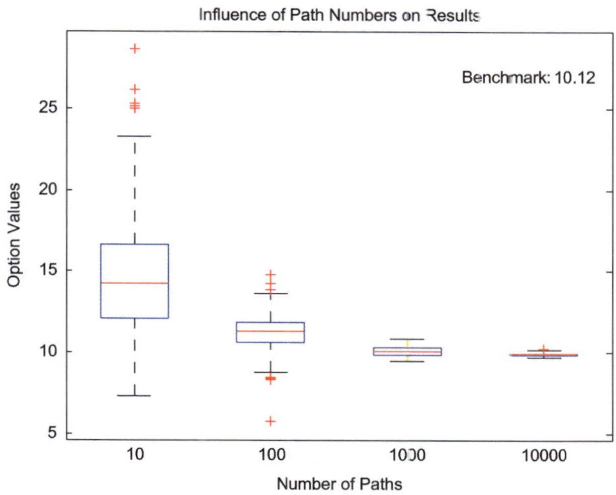

Fig. 7.2 Boxplots with the distribution of the results obtained by running the LS algorithm 100 times per number of paths: 10, 100, 1,000 and 10,000

1. *Mersenne Twister (MT)*: The MT is a widely-used pseudo-random number generator, whose MT19937 version is the one implemented in this work. It produces a sequence of 32-bit unsigned integer random numbers, and has a period of $2^{19937} - 1$. The algorithm code is explicitly shown in [10], and could be seen as split into two main parts [11]:

    - A set of 624 internal states used to generate the random numbers. This internal states are initialized through a seed that generates the initial values, and an actualization process modifies the states every 624 output numbers
    - A tempering function, a sequence of xor operations, that outputs the final number

    It is possible to pipeline this algorithm in order to achieve one output per clock cycle. In fact, the work is done on the actualization process itself, so that each state is actualized as soon as it has been used for the last time in the current cycle.
2. *Inverse Cumulative Distribution Function (ICDF)*: The MT module presented before generates uniformly distributed random numbers, whereas the BS model requires normally distributed ones. Previous work on this field has provided with an efficient implementation of the ICDF to obtain the desired standard normal distribution [13]. Furthermore, the mentioned implementation generates single-precision floating-point random numbers, which will match later with the setup for this work. A note is made, however, on the fact that the method does not precisely guarantee a valid output at every single clock cycle, but nevertheless presents a good tradeoff between hardware utilization and performance, as compared to more expensive approaches like the Box-Muller method [13].

3. *Antithetic Variates*: As mentioned before, the MC method suffers from slow convergence (high simulation runtime), which is overcome by attempting a faster reduction of its variance. In this regard, the easiest one is the method of antithetic variates, which works by introducing symmetry [8]. In this work, the antithetic method is implemented after the ICDF module, meaning that it works on normally-distributed random numbers. Under this condition, it can be proven that for a single random number z, then $-z$ is also a valid number, which reduces the overall number of generations by half. Furthermore, when using models based on Brownian motion to generate the paths, the payoffs of high-dimensional options can be typically written as:

$$P = h(Z_1, \ldots, Z_k). \tag{7.8}$$

   Under the assumption that h is monotonic on each variable, then it is possible to prove that Eqs. (7.9) and (7.10) are negatively correlated, which means that it can be used as a variance reduction technique. A similar approach on uniform random numbers is presented in [8].

$$P_1 = h(Z_1, \ldots, Z_k) \tag{7.9}$$

$$P_2 = h(-Z_1, \ldots, -Z_k) \tag{7.10}$$

4. *Correlation*: In the case of a two-dimensional option, the correlation process mentioned before is obtained in practice through the correlation of two independent random numbers, $y \sim N(0,1)$ and $z \sim N(0,1)$, and coefficient ρ, as in Eq. (7.11), delivering two correlated random numbers z and w as outputs [8].

$$w = \rho z + \sqrt{(1-\rho^2)}\, y \qquad \text{(correlation)} \tag{7.11}$$

The generated random number following the previous sequence are then fed into the BS model Eq. (7.6) in order to obtain the required paths.

7.2.5 LS Algorithm to Price American Options

The LS algorithm approximates the value of an American option by means of simulation [9]. The simulated MC paths represent the behaviour over time of the underlying assets (e.g. stocks), which compose the option to be priced. These paths could be obtained by different mathematical models with different degrees of complexity, for example BS. Once the paths have been generated, the option price is estimated by assessing which would be the best strategy the holder would follow that maximizes its profit. This strategy becomes, in turn, the worst-case scenario for the seller.

At the expiry date (maturity) the holder has only one choice, and that is to exercise the option only if it is ITM. However, at any other time, the holder can decide between:

- Exercising the option immediately
- Holding the option (called continuation)

The option should be exercised if the payoff of immediate exercise is higher than the continuation value. However, this continuation value is defined as the conditional expected value of continuing the option, assuming that the option is not exercised at or before the current time step. In general terms, the LS algorithm estimates this conditional expectation based on all generated paths at the current step, in order to derive the optimal excercise strategy.

In more detail, the LS algorithm uses least-squares linear regression to find the optimal exercise boundary. The basic steps are:

1. Generate N independent paths per underlying (stock) at all possible exercise dates, using a chosen Random Number Generator (RNG) and a chosen mathematical model (in our case with Eq. (7.6)). For multi-dimensional options, the Random Numbers (RNs) need to be correlated.
2. Initialize the cash-flow with the discounted payoffs at maturity.
3. Moving backwards one step in time, proceed as follows:

    - Linear regression: the goal is to find out whether to exercise the option or to hold it. For this purpose, the current discounted payoff (when exercised) is compared to the future expected return (for holding the option), approximated by regression. As an example, Fig. 7.3 plots the future return (cash-flow) over the current stock price for each path. Least-squares linear regression with user-defined basis functions is applied to obtain the expected future value, as shown in Fig. 7.3.
    - Cash-Flow update: For every path at the current time step compare the expected return in Fig. 7.3 with the current discounted payoff, take the larger one and update the corresponding value of the cash-flow.

    Repeat this process step by step until the initial day.
4. At the initial day, average all values in the cash-flow to obtain the option value.

The challenging part for LS is the choice of basis functions for regression. They highly depend on the exact form of the option being priced and need to be matched to the characteristics of the payoff function.

The flow described previously for the LS algorithm has been explained in simple terms aimed at giving a quick background on the topic. For the study case discussed in later sections pricing two-dimensional American maximum call options, a formal algorithm is given in the Appendix.

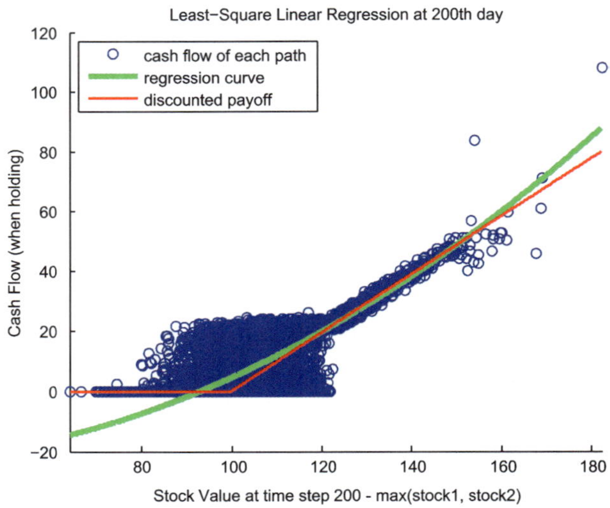

Fig. 7.3 Regression process example for a two-dimensional American max call option at time step $i = 200$. For each path the cash flow (holding value) is drawn over the current stock values (*circles*). The discounted payoff (exercising value) is shown, as well as the computed regression curve based on the drawn circles (expected mean future holding value)

7.2.6 Related Work

The use of FPGAs for accelerating financial simulations has become attractive with the first available devices. Many papers are available that propose efficient random number generation methods and paths generations. Most of the research focuses on the BS market model. For MC methods De Schryver et al. have shown that FPGAs are 33× more energy-efficient in the Heston market model [14]. For the GARCH model Thomas et al. could show speedups of 80× [16], for the Black Scholes model they showed speedups of 313× [19]. Sridharan et al. have extended this work to multi-asset options in the Black Scholes model [15], presenting speedups of up to 350× for one FPGA device. All four implementations are not able to price American options.

At the time this work was being carried out, there was only one publication of an architecture able to price American options by means of MC methods [18]. Their work is based on the LS algorithm and it has presented speedups of 20× in FPGA compared to CPU. It makes use of an efficient fully parallel architecture and an external memory chip to store the simulated MC paths. Some of the ideas presented in their work have been used as the basis of our new architecture. Nevertheless, their design makes use of 26/32-bit fixed-point arithmetic with a target resolution of 10^{-4} [17, 18].

However, their design presents several opportunities for improvement:

- Only 4K paths in MC simulations (compared to the minimum 10K paths suggested in the preceding sections)
- The use of an external memory chip, with its related power consumption and bandwidth limitation (imposed by technology)

The latter can be overcome by means of recomputation. A new approach, coined Reverse LS [2, 20], is based on this technique and is the subject of the following sections.

7.3 Reverse Longstaff-Schwartz

In the formulation of the LS algorithm in Sect. 7.2.5, first all paths are generated in step 1 and then traversed in reverse order in step 3. That means the value of each stock price at each time step for all paths has to be stored and communicated between these steps. A total of d.m.N values are generated, d being the dimension of our derivative, m the number of steps, and N the number of MC paths. We call this standard approach the *path storage solution*.

For FPGAs, with only limited internal storage of a few MB, this poses a huge design challenge and in general requires to use several external high-speed memory devices, making the design much more complex. We will now present a novel idea based on recomputation to avoid this massive storage of data.

Instead of storing the paths at each time step, we only store the final stock prices at maturity $\hat{S}_{t_m}$ and then recompute all the other alongside step 3 of the LS algorithm. For that to work we need to find a way to compute the stock price $\hat{S}_{t_i}$ based on the future price $\hat{S}_{t_{i+1}}$:

$$\hat{S}_{t_m} \to \hat{S}_{t_{m-1}} \dots \to \hat{S}_{t_1} \to \hat{S}_{t_0}.$$

The discretized BS equation in Eq. (7.6) is reversible provided we supply the same RNs, such that:

$$\hat{S}_{t_i} = \hat{S}_{t_{i+1}} \exp\left(\left(\frac{\sigma^2}{2} - r + q \right) \Delta t - \sigma \sqrt{\Delta t}\, \Delta W_i \right).$$

In this work, we are using the MT 19937 algorithm to generate a sequence of RNs. Instead of storing the RNs the idea is to build a RNG that generates exactly the opposite sequence, starting from the last one. Fortunately, the MT is a linear RNG, meaning that its state transition function is reversible. In fact, while the tempering function is kept unaltered, only the internal states are to be recomputed [3]. In general this works for all linear RNGs. Based on this a reversed MT can be built. As a result, the Reverse LS method only needs to store and communicate d.N values.

7.4 Architecture

In this section an overview of the whole operation is given, beginning with the paths generation, going through the LS algorithm and computing the final option value. The two proposed solutions are described and compared: *Paths Storage* in an external memory chip versus the novel approach coined *Reverse LS*. A more detail description of the main blocks is covered in the subsequent sections, concluding with notes on how the architecture achieves high-throughput operation.

7.4.1 General Architecture

In general terms, CPUs can be considered as a general purpose device with a fixed hardware structure, which run a program based on a set of predefined hardwired instructions. On the contrary, FPGAs provide with a flexible hardware that can be configured according to the application, enabling dedicated blocks to run more efficiently. There are, however, recent hybrid CPU/FPGA systems, like the Xilinx Zynq, which combine both worlds and provide enough resources to attempt an efficient hardware-software partitioning with low communication latency between both parts. By pipelining the design and fully exploiting the available FPGA resources through multiple parallel instances, the architecture is able to achieve high throughput. The efficiency in terms of energy consumption is the result of carefully implemented modules with minimum resources utilization.

As mentioned before, one particular characteristic of the LS algorithm is that it can only start working (backwards from maturity towards the initial day) once all MC paths have been generated. At this point, the modules in charge of generating this data return to idle, unnecessarily consuming valuable resources on the FPGA. This situation is overcome by exploiting a powerful feature available in Xilinx Zynq devices: the FPGA can be dynamically reprogrammed, either totally or partially.

The preceding explanation leads to an architecture divided into three steps:

- STEP 1: Forward paths generation until maturity
- STEP 2: FPGA reprogramming
- STEP 3: LS operation

Reprogramming in step 2 implies that the preceding and succeeding steps have access to the total amount of resources on FPGA. The time it takes to reprogram the FPGA could be amortized depending on the setup, as it will be explained in later chapters.

The general architecture is presented in Fig. 7.4. In step 1, multiple instances of the forward paths generation block increase bandwidth. In step 3, after reprogramming, the LS algorithm starts working in a pipelined fashion, in order to compute one value per clock cycle. Again, multiple parallel instances are possible in order to

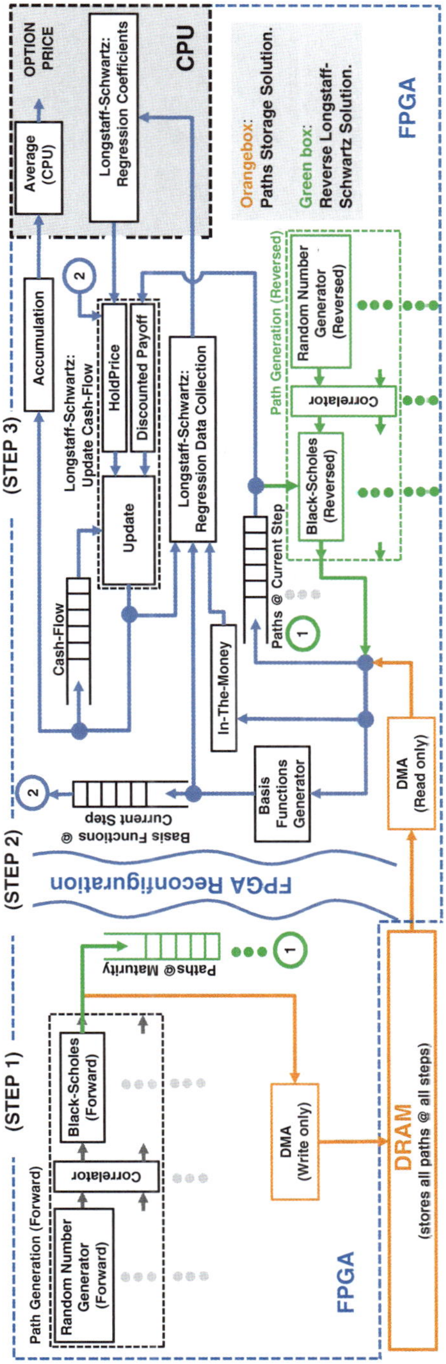

Fig. 7.4 Design architecture including both solutions: *Paths Storage* vs *Reverse LS*

increase bandwidth. Once the initial day is reached, the values from the cash-flow are averaged, which yields the option price.

The architecture in Fig. 7.4 is suited for high-dimensional options, where each instance of the path generation blocks (either forward of reversed) is capable of generating paths for each of the underlying assets (dimensions) simultaneously. Besides, the building blocks of the LS can also be adjusted accordingly.

7.4.2 Paths Storage vs Reverse LS

A straightforward approach is to store all generated paths in an external Dynamic Random-Access Memory (DRAM), as depicted in Fig. 7.4. First, there is a full write process to DRAM that takes place alongside the paths generation. Once the DRAM has been populated and the FPGA reprogrammed, the LS requests all paths, step by step, in a reverse sequence (from maturity towards the initial day). However, this approach presents three disadvantages:

- Data size: a large number of paths, steps, or dimensions, might be enough to exceed the available memory capacity
- Bandwidth: limited by technology based on the memory type (e.g. DDR3), data-bus width, and clock frequency
- Dynamic power consumption: while writing and reading data

Alternatively, the proposed Reverse LS solution overcomes the mentioned disadvantages by recomputing the paths backwards, from maturity, in parallel to the LS algorithm, as shown in Fig. 7.4. The forward paths generation process still computes all MC paths, but only needs to store the paths at maturity. A partial reconfiguration of the FPGA keeps this data on the FPGA, in order for the LS algorithm to start operation immediately in step 3.

7.4.3 Paths Generation: Forward and Reversed

The summarized forward paths generation block from Fig. 7.4 is presented in its full version in Fig. 7.5, and follows the same steps detailed in Sect. 7.2.4. Since paths belonging to each underlying are independent of each other, multiple parallel instances of the complete block are possible, as shown in Fig. 7.4 with dots. The block shown in Fig. 7.5 is configured for two-dimensional options, where each of the BS modules generates paths for one of the underlyings (dimensions). Therefore, this block can be easily extended to higher dimensions by adjusting the number of parallel internal modules, as shown in Fig. 7.4 with dots.

The reversed paths generation block is presented in Fig. 7.6, configured for two-dimensional American options, and following the explanation in Sect. 7.3. It is similar to its forward counterpart, with the exception that now the paths

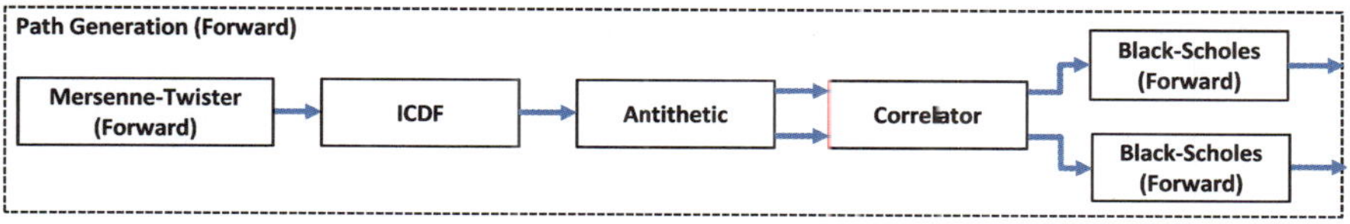

Fig. 7.5 Paths generation forward in detail, configured for two-dimensional American options

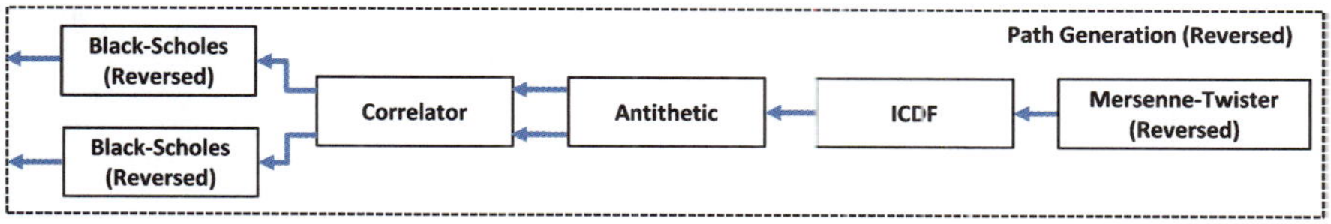

Fig. 7.6 Reversed paths generation in detail, configured for two-dimensional American options

are regenerated from maturity until the initial day (backwards), step by step. As presented before, the BS module is easily reversible. The backward operation of the MT module only needs to reverse the update process that modifies its internal states (the tempering function is kept unaltered). To obtain the same sequence of random numbers in reverse order, it only requires a copy of the last states and final index of its forward counterpart.

7.4.4 LS Implementation

The blocks required to update the cash-flow are fairly straightforward to implement following Sect. 7.2.5, and can be easily parallelized. However, the regression step presents a higher complexity in terms of implementation.

The core of the regression process consists of finding the regression coefficients required to generate the conditional expectation function at every step. These coefficients $\mathbf{b}$ are obtained by solving the system of linear equations:

$$X\mathbf{b} = \mathbf{y}, \tag{7.12}$$

where each row of X contains the values of the basis functions for every path that is *ITM*, and $\mathbf{y}$ contains the corresponding value in the cash-flow. The number of coefficients in $\mathbf{b}$ equals the number of basis functions.

Solving the regression process in hardware becomes either too expensive in terms of resources (fully parallel implementation) or requires a large latency (serialized version). It also becomes inflexible in terms of the method used and the number of coefficients to be calculated. To lift these restrictions, an intelligent hardware-software partitioning is introduced by calculating the coefficients on CPU. In order to reduce the communication overhead between FPGA and CPU, the size of the matrices is reduced, following [18], by rewriting Eq. (7.12) as:

$$(X^T X)\mathbf{b} = (X^T \mathbf{y}), \tag{7.13}$$

where for k basis functions, the size of $(X^T X)$ and $(X^T \mathbf{y})$ is $k \times k$ and $k \times 1$ respectively. It has already been proven that this process can be pipelined by means of accumulators [18]. For monomial-type basis functions x^0, x^1 and x^2, these accumulators become:

$$X^T X = \begin{pmatrix} \sum_n x_n^0 & \sum_n x_n^1 & \sum_n x_n^2 \\ \sum_n x_n^1 & \sum_n x_n^2 & \sum_n x_n^3 \\ \sum_n x_n^2 & \sum_n x_n^3 & \sum_n x_n^4 \end{pmatrix} \quad ; \quad X^T \mathbf{y} = \begin{pmatrix} \sum_n y_n \\ \sum_n y_n x_n \\ \sum_n y_n x_n^2 \end{pmatrix} \tag{7.14}$$

Different methods can be used to solve Eq. (7.13), such as Cholesky decomposition, or the direct method via matrix inverse Eq. (7.15). Although the latter is the one implemented in this work, the Cholesky decomposition is more efficient and can be also easily implemented in the proposed architecture since these operations are executed in software.

$$(X^T X)^{-1} = \frac{1}{det(X^T X)}(Adjoint(X^T X)) \tag{7.15}$$

7.4.5 High-Throughput Operation

It is possible to achieve high-throughput operation along the entire architecture presented in Fig. 7.4. In fact, every module is designed in a pipelined fashion in order to process one new value every clock cycle. Furthermore, several blocks work in parallel, with minimum latency between each other:

- Paths Generation Forward and Direct Memory Access (DMA) (full write): data is sent to DRAM as soon as it is available, with a minimum latency enough to prepare the first DMA burst
- LS and Paths Generation Reversed / DMA (full read): regression coefficients are computed in CPU and sent to the Update Cash-Flow module. As soon as the first path in the cash-flow is updated, two extra events happen:
  - This new value is available for the next Regression Data Collection
  - The value of the stock (path) already used at the current step is no longer required and is immediately updated by either the Paths Generation Reversed module or the DMA (full RD), depending on the implemented solution. In either case, at this point in time the new value of the path has been waiting

to be delivered. It is then not only sent to the corresponding vector, but it is also sent simultaneously to determine if it is ITM and to generate the basis functions

By means of the previous explanations, high-throughput operation for the overall architecture is possible.

7.5 Amortization of FPGA Reconfiguration

Reconfiguring the FPGA implies certain time and energy consumption which can easily exceed the runtime and energy consumption required when pricing a single option. However, when pricing a large set of options, the combination of the Paths Storage approach and the novel Reverse LS allows for easy amortization of the mentioned reconfiguration. In this case, all paths are generated for every option, but only the ones at maturity are stored in an external memory chip. Once the process is finished, the FPGA is reconfigured only once and the options are priced one by one, initializing the paths from the external memory and recomputing the paths backwards by means of the Reverse LS.

7.6 Setup

A comparison between the paths storage approach against their recomputation is only possible in a common setup. In this regard, there is a key observation to make: whereas DRAM chips have an upper limit on bandwidth (defined by the memory type, the clock frequency and the width of its data bus), the bandwidth in an FPGA is only dependent on the number of available resources (hence the number of parallel instances). However, FPGA resources vary considerably among devices and vendors. As a result, both implementations are set to run at the maximum DRAM bandwidth and compared in terms of the energy consumption. Up to the mentioned bandwidth, the lowest energy consumption determines the most profitable approach. Above it, the DRAM itself will not suffice the required bandwidth.

The complete setup, as well as the hardware resources, are detailed in Table 7.1. The FPGA clock is a submultiple of the one used in DRAM, and enough instances of all blocks are used in order to achieve the target bandwidth of 4,266 MB/s.

Although the chosen setup targets two-dimensional options as a testcase, the architecture proposed in Sect. 7.4.1 can be easily adapted for high-dimensional American options.

Our implementation has also been cross-verified with a binomial tree implementation: *Reverse LS*: $P = 9.92 \pm 0.24$; *Binomial Tree (Benchmark)*: $P = 10.12$; Setup:

Table 7.1 Setup table

Detail	Description
Option style	American
Option type	Call
Option characteristics	maximum
Dimensions	2
Basis functions type	Monomial
Basis functions detail	$1, max(S_1, S_2), max(S_1, S_2)^2$
Paths per dimension	10K
Steps	365
Data type	Single-precision floating point
Total data	27.85 MB
Platform	ZC702 evaluation kit
Operating system	Linux (Linaro)
DRAM type	DDR3
DRAM data-bus	32 bits
DRAM clock	533.33 MHz
DRAM bandwidth	4266.64 MB/s
FPGA clock	133.33 MHz
FPGA bandwidth	4266.64 MB/s

$S_{1,2}(0) = 100$, $K = 100$, $r = 0.05$, $q_{1,2} = 0.10$, $\sigma_{1,2} = 0.2$, $\rho = 0.1$. The chosen basis functions generally deliver good results for general options, however not the best result. This depends on the type and the number of basis functions, which need to be tried and tested.

7.7 Tools and Estimation Methodology

The different modules have been implemented in Vivado High-Level Synthesis (HLS) using C, and optimized for high-performance at a clock period of 7.5 ns (133.33 MHz) with a minimum number of FPGA resources. The place-and-route (P&R) report on resources utilization was then fed into the Xilinx Power Estimator [21] in order to obtain power estimations of individual blocks. The estimated values have been checked by means of testbenches on the Xilinx Zynq ZC702 Evaluation Kit. In a similar way, DRAM DDR3 power consumption is based on measured values at different bandwidths on the same board. The testbench followed the same access pattern used in the full architecture and achieved a maximum of 83 % and 87 % of the peak theoretical bandwidth for writing and reading respectively [20]. Then these values were extrapolated to the maximum theoretical bandwidth available (4,266 MB/s at 533.33 MHz and 32-bit data bus).

In terms of energy consumption, all values are derived from the obtained average power consumption and the required runtime.

7.8 Results

For the given setup, the resources utilization on the FPGA is detailed in Table 7.2, grouped by major blocks. A note is made on the fact that the minimum Zynq device on which the given configuration fits, with the required parallel instances, is the Z-7030 device.

As mentioned before, every single building block in the proposed architecture has been fully pipelined with an initiation interval of one clock cycle (II = 1). This means that every block starts processing a new data value in every clock cycle. At 4,266 MB/s, the total amount of data (27.85 MB) is processed in approximately 6.53 ms, as presented in Table 7.3. The total runtime in this case, including the communication overhead between CPU and FPGA and excluding the FPGA reconfiguration, adds up to 16.94 ms for one option pricing.

Table 7.2 FPGA resources breakdown

Step	Block	LUT	FF	DSP	BRAM
1	Path generation forward	18,404	17,376	188	88
	Paths @ Maturity	1,752	1,648	0	64
2	Reconfiguration	–	–	–	–
3	Longstaff-Schwartz	28,296	33,468	212	108
	Path generation reversed	24,048	23,932	164	88

Table 7.3 Power, runtime and energy consumption breakdown

Block	Dynamic power (mW)	Runtime (ms)	Energy (mJ)
Path generation forward	1,239	6.53	8.09
Paths @ Maturity	334	0.02	0.01
Paths storage full WR	1,265	6.53	8.26
MT communication overhead 1/2	160	1.21	0.19
Reprogramming [1]	1,860	50.00	93.00
MT communication overhead 2/2	160	0.62	0.09
Paths storage full RD	1,526	6.53	9.97
Path generation reversed	1,392	6.53	9.09
Regression data collection	795	6.53	5.19
Regression coefficients (CPU)	160	2.03	0.32
Update cash-flow	374	6.53	2.44
Cash-flow	174	6.53	1.14
Paths @ Current-Step	334	6.53	2.18
Accumulation (Average)	103	0.02	0.00

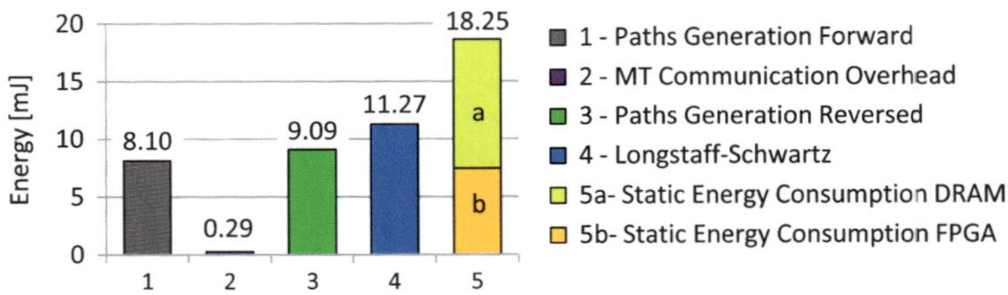

Fig. 7.7 Dynamic energy consumption breakdown

7.8.1 Dynamic Energy Consumption Breakdown

Based on the dynamic power and the runtime, it is possible to derive the dynamic energy consumption of every building block shown in Fig. 7.4, as detailed in Table 7.3.

Figure 7.7 presents the dynamic energy consumption breakdown of the whole architecture when the novel Reverse LS approach is implemented. *MT communication overhead* refers to the energy consumed to initialize the internal states of the forward MT modules, read the final states and index, and initializing the reversed MT modules. The *LS* column in Fig. 7.7 includes the energy consumption in FPGA (10.95 mJ) and in CPU (0.32 mJ). The latter includes the computation of the regression coefficients, as well as the communication overhead when reading the accumulated matrices and writing back the coefficients.

An optimized CPU implementation of the entire algorithm in Matlab on an Intel i5-2450M (2.50 GHz) core with, 6 GB of RAM, requires, for the given setup, 270 ms and an energy consumption of 12.70 J. The latter has been obtained at the power-plug with all unnecessary components in the computer disabled. In contrast, our implementation in Zynq requires 16.94 ms and consumes approximately 47 mJ, providing a speedup of 16× in runtime and 268× in energy consumption.

7.8.2 Reverse LS Versus Paths Storage

When comparing the regeneration of the paths in FPGA against the storage of all paths in DRAM (both when writing and reading data), there is a reduction in the energy consumption of 2×, as depicted in Fig. 7.8. All values shown are based on the given setup and methodology. To make a fair comparison, only the additional (dynamic) energy consumption is taken into account. This is due to the fact that in a hybrid CPU/FPGA device, like the Xilinx Zynq running Linux on the ARM cores, the DRAM is already being used by the operating system.

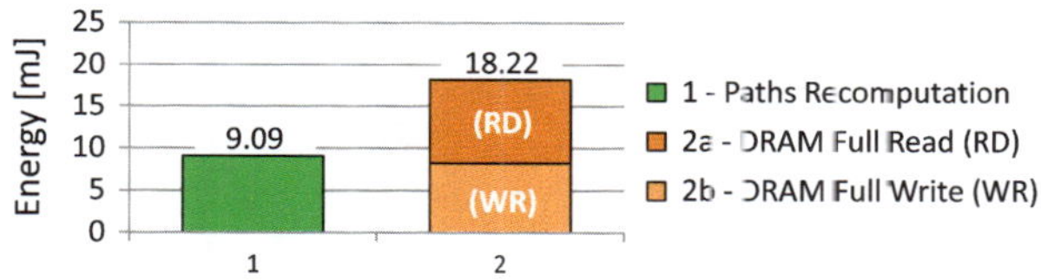

Fig. 7.8 Dynamic energy consumption when regenerating all paths in FPGA and when storing the paths in an external memory (DRAM)

7.8.3 Comparison to Related Work

The reference work [18] presented a dedicated FPGA implementation targeted for one specific option and setting. It further uses a number format specialized for this usecase based on 26/32-bit fixed-point operations. With our proposed architecture we show how it is possible to target high-dimensional options We further use single-precision floating-point operations, so that the user does not have to take care of the accuracy of the solution.

The main inconvenience in comparing our work to the reference resides in the fact that both architectures target different devices at different technology nodes. Under these circumstances, it was decided to run the comparison on the basis of energy efficiency, by porting their work [18] to the same Xilinx Zynq device based on their published resources utilization. Although this approach is just a coarse estimation, it could still be considered a valid setup for a comparison purposes. For their work, one option pricing consumes, according to XPE, 2.46 mJ dynamic power including one DRAM chip. Our downscaled architecture to one-dimension and the same number of paths and steps only requires 1.85 mJ dynamic power, being a 33 % improvement. This means that we achieve higher energy-efficiency while providing higher accuracy. We make this possible with FPGA reconfiguration in combination with an optimized scheduling, and our novel Reverse LS approach.

7.9 Conclusion

American option pricing is a computational challenge for financial institutions, which operate huge clusters. In this work a high-throughput and energy-efficient pricing system for American options has been presented, targeting hybrid CPU/FPGA devices. Compared to the state-of-the-art, this is the first FPGA-based implementation targeting the full range of high-dimensional American options.

Our main contribution is Reverse Longstaff-Schwartz, a bit-true algorithmic transformation where recomputation is exploited. Paths storage is minimized by means of recomputation, removing any bandwidth limitation and significantly improving energy-efficiency. By additionally making use of runtime reconfiguration and utilizing an optimized scheduling to amortize the reconfiguration times, we are able to deliver higher energy-efficiency. In this regard, the resulting architecture is 16× faster and 268× more energy-efficient than an optimized Intel i5 implementation in Matlab.

Acknowledgements We gratefully acknowledge the partial financial support from the Center of Mathematical and Computational Modelling $(\mathrm{CM})^2$ of the University of Kaiserslautern, from the German Federal Ministry of Education and Research under grant number 01LY1202D, and from the Deutsche Forschungsgemeinschaft (DFG) within the RTG GRK 1932 "Stochastic Models for Innovations in the Engineering Sciences", project area P2. The authors alone are responsible for the content of this work.

Appendix

Algorithm 1 Longstaff Schwartz MC method to price American maximum call option on two stocks

Input: discounted payoff $g(S)$
Output: option price V_N

1: Generate N independent paths for two stocks at all possible exercise dates: $\{S_1^n(t_0), S_1^n(t_1), \ldots, S_1^n(t_m)\}$ and $\{S_2^n(t_0), S_2^n(t_1), \ldots, S_2^n(t_m)\}$, with $n = 1, \ldots, N$, $t_i = \frac{T}{m} \times i$, $i = 1, \ldots, m$ and $S_1^n(t_0) \equiv S_1(0)$, $S_2^n(t_0) \equiv S_2(0)$ as follows:

$$S_1(t_i) = S_1(t_{i-1})e^{((r-q_1-\frac{1}{2}\sigma_1^2)\Delta t+\sigma_1\sqrt{\Delta t}Z_1)}$$

$$S_2(t_i) = S_2(t_{i-1})e^{((r-q_2-\frac{1}{2}\sigma_2^2)\Delta t+\sigma_2\sqrt{\Delta t}Z_2)}$$

with $Z_1 = u_1$ and $Z_2 = \rho u_1 + \sqrt{1-\rho^2}u_2$, where $u_1, u_2 \sim N(0,1)$.

2: At maturity $t_m = T$, fix the discounted terminal values of the American option for each path $n = 1, \ldots, N$:

$$V^n(t_m) = e^{-rT}(\max\{S_1^n(t_m), S_2^n(t_m)\} - K)^+$$

3: Compute backward at each potential exercise date t_i for $i = m-1, m-2, \ldots, 1$:

1. Choose k basis functions: $\{H_1, \ldots, H_k\}$.
2. Consider the subset of paths $\Theta_{\hat{N}} \subset \{1, \ldots, N\}$ for which the option is ITM, i.e. $\max\{S_1^n(t_m), S_2^n(t_m)\} > K$ holds for $n \in \Theta_{\hat{N}}$.
3. Solve the least-square linear regression problem:

$$\min_{a_l \in \mathbb{R}} \frac{1}{\hat{N}} \sum_{n=1}^{\hat{N}} (V^n(t_i) - \sum_{l=1}^{k} a_l H_l(S_{1,2}^n(t_i)))^2$$

$$S_{1,2}^n(t_i) := \{S_1^n(t_i), S_2^n(t_i)\} \quad \text{for simplicity}$$

and obtain the optimal coefficient a^*:

$$\begin{aligned} a^* &:= [a_1^*, \ldots, a_k^*]^\top \\ &= (X^\top X)^{-1} X^\top Y \in \mathbb{R}^{k \times 1} \end{aligned}$$

(continued)

Algorithm 1 (continued)

with $Y := [V^1(t_i),\ldots,V^{\hat{N}}(t_i)]^\top \in \mathbb{R}^{\hat{N}\times 1}$,

$$X := \begin{pmatrix} H_1(S^1_{1,2}(t_i)) & \ldots & H_k(S^1_{1,2}(t_i)) \\ \vdots & \ldots & \vdots \\ H_1(S^{\hat{N}}_{1,2}(t_i)) & \ldots & H_k(S^{\hat{N}}_{1,2}(t_i)) \end{pmatrix} \in \mathbb{R}^{\hat{N}\times k}$$

4. Calculate the approximation of the value for continuing the option $C^n(t_i)$ and the value for exercising the option $E^n(t_i)$ for each path $n \in \Theta_{\hat{N}}$:

$$C^n(t_i) = \sum_{l=1}^{k} a_l^* H_l(S^n_{1,2}(t_i))$$

$$E^n(t_i) = e^{-rt_i}(\max\{S^n_1(t_i), S^n_2(t_i)\} - K)^+$$

5. Compare the value of $C^n(t_i)$ and $E^n(t_i)$ to decide whether to exercise or to continue the option:

$$V^n(t_i) = \begin{cases} E^n(t_i), & \text{if } n \in \Theta_{\hat{N}} \text{ and } E^n(t_i) \geq C^n(t_i) \\ V^n(t_{i+1}), & \text{otherwise} \end{cases}$$

4: Compute $V_N = \left(\frac{1}{N}\sum_{i=1}^{N} V^n(t_1)\right)$ as an approximation for the American option price

References

1. Brugger, C., de Schryver, C., Wehn, N.: HyPER: a runtime reconfigurable architecture for Monte Carlo option pricing in the Heston model. In: 2014 IEEE 24th International Conference on Field Programmable Logic and Applications (FPL), Munich (2014). doi: 10.1109/FPL.2014.6927458
2. Brugger, C., Varela, J.A., Wehn, N., Tang, S., Korn, R.: Reverse Longstaff-Schwartz American option pricing on hybrid CPU/FPGA systems. In: Proceedings of the 2015 Design, Automation & Test in Europe Conference & Exhibition (DATE), Grenoble, pp. 1599–1602. ACM (2015)
3. Hagita, K., Takano, H., Nishimura, T., Matsumoto, M.: Reverse generator MT19937. Fortran source code (2000). http://www.math.sci.hiroshima-u.ac.jp/~m-mat/MT/VERSIONS/FORTRAN/REVmt19937b.f. Last access 16 Sept 2014
4. Hull, J.C.: Options, Futures, and other Derivatives, 8th edn. Pearson, Harlow (2012)
5. Jin, Q., Luk, W., Thomas, D.B.: On comparing financial option price solvers on FPGA. In: 2011 IEEE 19th Annual International Symposium on Field-Programmable Custom Computing Machines (FCCM), Salt Lake City, pp. 89–92 (2011). doi:10.1109/FCCM.2011.30
6. Kandemir, M., Li, F., Chen, G., Chen, G., Ozturk, O.: Studying storage-recomputation tradeoffs in memory-constrained embedded processing. In: 2005 Proceedings in Design, Automation and Test in Europe, Munich, pp. 1026–1031. IEEE (2005)
7. Koc, H., Kandemir, M., Ercanli, E., Ozturk, O.: Reducing off-chip memory access costs using data recomputation in embedded chip multi-processors. In: Proceedings of the 44th Annual Design Automation Conference, San Diego, pp. 224–229. ACM (2007)
8. Korn, R., Korn, E., Kroisandt, G.: Monte Carlo Methods and Models in Finance and Insurance. CRC, Boca Raton (2010)

9. Longstaff, F.A., Schwartz, E.S.: Valuing American options by simulation: a simple least-squares approach. Rev. Financ. stud. **14**(1), 113–147 (2001)
10. Matsumoto, M.: Mersenne twister home page (2007). http://www.math.sci.hiroshima-u.ac.jp/~m-mat/MT/emt.html. http://www.math.sci.hiroshima-u.ac.jp/~m-mat/MT/emt.html. Last access 02 July 2014
11. Matsumoto, M., Nishimura, T.: Mersenne twister: a 623-dimensionally equidistributed uniform pseudo-random number generator. ACM Trans. Model. Comput. Simul. **8**(1), 3–30 (1998). doi:http://doi.acm.org/10.1145/272991.272995
12. Monetary, Economic Department: Statistical release: OTC derivatives statistics at end-December 2013. Technical report, Bank for International Settlements (2014). http://www.bis.org/publ/otc_hy1405.pdf
13. de Schryver, C., Schmidt, D., Wehn, N., Korn, E., Marxen, H., Korn, R.: A new hardware efficient inversion based random number generator for non-uniform distributions. In: Proceedings of the 2010 International Conference on Reconfigurable Computing and FPGAs (ReConFig), Cancun, pp. 190–195 (2010). doi:10.1109/ReConFig.2010.20
14. de Schryver, C., Shcherbakov, I., Kienle, F., Wehn, N., Marxen, H., Kostiuk, A., Korn, R.: An energy efficient FPGA accelerator for Monte Carlo option pricing with the Heston model. In: Proceedings of the 2011 International Conference on Reconfigurable Computing and FPGAs (ReConFig), Cancun, pp. 468–474 (2011). doi:10.1109/ReConFig.2011.11
15. Sridharan, R., Cooke, G., Hill, K., Lam, H., George, A.: FPGA-based reconfigurable computing for pricing multi-asset barrier options. In: Proceedings of Symposium on Application Accelerators in High-Performance Computing PDF (SAAHPC), Argonne (2012)
16. Thomas, D.B., Bower, J.A., Luk, W.: Hardware architectures for Monte-Carlo based financial simulations. In: Proceedings of the IEEE International Conference on Field Programmable Technology FPT 2006, Bangkok, pp. 377–380 (2006). doi:10.1109/FPT.2006.270352
17. Tian, X., Benkrid, K.: Fixed-point arithmetic error estimation in Monte-Carlo simulations. In: 2010 International Conference on Reconfigurable Computing and FPGAs (ReConFig), Cancun, pp. 202–207 (2010). doi:10.1109/ReConFig.2010.14
18. Tian, X., Benkrid, K.: Implementation of the Longstaff and Schwartz American option pricing model on FPGA. J. Signal Process. Syst. **67**(1), 79–91 (2012). doi:10.1007/s11265-010-0550-1
19. Tse, A.H., Thomas, D.B., Tsoi, K.H., Luk, W.: Efficient reconfigurable design for pricing Asian options. SIGARCH Comput. Archit. News **38**(4), 14–20 (2011). doi:10.1145/1926367.1926371. http://doi.acm.org/10.1145/1926367.1926371
20. Varela, J.A.: Embedded architecture to value American options on the stock market. Master's thesis, Microelectronic Systems Design Reseach Group, Department of Electrical Engineering and Information Technology, University of Kaiserslautern (2014)
21. Xilinx: XPower estimator (XPE) (2014). http://www.xilinx.com/products/design_tools/logic_design/xpe.htm. Last access: 16 Sept 2014

Chapter 8
Bringing Flexibility to FPGA Based Pricing Systems

Christian Brugger, Christian De Schryver, and Norbert Wehn

Abstract High-speed and energy-efficient computations are mandatory in the financial and insurance industry to survive in competition and meet the federal reporting requirements. While FPGA based systems have demonstrated to provide huge speedups, they are perceived to be much harder to adapt to new products. In this chapter we introduce *HyPER*, a novel methodology for designing Monte Carlo based pricing engines for hybrid CPU/FPGA systems. Following this approach, we derive a high-performance and flexible system for exotic option pricing in the state-of-the-art Heston market model. Exemplarily, we show how to find an efficient implementation for barrier option pricing on the Xilinx Zynq 7020 All Programmable SoC with HyPER. The constructed system is nearly two orders of magnitude faster than high-end Intel CPUs, while consuming the same power.

8.1 Introduction

The recent advance in financial market models and products with ever increasing complexity, as well as the more stringent regulations on risk assessment from federal agencies have led to a steady growth of computational power. Additionally, increasing energy costs force finance and insurance institutes to consider new technologies for executing their computations. Graphics Processor Units (GPUs) have already demonstrated their benefit for speeding up financial simulations and are state-of-the-art in finance business nowadays [2, 21].

However, Field Programmable Gate Arrays (FPGAs) have been shown to outperform GPUs with respect to speed and energy efficiency by far for those tasks [6, 15, 17]. They are currently starting to emerge in finance institutes such as J.P. Morgan [1, 7] or Deutsche Bank [12]. Nevertheless, most problems cannot be efficiently ported to pure data path architectures, since they contain algorithmic steps that are executed best on a Central Processing Unit (CPU).

C. Brugger (✉) • C. De Schryver • N. Wehn
Microelectronic Systems Design Research Group, University of Kaiserslautern, Kaiserslautern, Germany
e-mail: brugger@eit.uni-kl.de; schryver@eit.uni-kl.de; wehn@eit.uni-kl.de

C. De Schryver (ed.), *FPGA Based Accelerators for Financial Applications*,
DOI 10.1007/978-3-319-15407-7_8

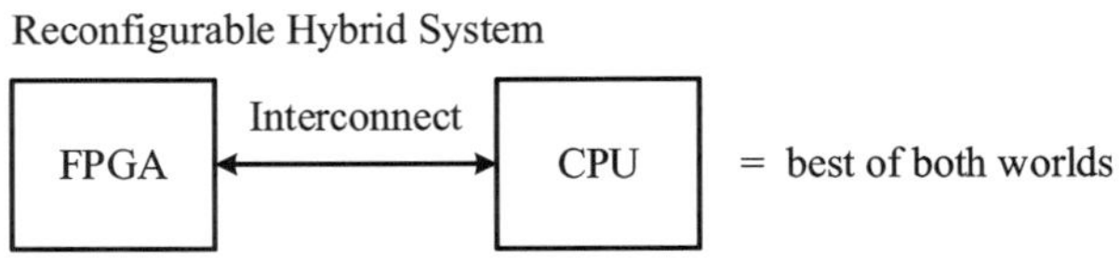

Fig. 8.1 In this work we target reconfigurable hybrid systems, i.e. heterogeneous FPGA/CPU platforms. With this setup we can exploit the efficiency of reconfigurable logic and the flexibility of a processor, having best of both worlds

Hybrid devices as shown in Fig. 8.1 combine standard CPU cores with a reconfigurable FPGA area, connected over multiple high-bandwidth channels. They allow running an Operating System (OS) that is able to (re-)configure the FPGA part at runtime, e.g. for instantiating problem specific accelerators. A prominent example is the recent Xilinx Zynq All Programmable System on Chip (SoC).

In addition to the technological improvements, there are advances in the algorithmic domain as well. Although classical Monte Carlo (MC) methods are still prevailing, for example Multilevel Monte Carlo (MLMC) methods are more and more called into action [8, 10]. They can help to reduce the computational effort in total, but require a higher complexity in the controlling and require a more flexible execution platform.

In this chapter, we illustrate how we can combine the benefits of dedicated hardware accelerators with high flexibility as required by many practical applications on hybrid CPU/FPGA systems. For this purpose we have combined the current trends both from technology and computational stochastics to an option pricing platform for reconfigurable hybrid architectures. The proposed *HyPER* framework can handle a wide range of option types, is based on the state-of-the-art Heston model, and extensively uses dynamic runtime reconfiguration during the simulations. To derive the architecture, we have applied a platform based design methodology including Hardware/Software (HW/SW) split and dynamic reconfiguration.

In particular, we focus on the following points:

- We propose an energy-efficient and modular option pricing framework called *HyPER* that is generically applicable to all kinds of hybrid CPU/FPGA platforms.
- We show how the special characteristics arising from reconfigurable hybrid systems can be included in a platform based design methodology.
- We have implemented HyPER configuration setup on the Xilinx Zynq-7000 All Programmable SoC relevant to practitioners. For this implementation we give detailed area, performance, and energy numbers.

8.2 Background and Related Work

The use of FPGAs for accelerating financial simulations has become attractive with the first available devices. Many papers are available that propose efficient random number generation methods and path generations [5, 13, 14, 18–20, 22]. Although most are focused on the Black-Scholes market model, there are a few publications on non-constant volatility models as well. Benkrid [20], Thomas, Tse, and Luk [18, 22] have thoroughly investigated the potentials of FPGAs and heterogeneous platforms for the Generalized Autoregressive Conditional Heteroskedasticity (GARCH) setting in particular. Thomas has come up with a Domain-Specific Language (DSL) for reconfigurable path-based MC simulations in 2007 [18] that supports GARCH as well. It allows to describe various path generation mechanisms and payoffs and can generate software and hardware implementations. That way, Thomas' DSL is similar to our proposed framework. However, it does neither incorporate MLMC simulations nor automatic HW/SW splitting.

For the Heston setting, Delivorias has demonstrated the enormous speedup potential of FPGAs for classical MC simulations compared to CPUs and GPUs in 2012 [6]. The results are included in Chap. 3, but do neither include energy nor synthesis numbers.

De Schryver et al. have shown in 2011 that Xilinx Virtex-5 FPGAs can save around 60 % of energy compared to a Tesla C2050 GPU [15]. Sridharan et al. have extended this work to multi-asset options in 2012 [17], showing speedups up to 350 for one FPGA device compared to an SSE reference model on a multi-core CPU. De Schryver et al. have enhanced their architecture further to support modern MLMC methods in 2013 [16]. Their architecture is the basis for our proposed implementation in this paper.

Our HyPER platform was first presented in [3]. A hardware prototype was exhibited at the ReConFig 2013 and the FPL 2014 conferences.

8.2.1 Heston Model

The Heston model is a mathematical model used to price products on the stock market [9]. Nowadays, it is widely used in the financial industry. One main reason is that the Heston model is complex enough to describe important market features, especially volatility clustering [10]. At the same time, closed-form solutions for simple products are available. This is crucial to enable calibrating the model against the market in realistic time.

In the Heston model the price S and the volatility ν of an economic resource are modeled as stochastic differential equations:

$$\begin{aligned} dS_t &= S_t r dt + S_t \sqrt{\nu_t} dW_t^S, \\ d\nu_t &= \kappa(\theta - \nu_t)\, dt + \sigma \sqrt{\nu_t} dW_t^\nu. \end{aligned} \tag{8.1}$$

The price S can reflect any economic resource like assets or indices as the S&P 500 or the Dow Jones Industrial Average. S can also be the stock price of a company. The volatility ν is a measure for the observable fluctuations of the price S. The fair price of a derivative today can be calculated as $P = \mathbb{E}[g(S_t)]$, where g is a corresponding discounted payoff function. Although closed-form solutions for simple payoffs like vanilla European call or put options exist, so-called *exotic* derivatives like barrier, lookback, or Asian options must be priced with compute-intensive numerical methods in the Heston model [10]. A very common and universal choice are Monte Carlo (MC) methods that we consider in this chapter.

8.2.2 Monte Carlo Methods for the Heston Model

Simulating the Heston model in Eq. (8.1) requires the application of an appropriate discretization scheme. In this work we have applied *Euler discretization* that has been shown to work well with in the MLMC Heston setting [11]. Discretizing Eq. (8.1) into k steps with equal step sizes $\Delta t = \frac{T}{k}$ leads to the discrete Heston equations given by:

$$\hat{S}_{t_{i+1}} = \hat{S}_{t_i} + r\hat{S}_{t_i}\Delta t + \hat{S}_{t_i}\sqrt{\hat{\nu}_{t_i}}\sqrt{\Delta t}\left(\sqrt{1-\rho^2}Z_i^S + \rho Z_i^\nu\right),$$

$$\hat{\nu}_{t_{i+1}} = \hat{\nu}_{t_i} + \kappa(\theta - \hat{\nu}_{t_i})\Delta t + \sigma\sqrt{\hat{\nu}_{t_i}}\sqrt{\Delta t}\, Z_i^\nu. \tag{8.2}$$

While the initial asset price $S_0 = \hat{S}_{t_0}$ and r can be observed directly at the market, the five Heston Parameters κ, θ, σ, ρ, and $\nu_0 = \hat{\nu}_{t_0}$ are obtained through calibration, compare Chaps. 2 and 10. Z_i^S and Z_i^ν are two independent normal distributed random variables with mean zero and variance one. With this method an approximated solution $\hat{S}_t$ can be obtained by linearly interpolating $\hat{S}_{t_0}, \ldots, \hat{S}_{t_k}$.

The classic MC algorithm estimates the price $P = \mathbb{E}[g(S_t)]$ of a European derivative with a final payoff function $g(S)$ as the sample mean of simulated instances of the discounted payoff values $g(\hat{S})$, i.e.,

$$\mathcal{A}^{\text{std}} = \frac{1}{N}\sum_{i=1}^{N} g(\hat{S}_i),$$

where $\hat{S}_1, \ldots, \hat{S}_N$ are independent identically distributed copies of $\hat{S}$.

For the implementation, we have used the same algorithmic refinements as in the data path presented in [16] (antithetic variates, full truncation, log price simulation).

8.2.3 The Multilevel Monte Carlo Method

The MLMC method as proposed by Giles in 2008 uses different discretization *levels* within one MC simulation [8]. It is based on an iterative result refinement strategy, starting from low levels with coarse discretizations and adding corrections

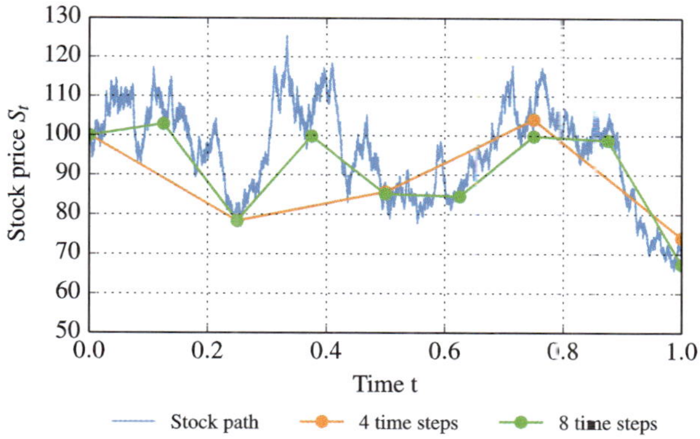

Fig. 8.2 MLMC approximates the real stockpath that has infinite information, with multiple levels of discretization. In this case the path is approximated with four and eight discretization points

from simulations on higher levels with finer discretizations. Figure 8.2 illustrates a continuous stock path with two different discretizations (4 and 8 steps). It is obvious that the computational effort required to compute one path increases for higher levels. For a predefined *accuracy* of the result, the MLMC method tries to balance the computational effort on all levels, therefore much more paths are computed on lower levels (with coarser discretizations). Since for finer discretizations the variances decrease, it is sufficient to simulate fewer paths on higher levels. In total, this leads to an asymptotically lower computational effort for the complete simulation [8]. For our investigated financial product "European barrier options", MLMC has explicitly shown to provide benefits also for practical constellations [11].

Let us formalize the idea. Without loss of generality one can assume that $k = M^{L-1}$ discretization points for a fixed M and some integer L are sufficient to obtain the desired accuracy. We define $\hat{S}^{(l)}$ for $l = l_0, \ldots, L$ as the approximated solution of Eq. (8.1) with M^{l-1} discretization points. Then, in contrast to the classic MC estimate where the "single" approximation $\hat{S}^{(l)}$ is used, one considers the sequence of approximation $\hat{S}^{(l_0)}, \ldots, \hat{S}^{(L)}$. With the telescoping sum

$$\hat{P} = \mathbb{E}\left[\hat{S}^{(L)}\right] = \mathbb{E}\underbrace{\left[g\left(\hat{S}^{(l_0)}\right)\right]}_{\hat{D}_{l_0}} + \sum_{l=l_0+1}^{L} \mathbb{E}\underbrace{\left[g\left(\hat{S}^{(l)}\right) - g\left(\hat{S}^{(l-1)}\right)\right]}_{\hat{L}_l}. \qquad (8.3)$$

the single expected value by expected values of differences. Each of the expectations on the right are called *levels*. The MLMC algorithm approximates each of these levels with independent classic MC algorithms. To get a convergent and efficient MLMC algorithm, it is important that the variances of the levels

$$V_l = \mathbb{V}\mathrm{ar}[D_l]$$

decay to zero fast enough. One way to achieve fast enough convergence of V_l is choose a suitable discretization scheme and to let $\hat{S}^{(l)}$ and $\hat{S}^{(l-1)}$ depend on the same Brownian path. At the end the MLMC algorithm aims at reducing the overall computational cost by optimally distributing the workload over all levels [8].

In our setup it has been explicitly shown that Euler discretization is sufficient [11]. Using the discretized Heston model, Eq. (8.2), the price P can be calculated according to Eq. (8.3) with L individual MC algorithms. To reach the target accuracy ε, N_l paths are evaluated on each level l, given by:

$$N_l = \left\lceil \varepsilon^{-2} \sqrt{V_l} \sum_{k=l_0}^{L} \sqrt{V_k} \right\rceil . \tag{8.4}$$

The level variances are estimated with initial $N_l = 10^4$ samples. To let $\hat{S}^{(l)}$ and $\hat{S}^{(l-1)}$ depend on the same Brownian path, the same random numbers of the fine path are also used to approximate the coarse path, by adding up the M previous random numbers of the fine path.

8.3 Methodology

The classical MC algorithm only uses one fixed discretization scheme and is very regular. MLMC methods as introduced in the previous section are more complicated and rely on an iterative scheme with high inherent dynamics. For both methods dedicated FPGA architectures have been proposed [15, 16] (also see Sect. 8.2). However, they are static architectures that use exactly one single generic FPGA configuration throughout the entire computation and for all products.

In this work we systematically approach the inherent dynamics of the MLMC algorithm and propose a pricing platform that incorporates them. The dynamics in particular are:

- The huge variety of the financial products and their different structure on how to calculate their price.
- The specialty of the first level, which calculates only one price path, while the higher levels calculate two paths simultaneously.
- The different number of discretization steps used in the iterative refinement strategy and the impact on the FPGA architecture.

Our goal is to design a pricing system that exploits the characteristics of the underlying hybrid CPU/FPGA execution platform efficiently for each part of the iterative algorithm and for all products traded on the market. A static design can never cover the complete range of those dynamics. Therefore we introduce a platform based design methodology that captures all the important characteristics of the problem and hybrid systems in general, but leaves enough flexibility to price arbitrary products and to target any specific hybrid device, see Fig. 8.3. It comes with three key features that address the dynamics:

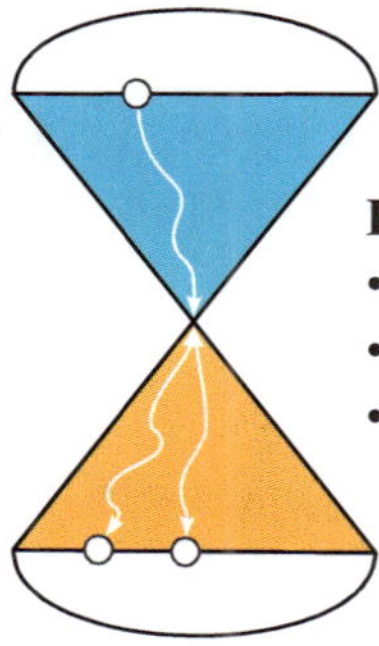

Fig. 8.3 The HyPER platform makes use of a platform based design methodology in which both the flexibility in the application and architectural space is captured by an automated approach. Once the user specifies the exact financial product and the target platform, the HyPER platform generates an optimal implementation for exactly this setup

- A modular pricing framework that is easily extensible, and consist of reusable building blocks with standardized ports to minimize the effort for adding new products.
- Extensive use of online reconfiguration of the FPGA to always have the best architecture available at any time, while still keeping the overhead of reconfiguration in mind.
- Use of static optimization to find the optimal configurations for a given financial product and specific hybrid device. The goal of the optimizer is to exploit all available degrees of freedom, including HW/SW splitting and the flexibility of the modular architecture.

With this new methodology it is possible to design a novel pricing system that is aware of the inherent dynamics of the problem. We introduce the resulting framework as the *HyPER pricing system* in the next section

8.4 The HyPER Pricing System

HyPER is a high-speed pricing system for option pricing in the Heston model. It uses the advanced Multilevel Monte Carlo (MLMC) method and targets hybrid CPU/FPGA systems. To be able to efficiently price the vast majority of exotic options traded on the market it is based on reusable building blocks. To adapt the FPGA architecture to the requirements of the multilevel simulation in each part of the algorithm, it exploits online dynamic reconfiguration

8.4.1 Modular Pricing Architecture

For each level l the main steps of the MLMC algorithm are:

1. Simulate N_l MC paths $\hat{S}^{(l)}$ and optionally $\hat{S}^{(l-1)}$ with $k = M^l$ time steps.
2. Calculate the coarse and fine payoff $g(.)$ for each path.
3. Calculate the mean $\mathbb{E}[\hat{D}_l]$ and variance $V_l = \mathbb{V}\text{ar}[\hat{D}_l]$ of the difference of all coarse and fine payoffs, according to Eq. (8.3).

This is done for $l = l_0, \ldots, L$. For practical problems the first level l_0 is typically equal to 1, the multilevel constant M equal to 4, and the maximum level L between 5 and 7. The number of MC steps NM^l is roughly the same on each level and in the order of 10^{12} [8, 11].

Step 1 is the computationally most intensive part of the multilevel algorithm since it requires solving Eq. (8.2). This involves Brownian increment generation (*Increment Generator*) and calculating the next step of each path, step by step, path by path (*Path Generator*). In HyPER we therefore implement it on the FPGA part of the hybrid architecture. While for the first level l_0 only one type of paths is calculated (*Single-Level Kernel*), for higher levels fine and coarse paths are required with the same Brownian increments. This makes the kernel more complicated and involves more logic resources (*Multilevel Kernel*). This covers the frontend of the HyPER architecture shown in Fig. 8.4.

The Brownian increments are generated with a uniform Random Number Generator (RNG) and transformed to normally distributed random numbers. We choose the Mersenne Twister MT19937 for the uniform RNG and an Inverse Cumulative Distribution Function (ICDF) approach for the transformation. We further use antithetic variates as a variance reduction technique [10].

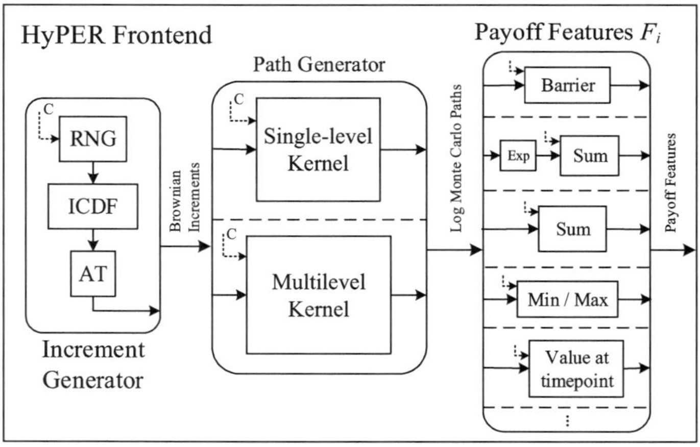

Fig. 8.4 The HyPER frontend is a modular pipeline in which each blocks are fully utilized in each cycle. Payoff features are user defined and can be extended to generate path dependent features as required for the financial product being prices. While often times only a small set of features are required for a specific product, only the necessary blocks are mapped to the system

8.4.1.1 Payoff Computation

Part 2 involves the payoff computation and is strongly dependent on the option being priced. With the HyPER architecture we cover arbitrary European options, including barrier options that depend on whether a barrier is hit or not, and Asian options for which the payoff depends on the average of the stock price. For such path dependent payoffs every price of the path has to be considered. This leads to the dilemma that on the one hand a high-throughput payoff computation is needed, since the prices are generated on the FPGA fabric with one value per clock cycle. On the other hand the payoff computation may involve complex arithmetics that are not used in each cycle. Considering the payoff procedure carefully in the HW/SW splitting process is therefore crucial.

One of the key insights of the HyPER pricing system is to split the discounted payoff function $g(\hat{S}_t)$ in two separate parts: A path dependent part F_i and a path independent part h. The idea is to put the path dependent part F_i on the FPGA and the independent part h on CPU. We express the payoff as:

$$g(\hat{S}_t) = h(F_1(\hat{S}_t), \ldots, F_n(\hat{S}_t)).$$

We call the path dependent functions F_i *features* and choose them such that they contain as little arithmetic operations as possible. h does not directly depend on $\hat{S}_t$. Let us look at an example: Asian Call options with strike K. Their payoff is given by:

$$g^{\text{Asian}}(\hat{S}_t) = e^{-rT} \max\left(\frac{1}{k}\sum_{i=1}^{k} \hat{S}_{t_i} - K, 0\right).$$

In this case the sum is path dependent and we can identify the result of this sum as feature F:

$$F(\hat{S}_t) = \sum_{i=1}^{k} \hat{S}_{t_i}, \quad \text{and } g^{\text{Asian}}(\hat{S}_t) = h(F(\hat{S}_t)),$$

$$\Rightarrow h(x) = e^{-rT} \max\left(k^{-1}x - K, 0\right).$$

For each MC path we now get one feature F instead of all prices from all the time steps. This dramatically reduces the bandwidth requirements for the backend, for example on level 5 from one value per cycle to one value every 1,024 cycles.

We have analyzed commonly traded European options[1] and extracted five general features with which it is possible to price all of them. They are given in Figs. 8.4

[1] Call and put options of type Vanilla, barrier (upper or lower, knock-in or knock-out, one barrier or multiple, unconditioned or windowed), Asian (geometric or arithmetic), Digital, and Lookback (fixed or floating strike) or any combinations of such types.

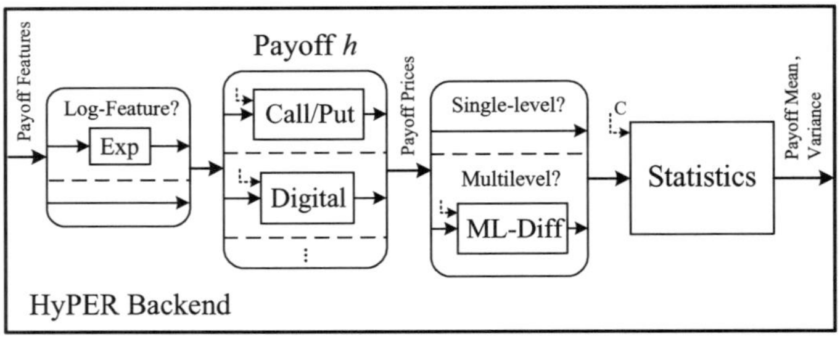

Fig. 8.5 The HyPER backend processes path features generated from the frontend and calculates statistics like the mean and variance of the payoff of the financial product being priced. Since features are generated once per path, the backend can process data from multiple frontends. Due to low demands on later blocks they can be mapped to the backend by the HyPER platform

and 8.5. Even highly exotic types like digital Asian barrier options are included. If a feature should not be present for a very specific option type, it can be easily identified and added to the list.

In general, only very few features are necessary to define the payoff g of an option. This shows the general usefulness of this payoff split and suggest to consider HW/SW partitions after all features have been generated. For the first part of the architecture starting from RNG and continuing with path simulation, a HW/SW split is normally not suggestive due to high bandwidth requirements inside. We call this part of the architecture the *HyPER frontend* as depicted in Fig. 8.4.

8.4.1.2 HyPER Backend

Everything following is called the *HyPER backend*. The stock prices in the frontend are calculated as $\log(\hat{S}_t)$. While some of the features like min / max can even be applied to them, for most of the features we have to go back to normal prices at some point. So the backend includes exponential transformations for log-features, the path independent parts of the payoff functions h (*Payoff*), and a statistic block that calculates Step 3 of the MLMC algorithm (see Fig. 8.5). The rest of the algorithm is handled on the CPU. On higher levels where fine and coarse paths are calculated, the statistic is evaluated for the differences. The rate of this differences is half the price rate, and we can always use the statistic core with an Initiation Interval (II) of 2, a core that takes one value every second clock cycles. For the first level l_0 we take the core with II $= 1$. Figure 8.6 shows the complete pricing system, the HyPER architecture.

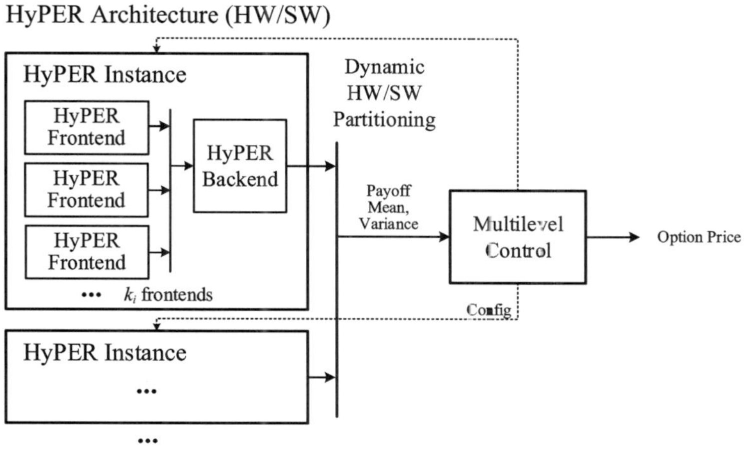

Fig. 8.6 For a given financial product and hybrid system the HyPER platform generates well-matched HyPER architectures for each level, including an optimal HW/SW partitioning. It is then used by the multilevel control to compute the option price in multiple iterations. In each iteration a different HyPER architecture might be used and is reconfigured as necessary by the system

8.4.2 *Runtime Reconfiguration*

The overall performance of the hybrid option pricing system obviously depends on the actual configuration of the platform. It is important to note that for a given payoff function g there are still some degrees of freedom in the architecture, for example:

- The number of HyPER instances on the FPGA part,
- For each HyPER instance the number of frontends and where to make the HW/SW split in the backend, or
- The type of communication core for CPU/FPGA communication.

When running the MLMC algorithm, the backend processes the payoff features F_i from the frontend, one feature set F_i per path. For level one, new features are generated every 4th clock cycle, which suggests no HW/SW split inside or after the backend. For level $l = 5$, features are generated only every 1,024th clock cycle, which suggests an early HW/SW split right after the frontend.

To account for these changing requirements for different levels, we propose an algorithmic extension in which we reconfigure the hybrid system for each level, see Algorithm 1.

This leaves the question on how to find the optimal HyPER configuration $\mathcal{H}_l{}^*$ on each level, especially for the middle levels $l = 2, \ldots, 4$. This issue is addressed in the next sections.

Algorithm 1 Reconfigurable multilevel

Input: target accuracy ε, first level l_0 and last level L
Output: Approximated price of the option $\hat{P}$
  load $\mathcal{H}_{l_0+1}{}^*$, the optimal configuration for level l_0+1.
  for $l = l_0, \ldots, L$ **do**
    Estimate the level variances $V_l = \mathbb{V}\mathrm{ar}\big[\hat{D}_l\big]$, using an initial $N_l = 10^4$ samples.
  end for
  Calculate $N_{l_0}, \ldots, N_L$ according to Eq. (8.4).
  for all l in $\{l_0, \ldots, L\}$ **do**
    load $\mathcal{H}_l{}^*$, the optimal configuration for level l.
    Evaluate extra paths at each level up to N_l.
  end for
  Calculate the approximated price of the option $\hat{P}$ according to Eq. (8.3).

8.4.3 Static Optimizer

Based on a given platform $\mathcal{F}$ and payoff function g the static optimizer finds the set of optimal HyPER configurations used in the reconfigurable MLMC algorithm (Algorithm 1). This set is used to reconfigure the FPGA several times during the execution to boost the overall performance.

The optimizer maximizes the performance of HyPER by exploiting all degrees of freedom in the architecture. These are in particular:

- The number of HyPER instances N,
- The communication core Ψ, and
- For each HyPER instance $n \in \{1, \ldots, N\}$:
  - The number of frontends k_n,
  - The utilization factor of the frontend β_n, and
  - The HW/SW split Ω_k.

We express this freedom as $\mathcal{H}_l(\mathcal{F}, g; N, k_1, \ldots, k_N, \beta_1, \ldots, \beta_N, \Omega_1, \ldots, \Omega_N, \Psi)$ and from now on only write $\mathcal{H}_l(N, k_n, \beta_n, \Omega_n, \Psi)$ for brevity. The best architectures are therefore defined by:

$$
\begin{aligned}
\underset{N,k_n,\beta_n,\Omega_n,\Psi}{\text{maximize}} \quad & \mathbb{P}\mathrm{erformance}\big(\mathcal{H}_l(N, k_n, \beta_n, \Omega_n, \Psi)\big), \\
\text{subject to} \quad & \mathbb{A}\mathrm{rea}^{\varphi}\big(\mathcal{H}_l(\ldots)\big) \leq \alpha^{\varphi}\, \mathbb{A}\mathrm{rea}^{\varphi}\big(\mathcal{F}\big) \quad \forall \varphi, \\
& \mathbb{L}\mathrm{oad}\big(\mathcal{H}_l(\ldots)\big) \leq 1, \\
& \mathbb{B}\mathrm{andwidth}\big(\mathcal{H}_l(\ldots)\big) \leq \mathbb{B}\mathrm{andwidth}\big(\Psi\big),
\end{aligned}
$$

$\mathcal{F}$:	target reconfigurable hybrid system
g :	given payoff function
$N \in \mathbb{N}$:	number of HyPER instances,
$\Psi \in \{\text{available cores of } \mathcal{F}\}$:	communication core,
$k_n \in \mathbb{N}$:	number of frontends,
$\beta_n \in [0,1]$:	utilization factor of the frontends,
$\Omega_n \in \{\text{Ser., Exp, Payoff, ML-Diff, Stats}\}$:	HW/SW split,
$\varphi \in \{\text{LUT, FF, BRAM, DSP}\}$:	FGPA resource type,
α^{φ}	synthesis weight,
$\forall n \in \{1, \ldots, N\}$	(for each HyPER instance).

This concludes the HyPER platform, the whole methodology is shown in Fig. 8.7.

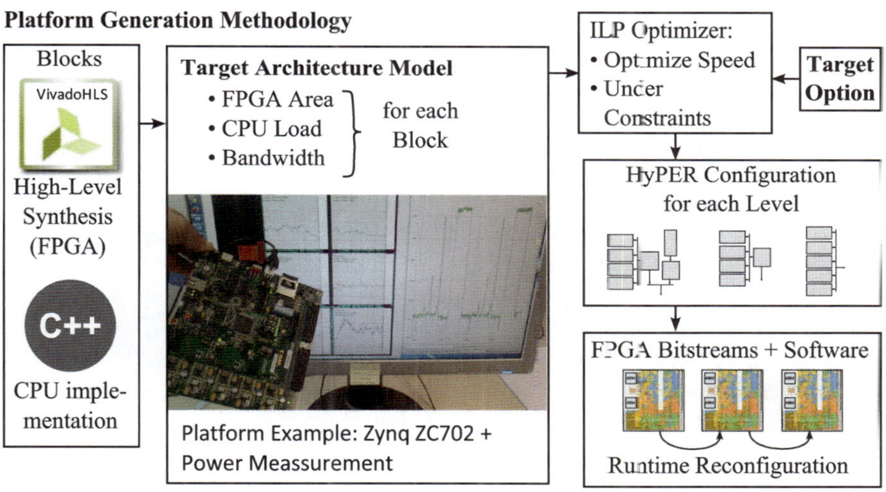

Fig. 8.7 The HyPER platform generation methodology requires CPU and FPGA implementations for each of the building blocks of the HyPER architecture (Fig. 8.6). Based on these designs that might e.g. written in HLS and C++, architecture models are derived specifying area, CPU, and bandwidth usage of each of the blocks for a specific target architecture (*middle*). The ILP optimizer uses those models for determining the HyPER architecture with the highest speed for a specific target options. In this process it generates different HyPER architecture (*right*) configurations for each of the levels, which are used synthesized to bitstreams and that reconfigured during runtime (*right bottom*)

8.5 HyPER on Zynq

In this section we thoroughly investigate the HyPER architecture for the Xilinx Zynq 7020 platform. It is a SoC that integrates a dual-core ARM Cortex-A9 processor and an FPGA into a tightly coupled hybrid system. For the financial product we choose barrier call options as a practical example.

In order to solve the static optimization we need to know how big the building blocks of the HyPER architecture from Fig. 8.6 are on our device $\mathcal{F}$ in Fig. 8.9. For that, we have implemented all the building blocks for the FPGA with Xilinx Vivado HLS for $f = 100\,\text{MHz}$ and single precision floating-point arithmetic. To implement the ICDF we followed [14]. We have run a complete place & route synthesis for each core and extracted the resource usage numbers from Xilinx Vivado. As the cores include the full AXI interfaces, these are accurate numbers and they do not change much for composed designs. The obtained numbers are shown in Table 8.1.

Furthermore, we have to know how much CPU load the blocks generate when they are mapped to the ARM processors. We estimated them by implemented the blocks as C++ functions and measuring the time per input value.

Additionally, we need to determine the speed and area of all available communication cores. We have used simple continuous streaming cores and measured the raw speed on the ARM cores. Finally, we have to specify how big our FPGA is and how much resources we want to use, as fully mapped devices cause routing congestions. The numbers of our complete analysis are given in Table 8.1.

We formulated the optimization problem, introduced in Sect. 8.4.3, as an Integer Linear Programming (ILP) problem and solved it with an ILP solver. As a result we got four unique architectures. The optimal parameters for each architecture $\mathcal{H}_l{}^*$ are listed in Table 8.2. Their metrics area, load, bandwidth, and performance are given in Table 8.3. Section 8.5 visualizes the found architectures.

Based on the configurations found by the ILP problem we can find the minimum set of configurations that still reach the maximum performance. For each level we estimate the performance for each configuration and keep the ones with maximum performance. The result is Table 8.4. We see that configurations $\mathcal{H}_1{}^*$, $\mathcal{H}_2{}^*$, and $\mathcal{H}_3{}^*$ are sufficient to reach maximum performance. $\mathcal{H}_l{}^*$ for $l \geq 4$ looks similar to $\mathcal{H}_3{}^*$, just instead of a DMA it has a streaming First in, First Out (FIFO) for the interface to the CPU. Therefore we save one reconfiguration (Fig. 8.8).

In the next section we evaluate these configurations in detail.

8.5.1 Results and Comparison

We have synthesized the optimal HyPER architectures $\mathcal{H}_l{}^*$ as defined in Table 8.2 and implemented the complete multilevel algorithm. As an example, the floorplan of $\mathcal{H}_3{}^*$ is shown in Fig. 8.9. On the ARM cores we boot a full Linaro Ubuntu. The Zynq platform supports online dynamic reconfiguration from the OS level in about

Table 8.1 Building blocks of HyPER on the Zynq-7000 series. Based on their area, throughput, and CPU timing numbers the HyPER platform can find the optimal HyPER architecture

Building blocks	LUT	FF	BRAM	DSP	CPU ns/val.
Increment generator:					
Mersenne Twister	301	323	4	0	–
ICDF	451	592	4	1	–
Antithetic core	228	258	0	0	–
Path generators:					
Single-level kernel	4,153	4,241	2	38	–
Multilevel kernel	5,607	5,326	6	43	–
Payoff features F_i:					
Barrier	180	158	0	0	–
Payoff h:					
Call/put	440	396	0	2	6
Backend:					
Feature					
Serializer $k\times 1$	$30k+65$	$65k+45$	0	0	–
Exponential	900	384	0	7	250
Multilevel difference	372	355	0	2	5
Statistics II=1	2,170	1,612	4	9	6
Statistics II=2	1,454	1,164	2	6	3

Com. interface Ψ FPGA → CPU	LUT	FF	BRAM	Bandwidth in MB/s
Config-Bus $1\times k$	$30k+50$	$2k+40$	0	<1
Streaming-Fifo	654	611	4	20
DMA-Core	1,864	3,122	4	250

Hybrid chip $\mathcal{F}$	LUT	FF	BRAM	DSP	ARM
Xilinx Zynq 7020	53,200	106,400	280	220	2 cores
Synthesis weight α	0.8	0.5	1	1	

50 ms. The running system is visible in the picture in Fig. 8.7, with the ZC706 board on the left, the generated paths in the middle, and the power measurements on the right of the picture.

To quantify the quality of our implementation, we have implemented a sophisticated CPU Heston pricer as a reference model. While Gaussian increment generation is only a small part of the HyPER architecture on FPGAs, it takes a significant time on CPUs (about 40 % of the overall runtime). We have compared several advanced libraries and selected the fastest Mersenne Twister RNG from the C++11 standard library and the Ziggurat method from the GNU Scientific Library (GSL), which we adapted to use "single" precision floating-point. We

Table 8.2 Optimal HyPER architectures for barrier option pricing on the Zynq 7020

Optimal HyPER Architectures $\mathcal{H}_l{}^*$
for $\mathcal{F}$ = Xilinx Zynq 7020, g = Barrier Call Option
$\mathcal{H}_1{}^* = \mathcal{H}_1\big(N = 2, \Psi = \text{DMA}$ $k_1 = 4, \beta_1 = 1, \Omega_1 = \text{Stats},$ $k_2 = 1, \beta_2 = 1, \Omega_2 = \text{Exp}\big)$
$\mathcal{H}_2{}^* = \mathcal{H}_2\big(N = 1, \Psi = \text{Config-Bus}$ $k_1 = 4, \beta_1 = 1, \Omega_1 = \text{Stats}\big)$
$\mathcal{H}_3{}^* = \mathcal{H}_3\big(N = 1, \Psi = \text{DMA}$ $k_1 = 5, \beta_1 = 0.966, \Omega_1 = \text{Serializer}\big)$
$\mathcal{H}_l{}^* = \mathcal{H}_l\big(N = 1, \Psi = \text{Streaming-Fifo}$ $k_1 = 5, \beta_1 = 1, \Omega_1 = \text{Serializer}\big) \quad \forall\, l \geq 4$

Table 8.3 Showing the area, CPU, and bandwidth requirements and their performance of each of the optimal HyPER configurations $\mathcal{H}_l{}^*$ given in Table 8.2

Optim.	Area in %				CUP	Bandw. (MB/s)		Perform.
HyPER	LUT	FF	BRAM	DSP	Load	Used	Available	MC step/s
$\mathcal{H}_1{}^*$	63	32	13	99	0.19	95	350	500 M
$\mathcal{H}_2{}^*$	58	27	15	87	0.00	0	<1	400 M
$\mathcal{H}_3{}^*$	69	35	19	100	1.00	30	350	483 M
$\mathcal{H}_4{}^*$	67	33	19	100	0.26	7	20	500 M
$\mathcal{H}_5{}^*$	67	33	19	100	0.07	2	20	500 M

Table 8.4 List of optimal configurations for each levels, used to find minimal set of configurations

Level	Configurations providing maximum performance
1	$\mathcal{H}_1{}^*$
2	$\mathcal{H}_2{}^*$
3	$\mathcal{H}_3{}^*$
≥4	$\mathcal{H}_3{}^*, \mathcal{H}_4{}^*$

have written the Monte Carlo step generation by hand and tuned its loop structure to support Advanced Vector Extensions (AVX). Additionally, we parallelized the whole program such that it uses all available cores. We have employed the Microsoft Visual C++ (MSVC) 2012 compiler, which has excellent auto-vectorization support, with compiler flags: "/O2 /arch:AVX /fp:fast /GL". Profile-guided optimization gave an additional 10 % speedup. The result is a high-speed reference implementation that has received as much care as HyPER itself.

As an execution platform, we had several choices between servers, desktops, and laptops. Among all of them, the laptop proved to be the most energy efficient platform. It is a Dell Latitude E6430 with an Intel Core i5-3320M manufactured in 22 nm and supporting the latest AVX instructions. The Zynq 7020 is fabricated with a 28 nm process. Both chips are the most recent generations available today.

For measuring the speed, we have calculated the price for barrier call options for the Heston benchmark parameters [4] in Table 8.5 with a target precision of $\varepsilon = 0.005$, start level $l_0 = 1$, last level $L = 5$, and multilevel constant $M = 4$ (compare

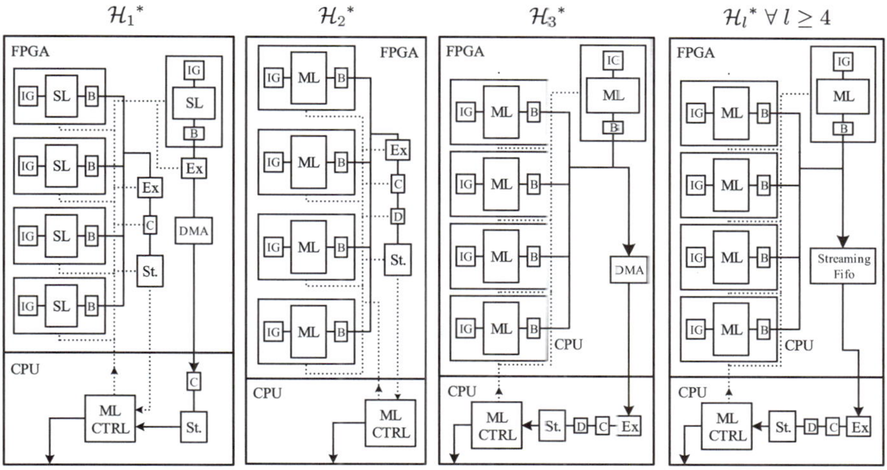

Fig. 8.8 Optimal HyPER architectures for barrier option pricing on the Xilinx Zynq 7020, as given in Table 8.2. They are specific configurations of the architecture in Fig. 8.6 with abbreviations (*IG* – Increment Generator, *SL* – Singlelevel Path Generator, *B* – Barrier, *Ex* – Exponential, *C* – Call, *St.* – Statistics, *ML CTRL* – Multilevel Control, *ML* – Multilevel Path Generator, *D* – Multilevel Difference). Note that configuration $\mathcal{H}_1^*$ contains two HyPER instances with different HW/SW partitioning

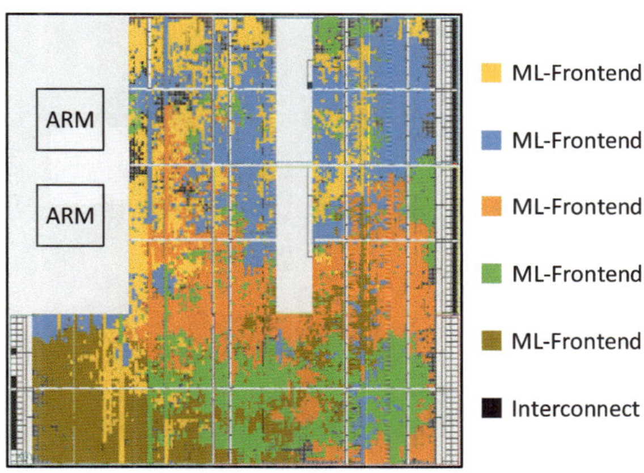

Fig. 8.9 Floorplan of the optimal HyPER Architecture $\mathcal{H}_3^*$ for level 3, as defined in Table 8.2, highlighting the five frontends and the interconnect Ψ

Chap. 4). We have validated that both implementations are correct and calculate the same number of MC paths N_l on each level as given in Table 8.7. We have measured the overall execution time and the power consumption. For the laptop we kept the power consumption to a minimum by turning of the display and Wi-Fi and have removed all USB devices. We have run the simulation in a loop and measured the average power at the power plug.

To measure the power of the hybrid platform, we have used the Xilinx ZC702 evaluation board. It is possible to measure all power lanes on a 50 ms basis. We have run the simulation in a loop and added up the average power consumption of each power lane, except the 3.3 V lane with about 0.7 W. The measured power includes

Table 8.5 Benchmark Heston parameters [4]

S_0	κ	θ	σ	r	v_0	ρ	K	T	Barrier	ε
100	3	0.16	0.4	0.02	0.1	−0.8	100	1	150	0.005

Table 8.6 Execution time and energy consumption

	Time (S)	Power (W)	Energy (J)
MC on CPU[a]	111	31.3	**3,460**
MC on Zynq[b]	26.1	2.77	**72**
MLMC on CPU[a]	29.9	30.6	**916**
Level 1	2.9	30.6	88
Level 2	4.2	30.6	130
Level 3	5.2	30.6	158
Level 4	7.0	30.6	214
Level 5	10.7	30.6	327
Reconf.	–	–	–
HyPER on Zynq[b]	8.83	2.87	**25.3**
Level 1	0.58	3.05	1.77
Level 2	1.39	2.41	3.35
Level 3	1.52	3.38	5.14
Level 4	2.07	2.96	6.11
Level 5	3.03	2.80	8.48
Reconf.	0.25	1.86	0.47

[a]Intel Core i5-3320M
[b]Zynq 7020

the Zynq 7020, Dynamic Random-Access Memory (DRAM), and oscillators, but not the peripherals like LEDs, USB, or HDMI controllers that have not been in use at all. To account for a power supply with 90 % efficiency, we have multiplied all measurements by 1.11.

The measured numbers are presented in Table 8.6. The CPU takes 30 s and 916 J, while HyPER takes 8.6 s and 25 J to price the product. This means the HyPER architecture on the Zynq is 3.4× faster and 36× more power efficient than the reference system. As option pricing is perfectly scalable over multiple instances, HyPER is 36× faster than the CPU for a fixed power budget.

Without reconfiguration, the best architecture for all levels would be $\mathcal{H}_2{}^*$. Pricing the same benchmark on this static architecture would take 10.5 s, which would be 19 % slower than the HyPER architecture with online reconfiguration.

8.5.2 *Comparison with Related Work*

In this section we compare HyPER on Zynq to related work [15] and [16], introduced in Sect. 8.2. Although the architectures [15, 16] are limited to barrier options while HyPER supports the whole spectrum of traded options, we evaluate them in this specific setting.

Table 8.7 Choosen MC path count N_l for each level when pricing our benchmark Barrier call options with MLMC for the Heston parameters in Table 8.5

Level l	Time steps $k = M^l$	MC paths N_l [$\times 10^6$]	Fine MC steps NM^l [$\times 10^8$]	Coarse MC steps NM^{l-1} [$\times 10^9$]
\protect \newucase {m}ultilevel Monte Carlo ($l_0 = 1, L = 5, M = 4$):				
1	4	72.5	0.29	–
2	16	27.8	0.44	1.11
3	64	9.2	0.59	1.47
4	256	3.2	0.83	2.07
5	1,024	1.18	1.21	3.03
Classical Monte Carlo:				
–	1,024	15.3	15.62	–

Reference [15] is a classical MC implementation on a hybrid system containing a Virtex 5 and a laptop. The HyPER architecture is superior on both the algorithmic and implementation level:

1. On algorithmic level, HyPER uses the faster MLMC algorithm. In our setup (Table 8.5) MLMC needs to evaluate 3.8× less steps than classical MC (see Table 8.7). A more elaborate numerical comparison between both algorithms can be found in [8], where Giles shows speedups from 3 to 100×, mainly depending on the option types considered.
2. While [15] uses a Virtex 5 with a static configuration and a laptop, we present a runtime reconfigurable architecture on a tightly coupled hybrid architecture.

Based on the numbers given in [15], it would take 110 s and 3,861 J to run the benchmark. That means HyPER is 12.5× faster and 153× more power efficient than [15] due to improvements on algorithmic and implementation level, see Table 8.8 for more details.

The MLMC architecture in [16] is a partial implementation only and no time or energy numbers are given for a complete pricing system. Specifically only synthesis results are given for parts of the architecture, mainly what we call HyPER frontend. The payoff computation has not been implemented. That is why no complete comparison can be made. Section IV of [16] suggests to do the payoff computations on an embedded CPU. We have shown in Sect. 8.4.2 that such a HW/SW split leads to high CPU speed and bandwidth requirement for small levels. The work of [16] would therefore require a powerful CPU. With HyPER we have solved this issue by dynamically changing the HW/SW partitioning during runtime. As a result, we expect our architecture to be far superior in power efficiency compared to [16].

We can compare the synthesis results in [16] with our implementation of the HyPER frontend, including increment generator, multilevel path generator, and barrier checker (see Table 8.8). While the two devices have almost the same FPGA

Table 8.8 Comparison HyPER on Zynq with related work for Heston MC Barrier option pricing

Architecture	De Schryver et al. [15]	De Schryver et al. [16]	HyPER on Zynq (this work)
Algorithm	Classical MC	Multilevel MC	Multilevel MC
Total MC steps ($\times 10^9$)	15.62	4.13	4.13
Time (s)	110	–	9
Energy (J)	**3,861**	–	**25**
Monte Carlo barrier frontend:			
LUT	5,480	10,300	6,770
FF	6,950	11,900	6,660
DSP	43	68	44
BRAM	10	128	22
Frequency (MHz)	102	120	100
Setup	Virtex 5 + Laptop	Virtex 6, synthesis only	Zynq-7000

fabric and both implementations use single-precision floating-point as calculation formats, we see that our implementation is significantly (>35 %) smaller. This difference might come from the way [16] models what we call *path generator*. They have split this part of the architecture in more than 10 pieces, each modeled individually with HLS and connected by Advanced eXtensible Interface (AXI) Stream components. In contrast to this approach, we have modeled everything in one HLS component with no internal buffers, making the design efficient and compact, with just 145 lines of code.

8.5.3 Flexibility Performance Tradeoff

In the last two chapters we have clearly shown that HyPER is far superior in energy-efficiency compared to CPU solutions and related work architectures. In this chapter we will comment on flexibility, see Fig. 8.10.

We define flexibility as how easy it is to add new financial products to the implementation. The MC CPU solution has the highest flexibility. While in general it as easy to add new payoff functions to a MLMC implementation as it is for a classical MC implementation, some payoffs need special treatments for MLMC algorithms. That is why we have a slightly less flexibility for MLMC for CPUs.

The classic MC architecture [15] is basically equivalent to a HyPER configuration with six SL frontends on the Zynq 7020. Adding new products to this architecture, requires the user to redesign new payoff blocks and change the whole FPGA architecture, which is extremely difficult. That is why this solution has the lowest flexibility.

With HyPER we get both another speedup and most off the flexibility back. The systematic approach of splitting payoffs into features and payoff functions makes it

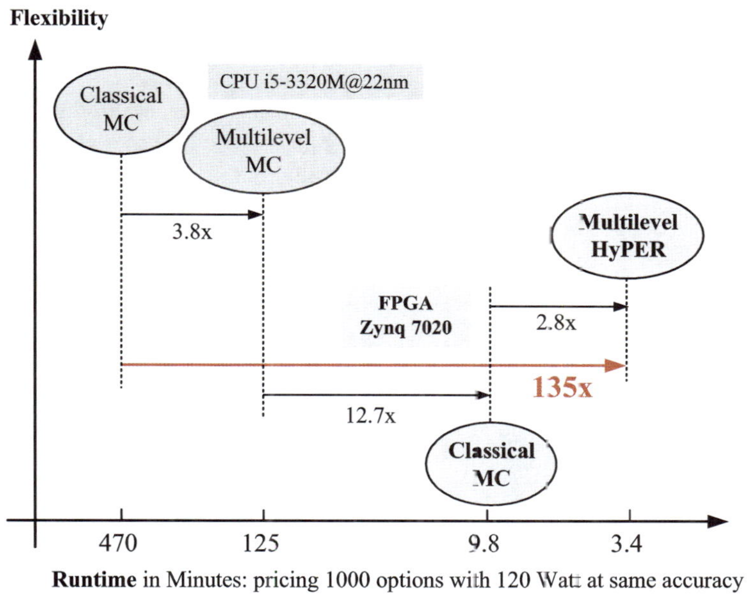

Fig. 8.10 Showing the runtime and flexibility of the presented architectures. HyPER is a clear winner in both flexibility and efficiency, being 135× faster than the classic MC CPUs solutions

very easy to find a good FPGA implementation for new products. In addition, it is completely clear where to put such new blocks, while reusing the same interfaces and all other blocks. Furthermore, it is possible to create HLS templates where a user just has to add the mathematical formulas in C++ syntax, so in general no FPGA knowledge is necessary. Once the new blocks are written in HLS new models can be derived automatically from them and fed into the static optimizer that will then generate the most efficient architectures.

With HyPER, adding new products is both easy and efficient, giving us back most of the flexibility a CPU solution has.

8.6 Block Modeling Extensions

In Sect. 8.4.3 we introduced a simple formalism to model the HyPER architecture. However, the formalism can be further extended to incorporate even more flexibility of the architecture.

1. While we synthesized all our building blocks for a fixed frequency, it might be possible to have a different frequency for each building block, or at least each FPGA configuration. For that we would synthesize each block for a set of frequencies. Here, the expected tendency is that faster cores consume more FPGA resources, and we might find a more balanced configuration this way.

2. For more complex Payoff functions it is beneficial to consider all possible pipeline IIs, what leads to smaller designs due to operator reuse.
3. In the HLS tool it is easy to trade off Digital Signal Processor (DSP) blocks agains Lookup Table (LUT)/Flip-Flop (FF) usage. By compiling all the cores for different DSP usage factors it should be possible to find more balanced chip configurations with possibly even higher throughput.

8.7 Conclusions

The HyPER platform is a novel option pricing system for hybrid reconfigurable platforms. It is based on state-of-the-art Multilevel Monte Carlo (MLMC) methods, the Heston market model, and covers a wide range of option types. As a platform, HyPER captures all essential aspects of the problem and implementation space in a systematic way to generate efficient implementations. It provides a formalism to describe options in a way that they can be optimally mapped to a hybrid system. In this formalism payoff functions are systematically split in two parts, one targeting the FPGA and the other one the CPU. Furthermore, it provides a reconfigurable multilevel algorithm enabling the platform to adapt itself to the changing requirements for different parts of the algorithm. With specific information of the implementation platform including area, runtime, and bandwidth information HyPER is able to yield the optimal implementation to price a financial product.

We have used the HyPER platform to find an efficient implementation for barrier options on the Xilinx Zynq 7020 All Programmable SoC. The implementation is $3.4\times$ faster and $36\times$ more power-efficient than a highly tuned software reference on an Intel Core i5 CPU.

As far as the authors know, HyPER is the first flexible FPGA based Heston pricing system supporting a wide range of traded options, while clearly outperforming previous specialized Heston Monte Carlo implementations at the same time.

Acknowledgements We gratefully acknowledge the partial financial support from the Center of Mathematical and Computational Modelling $(CM)^2$ of the University of Kaiserslautern, from the German Federal Ministry of Education and Research under grant number 01LY1202D, and from the Deutsche Forschungsgemeinschaft (DFG) within the RTG GRK 1932 “Stochastic Models for Innovations in the Engineering Sciences”, project area P2. The authors alone are responsible for the content of this work.

References

1. Advances and Innovations – Field Programmable Gate Arrays (FPGAs) (2015). http://careers.jpmorgan.com/experienced/jpmorgan/jobs/businesses/ib/technology/advances#Field_Programmable_Gate_Arrays__FPGAs_. Last access: 09 Feb 2015
2. Bernemann, A., Schreyer, R., Spanderen, K.: Accelerating Exotic Option Pricing and Model Calibration Using GPUs. Available at SSRN 1753596 (2011)

3. Brugger, C., de Schryver, C., Wehn, N.: HyPER: a runtime reconfigurable architecture for Monte Carlo option pricing in the Heston model. In: Proceedings of the 24th IEEE International Conference of Field Programmable Logic and Applications (FPL), Munich, pp. 1–8 (2014). doi:10.1109/FPL.2014.6927458
4. Brugger, C., de Schryver, C., Wehn, N., Omland, S., Hefter, M., Ritter, K., Kostiuk, A., Korn, R.: Mixed precision multilevel Monte Carlo on hybrid computing systems. In: Proceedings of the 2014 IEEE Conference on Computational Intelligence for Financial Engineering Economics (CIFEr), London, pp. 215–222 (2014). doi:10.1109/CIFEr.2014.6924076
5. Brugger, C., Weithoffer, S., de Schryver, C., Wasenmüller, U., Wehn, N.: On parallel random number generation for accelerating simulations of communication systems. Adv. Radio Sci. **12**, 75–81 (2014). doi:10.5194/ars-12-75-2014. http://www.adv-radio-sci.net/12/75/2014/
6. Delivorias, C.: Case studies in acceleration of Heston's stochastic volatility financial engineering model: GPU, cloud and FPGA implementations. Master's thesis, The University of Edinburgh (2012). http://www.hpcfinance.eu/sites/www.hpcfinance.eu/files/Christos_Delivorias_0 pdf
7. Feldman, M.: JP Morgan Buys Into FPGA Supercomputing. http://www.hpcwire.com/2011/07/13/jp_morgan_buys_into_fpga_supercomputing/ (2011). http://www.hpcwire.com/2011/07/13/jp_morgan_buys_into_fpga_supercomputing/. Last access: 09 Feb 2015
8. Giles, M.B.: Multilevel Monte Carlo path simulation. Oper. Res.-Baltim. **56**(3), 607–617 (2008). doi:10.1.1.121.713
9. Heston, S.L.: A closed-form solution for options with stochastic volatility with applications to bond and currency options. Rev. Financ. Stud. **6**(2), 327 (1993). doi:10.1093/rfs/6.2.327
10. Korn, R., Korn, E., Kroisandt, G.: Monte Carlo Methods and Models in Finance and Insurance. CRC, Boca Raton (2010)
11. Marxen, H.: Aspects of the application of multilevel Monte Carlo methods in the Heston model and in a Lévy process framework. Ph.D. thesis, University of Kaiserslautern (2012)
12. Schmerken, I.: Deutsche Bank Shaves Trade Latency Down to 1.25 Microseconds. http://www.advancedtrading.com/infrastructure/229300997 (2011). http://www.advancedtrading.com/infrastructure/229300997. Last access: 09 Feb 2015
13. de Schryver, C., Schmidt, D., Wehn, N., Korn, E., Marxen, H., Korn, R.: A new hardware efficient inversion based random number generator for non-uniform distributions. In: Proceedings of the 2010 International Conference on Reconfigurable Computing and FPGAs (ReConFig), Cancun, pp. 190–195 (2010). doi:10.1109/ReConFig.2010.20
14. de Schryver, C., Schmidt, D., Wehn, N., Korn, E., Marxen, H., Kostiuk, A., Korn, R.: A hardware efficient random number generator for nonuniform distributions with arbitrary precision. Int. J. Reconfigurable Comput. (IJRC) **2012**, 1–11 (2012). doi:10.1155/2012/675130. Article ID 675130
15. de Schryver, C., Shcherbakov, I., Kienle, F., Wehn, N., Marxen, H., Kostiuk, A., Korn, R.: An energy efficient FPGA accelerator for Monte Carlo option pricing with the Heston model. In: Proceedings of the 2011 International Conference on Reconfigurable Computing and FPGAs (ReConFig), Cancun, pp. 468–474 (2011). doi:10.1109/ReConFig.2011.11
16. de Schryver, C., Torruella, P., Wehn, N.: A multi-level Monte Carlo FPGA accelerator for option pricing in the Heston model. In: Proceedings of the IEEE Conference on Design, Automation and Test in Europe (DATE), Grenoble, pp. 248–253 (2013)
17. Sridharan, R., Cooke, G., Hill, K., Lam, H., George, A.: FPGA-based reconfigurable computing for pricing multi-asset barrier options. In: Proceedings of Symposium on Application Accelerators in High-Performance Computing PDF (SAAHPC), Chicago, IL (2012)
18. Thomas, D.B., Luk, W.: A domain specific language for reconfigurable path-based Monte Carlo simulations. In: International Conference on Field-Programmable Technology, 2007, ICFPT 2007, Kitakyushu, pp. 97–104 (2007). doi:10.1109/FPT.2007.4439237
19. Thomas, D.B., Luk, W., Leong, P.H., Villasenor, J.D.: Gaussian random number generators. ACM Comput. Surv. **39**(4), 11 (2007). doi:http://doi.acm.org/10.1145/1287620.1287622
20. Tian, X., Benkrid, K., Gu, X.: High performance Monte-Carlo based option pricing on FPGAs. Eng. Lett. **16**(3), 434–442 (2008)

21. du Toit, J., Ehrlich, I.: Local volatility FX basket option on CPU and GPU. Technical report, The Numerical Algorithms Group Ltd (2013). http://www.nag.co.uk/numeric/gpus/local-volatility-fx-basket-option-on-cpu-and-gpu.pdf. Last access: 09 Feb 2015
22. Tse, A., Thomas, D., Tsoi, K., Luk, W.: Dynamic scheduling Monte-Carlo framework for multi-accelerator heterogeneous clusters. In: 2010 International Conference on Field-Programmable Technology (FPT), pp. 233–240 (2010). doi:10.1109/FPT.2010.5681495

Chapter 9
Exploiting Mixed-Precision Arithmetics in a Multilevel Monte Carlo Approach on FPGAs

Steffen Omland, Mario Hefter, Klaus Ritter, Christian Brugger, Christian De Schryver, Norbert Wehn, and Anton Kostiuk

Abstract Nowadays, high-speed computations are mandatory for financial and insurance institutes to survive in competition and to fulfill the regulatory reporting requirements that have just toughened over the last years. A majority of these computations are carried out on huge computing clusters, which are an ever increasing cost burden for the financial industry. There, state-of-the-art CPU and GPU architectures execute arithmetic operations with predefined precisions only, that may not meet the actual requirements for a specific application. Reconfigurable architectures like Field Programmable Gate Arrays (FPGAs) have a huge potential to accelerate financial simulations while consuming only very low energy by exploiting dedicated precisions in optimal ways.

In this work we present a novel methodology to speed up Multilevel Monte Carlo (MLMC) simulations on reconfigurable architectures. The idea is to aggressively lower the precisions for different parts of the algorithm without loosing any accuracy at the end. For this, we have developed a novel heuristic for selecting an appropriate precision at each stage of the simulation that can be executed with low costs at runtime. Further, we introduce a cost model for reconfigurable architectures and minimize the cost of our algorithm without changing the overall error.

S. Omland (✉) • M. Hefter • K. Ritter
Computational Stochastics Research Group, University of Kaiserslautern, Kaiserslautern, Germany
e-mail: omland@mathematik.uni-kl.de; hefter@mathematik.uni-kl.de; ritter@mathematik.uni-kl.de

C. Brugger • C. De Schryver • N. Wehn
Microelectronic Systems Design Research Group, University of Kaiserslautern, Kaiserslautern, Germany
e-mail: brugger@eit.uni-kl.de; schryver@eit.uni-kl.de; wehn@eit.uni-kl.de

A. Kostiuk
Stochastic Control and Financial Mathematics Group, University of Kaiserslautern, Kaiserslautern, Germany
e-mail: kostiuk@mathematik.uni-kl.de

C. De Schryver (ed.), *FPGA Based Accelerators for Financial Applications*,
DOI 10.1007/978-3-319-15407-7_9

We consider the showcase of pricing Asian options in the Heston model. For this setup we improve one of the most advanced simulation methods by a factor of 3–9$\times$ on the same platform.

9.1 Introduction

Realistic pricing of exotic derivatives requires complex market models like the Heston model with its stochastic volatility process. For those models, in general there are no closed-form solutions for pricing various derivative classes, and compute-intensive numerical methods have to be used. In addition, financial institutes are forced to generate risk measures more frequently than in the past due to recent regulatory requirements. As a third point, more precise and faster pricing routines can increase the competitive advantage of financial companies when combined with reliable calibration procedures for input parameters.

The computed derivatives prices shall be available with a specific *accuracy*. It is important to strictly distinguish between the terms *accuracy* and *precision*, as introduced by Higham [35]. *Accuracy* means the absolute or relative error of an approximated quantity, in particular a result of a complete computation. *Precision* is the accuracy of the atomic arithmetic operations (e.g. such as $+$, $-$, $\cdot$, $/$) used in this computation. For example, accuracy and precision are the same if the final goal is to compute $z = x + y$, but the accuracy can be much worse than the actual precision for complex tasks like solving a system of linear equations.

Common simulation architectures like Central Processing Units (CPUs) or Graphics Processor Units (GPUs) only support very few standard number formats with fixed precision. The usually available predefined precisions are 32 or 64 bit integers, "single", and "double" precision floating point. Up to now, most quantitative analysts and other developers of executable algorithm implementations just rely on the available precisions of the target platforms. Custom precision formats are normally not considered on CPUs and GPUs since they have to be costly emulated on the available hardware. This leads to longer execution times and higher energy demands in total. The common algorithm design paradigm is:

> *Use the available precisions of your computing hardware and tune your algorithm to achieve the desired result accuracy as fast as possible.*

While CPUs and GPUs work with predefined internal number formats with fixed precision, reconfigurable architectures like FPGAs allow to use number formats with any custom precisions. Energy, speed, and accuracy of the computation can therefore be traded-off against each other effectively for the first time in real business. It has already been shown that FPGAs can be 10 times faster compared to state-of-the-art CPU or GPU clusters for the same energy budget and accuracy [16].

Nevertheless, the integration of new hardware architectures in pricing systems require considering platform-given restrictions already on the algorithm development level to achieve optimal implementations in the end. This postulates the coalescing of formerly distinct disciplines like quantitative finance, computer science, and hardware engineering.

As an example, computational finance theory typically does not consider errors introduced by limited precision computations. It is rather assumed, at least implicitly, that the algorithm is executed with *infinite* precision in the real number model, see [49]. However, limited and mixed precision calculations have always been in the focus of hardware engineers, especially in the power-limited embedded systems domain. In this chapter, we make one step towards closing the gap between high-level simulations in finance and the underlying finite precision execution platform. The new design paradigm is now:

Carefully select the needed precisions in your computation for a defined result accuracy.

Following this paradigm, we present a novel methodology to speed up Multilevel Monte Carlo (MLMC) simulations on reconfigurable architectures. It is based on exploiting custom number formats on FPGAs. By aggressively using reduced precisions for most stages of the algorithm, we speed up the computations, while maintaining the overall precision at the same time. We introduce a novel heuristic to choose the precisions that can be executed with low costs at runtime. Furthermore, to reflect the characteristics of reconfigurable hardware, we develop an analytical cost model that is in line with the real runtimes. It is used to minimize the costs of the algorithm without changing the overall error. We present the benefits of our methodology by speeding up Asian option pricing in the Heston model. Assuming a similar power consumption, reducing the execution time directly results in saving energy for one computation.

Our novel contributions are:

- An extended version of the state-of-the-art MLMC algorithm that integrates different precisions for different parts of the algorithm (Sect. 9.4.1).
- An analytic cost model for reconfigurable architectures like FPGAs (Sect. 9.4.4) and a heuristic for determining appropriate precisions for different levels of the MLMC simulation at runtime (Sect. 9.4.5).
- A comprehensive analysis that confirms that the heuristic is stable (Sect. 9.4.5) and that our proposed methodology leads to significant (3–9×) speedups (Sect. 9.5).

This chapter is an extended version of [10] presented at the CIFEr conference in London.

9.2 Background and Related Work

This section summarizes the necessary background from finance, computational stochastics, and hardware engineering that is needed to follow this chapter. It also includes the relevant state-of-the-art.

9.2.1 Pricing Under Heston's Stochastic Volatility Model

Market simulations always rely on underlying *models* that incorporate the important characteristics for specific markets. The Heston model has been introduced by Steven L. Heston in 1993 [33]. It extends the popular Black-Scholes (BS) model from 1973 [8] by a stochastic volatility process. Nowadays it is used as a standard model in many productive environments in the finance and insurance domain [41]. It is sophisticated enough to reflect important market characteristics and (semi-)closed-form pricing formulas exist for plain vanilla options. This is especially important to allow fast calibration, and therefore for the acceptance of the model in real business [45].

Under the risk-neutral measure, the Heston model consists of two correlated Stochastic Differential Equations (SDEs). Equation (9.1a) describes the asset price process S that matches the Black-Scholes SDE, except for the non-constant volatility $\sqrt{\nu}$ driven by its own SDE (9.1b).

$$dS_t = S_t r dt + S_t \sqrt{\nu_t} dW_t^S, \tag{9.1a}$$

$$d\nu_t = \kappa(\theta - \nu_t)\, dt + \sigma \sqrt{\nu_t} dW_t^\nu. \tag{9.1b}$$

At time $t = 0$, the asset spot price is S_0 and the volatility is given by $\sqrt{\nu_0}$. The two driving Brownian motions W^S and W^ν are correlated with the correlation coefficient ρ to reflect the observable volatility clustering effect of the market. All constants are positive except the correlation coefficient for which $|\rho| \leq 1$ holds. The variance process ν is given by a so-called square root diffusion or Feller diffusion. This type of process was also used by Cox, Ingersoll and Ross for a short rate model, see [14]. Thus, (9.1b) is sometimes called CIR-process.

The present price of an European derivative can be calculated as the expectation $a = \mathbb{E}[H(S)]$ where H is the corresponding discounted payoff function. We will consider two different kinds of payoff functions: Path-independent payoffs which depend only on a fixed time of the asset price process S, like put and call options, and path-dependent options which depend on the complete path of S. Options with these type of payoffs are often called exotic. Since path-dependent payoffs are a special case of path-independent payoffs, we will denote any discounted payoff function with H whenever it is clear from the context which case is meant.

For exotic derivatives like barrier, lookback, or Asian options, no closed-form solutions exist in general, and numerical methods have to be used to price those products in the Heston setting. In this chapter we focus on Monte Carlo (MC) methods, one of the most common choices in this setting. MC methods are very popular in a wide range of application domains due to their universalism. Especially in finance, they are often the only suitable numerical approach for pricing complex products [41].

9.2.2 *The Classic Monte Carlo Method*

Let us first consider the following problem: We aim at approximating the quantity

$$a = \mathbb{E}[H(S)],$$

where

- S is a random variable taking values in $\mathbb{R}$ and
- H is a measurable mapping from $\mathbb{R}$ to $\mathbb{R}$.

We assume that direct sampling from S is possible and that the second moment of $H(S)$ exists.

One famous method to approximate a is the *classic MC method* in which a is approximated by

$$\mathcal{A}^{\mathrm{MC}} = \frac{1}{N}\sum_{i=1}^{N} H(S_i), \tag{9.2}$$

where $S_1, \ldots, S_N$ are independent copies of S. The MC method is based on random experiments and therefore introduces an error, which can measured by the Mean Squared Error (MSE) defined as

$$\mathbb{MSE}\left(\mathcal{A}^{\mathrm{MC}}\right) = \mathbb{E}\left[\left(\mathcal{A}^{\mathrm{MC}} - a\right)^2\right]. \tag{9.3}$$

Since $S_1, \ldots, S_N$ are independent and have the same distribution as S, the MSE can be written as

$$\mathbb{MSE}\left(\mathcal{A}^{\mathrm{MC}}\right) = \mathbb{V}\mathrm{ar}\left(\mathcal{A}^{\mathrm{MC}}\right) = \frac{1}{N}\mathbb{V}\mathrm{ar}(H(S)), \tag{9.4}$$

i.e., the MSE of the classic MC is given by the variance of the estimator which is the variance of the basic experiment $H(S)$ divided by the number of replications N. Since we have assumed that the second moment of $H(S)$ exists, this error, sometimes called *statistical* error, is decreasing with increasing N.

As an example, consider S given by

$$S = S_0 e^{\left(r - \frac{\sigma^2}{2}\right)T + \sigma\sqrt{T}Z}$$

where Z has the standard normal distribution and H is given by

$$H(x) = e^{-rT} \max(x - K, 0).$$

Then, $a = \mathbb{E}[H(S)]$ is the fair price of an European call option with strike K and time to maturity T in the Black-Scholes model with spot price S_0, risk-free interest rate r and volatility σ. In this case, we have to sample N times from the standard normal distribution to calculate the MC estimate (9.2) for a. Note, that this example is only given as an illustration, since for the calculation of European call options in the Black-Scholes model closed-form solutions are available.

For path-dependent options and more complex models, like the Heston model, this is often not the case. Furthermore direct sampling is not possible or not feasible. In the following sections we will describe how the MC method can be used in these cases.

9.2.3 Monte Carlo Methods for SDEs

Models for pricing derivatives are often given as a multi-dimensional SDE driven by a Brownian motion W

$$\begin{aligned} X_0 &= x_0, \\ dX_t &= a(t, X_t)dt + b(t, X_t)dW_t, \quad 0 < t \leq T, \end{aligned} \tag{9.5}$$

with some specific drift function a, volatility function b and a (deterministic) initial value x_0. Here, the asset price process S is one of the components of the multidimensional stochastic process X.

In the Heston model,

$$X = \begin{pmatrix} X_1 \\ X_2 \end{pmatrix} = \begin{pmatrix} S \\ \nu \end{pmatrix}$$

and

$$\begin{aligned} a(t, x) &= \begin{pmatrix} x_1 r \\ \kappa(\theta - x_2) \end{pmatrix}, \\ b(t, x) &= \begin{pmatrix} x_1\sqrt{x_2} & 0 \\ 0 & \sigma\sqrt{x_2} \end{pmatrix} \begin{pmatrix} \rho & \sqrt{1-\rho^2} \\ 1 & 0 \end{pmatrix}, \end{aligned}$$

with $x = (x_1, x_2)$.

For advanced models we often do not know an analytic solution of the SDE or it is difficult to simulate from it. One way out is to approximate the solution of the SDE using a discretization scheme which will be described in the next section.

Due to Broadie and Kaya [9] there exists a method to simulate the Heston model exactly at finitely many discretization points. Since it is numerically very challenging and computationally very expensive, discretization schemes are often superior for path-dependent options (and options depending on many discrete observation dates), see e.g. [43].

9.2.3.1 Discretization Schemes for SDEs

For notational convenience, we will in the following consider some one-dimensional model for the asset prices process, i.e., $X = S$ and $x_0 = s_0$ in (9.5).

Fix integers $M \geq 2$ and $l \geq 1$ and define $D_l = M^{l-1}$ and $h_l = T/D_l$. Then, for example, the *Euler discretization scheme* with D_l steps and discretization points $t_m^{(l)} = mh_l$ for $m = 0, \ldots, D_l$ is given by

$$\hat{S}_0^{(l)} = s_0$$

and

$$\hat{S}_{m+1}^{(l)} = \hat{S}_m^{(l)} + a\left(t_m^{(l)}, \hat{S}_m^{(l)}\right)h_l + b\left(t_m^{(l)}, \hat{S}_m^{(l)}\right)\Delta W_{m+1}^{(l)}, \tag{9.6}$$

where $\Delta W_{m+1}^{(l)} = W\left(t_{m+1}^{(l)}\right) - W\left(t_m^{(l)}\right)$ with $m = 0, \ldots, D_l - 1$. An approximation $\hat{S}^{(l)}$ of the solution of SDE (9.5) is obtained by linear interpolation of $\hat{S}_0^{(l)}, \ldots, \hat{S}_{D_l}^{(l)}$. For more information on discretization schemes see, e.g., [40].

Abstractly speaking, a discretization scheme is a sequence of measurable functions $\varphi^{(l)}$ which map the Brownian motion W to approximations $\hat{S}^{(l)} = \varphi^{(l)}(W)$ of S.

Note that the Euler discretization does not depend on the whole path of the Brownian motion, but only on finitely many increments $\Delta W_1^{(l)}, \ldots, \Delta W_{D_l}^{(l)}$ which are independently normally distributed. Hence, Eq. (9.6) can be easily simulated, i.e. samples of $\hat{S}^{(l)} = \varphi^{(l)}(W)$ can be generated on a computer with a pseudorandom number generator. For the simulation of normally distributed random variables see, e.g. [29].

As already mentioned, direct sampling of the Heston model is possible, but not reasonable for path-dependent options. Besides Broadie and Kaya [9], approaches for exact sampling have been shown in [24, 30] and [44].

Numerically discretization of the Heston model is a challenging topic since the CIR-process used to model the volatility does not fulfill the standard textbook assumptions (like globally Lipschitz diffusion coefficient, see [40]) due to the square root in the diffusion coefficient. It is well known that the inequality

$$\sigma^2 \leq 2\kappa\theta,$$

known as the *Feller condition*, plays a crucial role for the quality of discretization schemes [3, 19] for the volatility process ν. We comment on the impact of this condition on the results of our numerical experiments in Sect. 9.5.

If the Feller condition is fulfilled, ν is strictly positive, otherwise ν might reach zero, but does not spend time there. Discretization schemes with strong rates (which is important for the multilevel method presented later) given the Feller condition (or even stronger conditions) have been shown in [1, 3, 7, 19] and [39]. Recently, Hutzenthaler et al. [37] have proven a strong rate under the condition that $0.5\sigma^2 < 2\kappa\theta$. Weak rates have been shown in [2] and [47].

Unfortunately the Feller condition is rarely satisfied in practice if the CIR-process is used as a volatility process as in the Heston model, see [4] and [13] and the calibration results in [5, 25] and [22]. There are two different approaches for discretization schemes which work for the full parameter space of the CIR-process: The first approach is based on the technique of Brodie and Kaya [9], but instead of sampling from the correct distribution of the CIR-process and its integral, one samples of easier distributions which match the first two moments and/or discretize the integrals. Methods of this kind with numerical results are presented in [4] and [31]. The second approach of discretization schemes, which are defined if the Feller condition is violated, is given by Euler-type discretizations with ad-hoc fixes for negative volatilities, see [7, 17, 18, 36, 38] and [43] where additional to a new scheme all the previous schemes have been compared. Alfonsi [2] combines a discretization scheme and an idea similar to [4] to obtain a discretization scheme that works whether the Feller condition holds or not.

9.2.3.2 The Classic Monte Carlo Method for SDEs

We aim at approximating the price $a = \mathbb{E}[H(S)]$ of a European derivative with path-dependent payoff function H where S is given by a one-dimensional SDE as in the previous section. Therefor, fix some integer L and consider an approximation $\hat{S}^{(L)} = \varphi^{(L)}(W)$ for some discretization scheme $\varphi^{(L)}$ with $D_L = M^{L-1}$ steps. If direct sampling of $H(S)$ is not possible,

$$\mathbb{E}\left[H\left(\hat{S}^{(L)}\right)\right] \tag{9.7}$$

may serve as an approximation to a. If the discretization $\hat{S}^{(L)}$ is based on finitely many Brownian increments (as it is the case for the Euler discretization), direct sampling of $\hat{S}^{(L)}$ is possible and we can apply the classic MC method to (9.7).

Overall, the *classic MC algorithm for SDEs* estimates a by the sample mean of simulated instances of the discounted payoff values $H\left(\hat{S}^{(L)}\right)$, i.e.,

$$\mathcal{A}^{\mathrm{MC}_{\mathrm{SDE}}} = \frac{1}{N}\sum_{i=1}^{N} H(\hat{S}_i^{(L)}), \tag{9.8}$$

where $\hat{S}_1^{(L)}, \ldots, \hat{S}_N^{(L)}$ are independent copies of $\hat{S}^{(L)}$. If $\hat{S}^{(L)}$ is the Euler discretization (9.6) then we need D_L Independent and Identically Distributed (i.i.d.) normal random variables for $\mathcal{A}^{\mathrm{MC}_{\mathrm{SDE}}}$ and thus (9.8) can be simulated easily.

Since the MC method (9.8) is based on random experiments, it has a *statistical error* that is measured by the *variance* of the estimator. Moreover, the application of a discretization scheme introduces a second type of error, the so-called *bias*. The overall MSE of $\mathcal{A}^{\mathrm{MC}_{\mathrm{SDE}}}$ can be decomposed into

$$\begin{aligned}
\mathbb{MSE}\left(\mathcal{A}^{\mathrm{MC}_{\mathrm{SDE}}}\right) &= \mathbb{E}\left[\left(\mathcal{A}^{\mathrm{MC}_{\mathrm{SDE}}} - a\right)^2\right] \\
&= \underbrace{\mathbb{V}\mathrm{ar}\left[\mathcal{A}^{\mathrm{MC}_{\mathrm{SDE}}}\right]}_{\text{stat. error}} + \Bigg(\underbrace{\mathbb{E}\left[H\left(\hat{S}^{(L)}\right)\right] - a}_{=b_L \text{ (bias)}}\Bigg)^2 \\
&= \frac{1}{N}\mathbb{V}\mathrm{ar}\left[H\left(\hat{S}^{(L)}\right)\right] + b_L^2.
\end{aligned} \tag{9.9}$$

The last equality follows from the independence of $\hat{S}_1^{(L)}, \ldots, \hat{S}_N^{(L)}$.

Typically, the variance of $H\left(\hat{S}^{(L)}\right)$ converges to $\mathbb{V}\mathrm{ar}(H(S)) > 0$ and thus, the statistical error can only be controlled by the increase of the simulated instances. Additionally, under suitable assumptions, the bias decreases with finer discretization steps. Hence, to obtain a MC estimate with sufficiently small MSE, one typically chooses L (and hence D_L) to be big enough to reach a certain bound for the bias and adjusts N such that the estimator has the desired statistical error. Smaller MSE induce higher cost: The computational costs of the classic MC algorithm are increasing in D_L and N. Optimal choices of D_L and N to reach a given error depend on the setting, see, e.g., [15].

9.2.3.3 The Multilevel Monte Carlo Method

The *MLMC method* was introduced by Heinrich in 1998 [32] and in the context of SDEs by Giles in 2008 [26]. It is an optimal algorithm in the context of SDEs under suitable assumptions [15]. A comprehensive introduction as well as pointers to recent developments of multilevel algorithm is given in [28]. The approximation of $a = \mathbb{E}[H(S)]$ for a path-dependent payoff H is an infinite dimensional integration problem. Problems of these kind have been studied in the framework of integration on sequence spaces since 5 years in [34, 42, 48] and recently in [6, 20, 21].

The subsequent presentation of the MLMC method will follow [46]. We assume that $D_L = M^{L-1}$ discretization points for a fixed M and some integer L are sufficient to obtain the desired bias accuracy. Denote by

$$\hat{S}^{(l)} = \varphi^{(l)}(W)$$

for $l=1,\dots,L$ the approximated solution of (9.5) by some discretization scheme $\varphi^{(l)}$ (e.g. the Euler scheme, see (9.6)) with $D_l = M^{l-1}$ equidistantly spaced discretization points. Then, in contrast to the classic MC estimate where the "single" approximation $\hat{S}^{(L)}$ is used, one employs the sequence of approximations $\hat{S}^{(1)},\dots,\hat{S}^{(L)}$. Consider the telescoping sum

$$\mathbb{E}\left[H\left(\hat{S}^{(L)}\right)\right] = \mathbb{E}\left[H\left(\hat{S}^{(1)}\right)\right] + \sum_{l=2}^{L} \mathbb{E}\underbrace{\left[H\left(\hat{S}^{(l)}\right) - H\left(\hat{S}^{(l-1)}\right)\right]}_{=\delta_l}, \tag{9.10}$$

where $H\left(\hat{S}^{(l)}\right) = H\left(\varphi^{(l)}(W)\right)$ and $H\left(\hat{S}^{(l-1)}\right) = H\left(\varphi^{(l-1)}(W)\right)$ are coupled via the same Brownian motion. With Eq. (9.10), one represents the target expected value of $H\left(\hat{S}^{(L)}\right)$ by expected values of differences of finer and coarser approximations. The expectations on the right hand side of Eq. (9.10) are called *levels*.

Note that typically the computational cost for generating independent samples of δ_l is increasing, but the variance of δ_l is decreasing for increasing l. This leads to the idea of multilevel algorithms, where each of the levels of (9.10) are approximated with independent classic MC algorithms: Let $N_1,\dots,N_L \in \mathbb{N}$ be given replication numbers and consider independent copies

$$W_{l,1},\dots,W_{l,N_l} \text{ for } l=1,\dots,L \tag{9.11}$$

of W. Then, the MLMC algorithm is given by

$$\mathscr{A}^{\text{MLMC}} = \mathscr{A}^{(1)} + \sum_{l=2}^{L} \mathscr{A}^{(l)}, \tag{9.12}$$

where

$$\mathscr{A}^{(1)} = \frac{1}{N_1}\sum_{n=1}^{N_1} H\left(\varphi^{(1)}(W_{1,n})\right) \text{ and}$$

$$\mathscr{A}^{(l)} = \frac{1}{N_l}\sum_{n=1}^{N_l} \left[H\left(\varphi^{(l)}(W_{l,n})\right) - H\left(\varphi^{(l-1)}(W_{l,n})\right)\right] \text{ for } l=2,\dots,L.$$

Using the independence in (9.11), the MSE of the MLMC algorithm can be decomposed as

$$\text{MSE}\left(\mathcal{A}^{\text{MLMC}}\right) = \sum_{l=1}^{L} \frac{V_l}{N_l} + b_L^2, \tag{9.13}$$

where

$$V_1 = \mathbb{V}\mathrm{ar}\left[H\left(\hat{S}^{(1)}\right)\right] \text{ and}$$
$$V_l = \mathbb{V}\mathrm{ar}\left[H\left(\hat{S}^{(l)}\right) - H\left(\hat{S}^{(l-1)}\right)\right] \text{ for } l = 2 \ldots, L.$$

Under suitable conditions, the variances V_l and the bias b_L are decreasing with increasing l respectively L while the computational cost is increasing. This behavior is exploited by the MLMC idea: Smaller variance makes less repetitions necessary to reach the same error bound for the statistical error. To get a convergent and efficient MLMC algorithm, it is important that the variances of the levels decay to zero fast enough. At the end the MLMC algorithm aims at reducing the overall computational costs by optimally distributing the workload over all levels [26].

Note that the error introduced by the bias is the same for the classic MC and the MLMC algorithm if the number of discretization points used for the last level of the multilevel algorithm agrees with the number of discretization points use in the classic MC algorithm, cf. Eqs. (9.9) and (9.13).

In the following we will explain how samples of $\mathcal{A}^{\mathrm{MLMC}}$ can be generated and especially how coupled paths $\left(\varphi^{(l)}(W), \varphi^{(l-1)}(W)\right)$ can be simulated in the case of the Euler discretization. For each level, the main idea is to generate increments of the Brownian motion for the fine discretization and reuse these increments in a proper way for the coarse discretization.

Let $Z_{m,n}^{(l)}$ for $l = 1, \ldots, L$, $m = 1, \ldots, M^{l-1}$ and $n = 1, \ldots, N_l$ be i.i.d. standard normally distributed random variables. Set

$$Y_{0,n}^{(l),f} = s_0$$

and

$$Y_{m+1,n}^{(l),f} = Y_{m,n}^{(l),f} + a\left(t_m^{(l)}, Y_{m,n}^{(l),f}\right) h_l + b\left(t_m^{(l)}, Y_{m,n}^{(l),f}\right) \sqrt{h_l} Z_{m+1,n}^{(l)},$$

for $m = 0, \ldots, M^{l-1} - 1$. Here, the random variables $Z_{1,n}^{(l)}, \ldots, Z_{M^{l-1},n}^{(l)}$ are scaled and use for the Brownian increments in the Euler scheme to obtain a fine approximation $Y_n^{(l),f}$. For $l > 1$ set

$$Y_{0,n}^{(l),c} = s_0$$

and

$$Y_{m+1,n}^{(l),c} = Y_{m,n}^{(l),c} + a\left(t_m^{(l-1)}, Y_{m,n}^{(l),c}\right) h_{l-1} + b\left(t_m^{(l-1)}, Y_{m,n}^{(l),c}\right) \sqrt{\frac{h_{l-1}}{M}} \sum_{i=1}^{M} Z_{Mm+i,n}^{(l)}$$

for $m = 0, \dots, M^{l-2} - 1$. Note, that for the coarse approximation $Y_n^{(l),c}$ the same random variables are used as for the fine approximation $Y_n^{(l),f}$ on the same level: $Z_{1,n}^{(l)}, \dots, Z_{M^{l-1},n}^{(l)}$ are summed up in groups of length M and scaled to obtain Brownian increments for the coarse path. In this way, $Y_n^{(l),f}$ and $Y_n^{(l),c}$ are coupled. On the other hand, $Y_n^{(l),f}$ and $Y_{n'}^{(l+1),c}$ for $n \neq n'$ are independent and distributed like $\hat{S}^{(l)}$. The last condition is needed for the telescope sum (9.15) to be true, whereas the coupling is necessary for the variance reduction effect of the MLMC algorithm. See Fig. 9.1 for a visualization of the coupling.

Finally set

$$A^{\mathrm{MLMC}} = \frac{1}{N_1} \sum_{n=1}^{N_1} H\left(Y_n^{(1),f}\right) + \sum_{l=2}^{L} \frac{1}{N_l} \sum_{n=1}^{N_l} \left[H\left(Y_n^{(l),f}\right) - H\left(Y_n^{(l),c}\right)\right].$$

Then, A^{MLMC} and $\mathcal{A}^{\mathrm{MLMC}}$ have the same distribution and A^{MLMC} is an implementable version of the MLMC algorithm.

9.2.4 Mixed Precision Architectures

While the former sections have focused on financial mathematics and computational stochastics background, this section is about mixed precision computing systems.

In contrast to standard CPU or GPU architectures where data formats are predefined, reconfigurable architectures like FPGAs or dedicated Application Specific Integrated Circuits (ASICs) are more flexible. They allow the customization of the data formats and widths in all points of the data path. This means that for every stage in the data path the necessary *precision* can be traded off against the size of the circuit and the energy demand for a given overall *accuracy*.

To get the best of both worlds we consider hybrid CPU/FPGA architectures. The high precision simulations and the precision selection procedure can be performed on the CPU, while the low-precision computations are done on the FPGA accelerator.

Although fixed point simulations would work as well with our methodology, we have observed that the overhead to represent the necessary value ranges is too high for our application. In this chapter, we therefore only consider floating point number formats.

Floating point numbers in general consist of an exponent and a significand. The represented number is defined as significand $\times \exp(\text{exponent})$. In binary form the significand and exponent are given in limited number of digits or bits. The number of digits for the significand, p, determines the numerical precision, while the exponent digits define the overall number range. For standard architectures the available number formats are "single" (8 exponent and 23 significand bits) and "double" (11 exponent and 52 significand bits). Later in this chapter we refer to "single" or

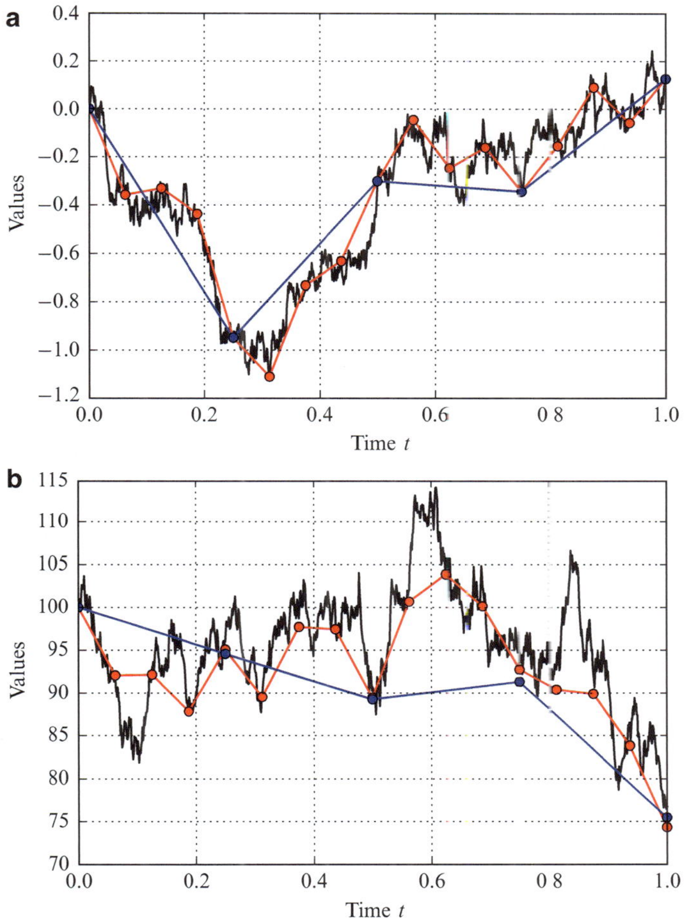

Fig. 9.1 (**a**) Sample paths of coupled fine and coarse discretization of a Brownian motion on a level with 16 respectively 4 discretization and (as an approximation to exact path) 1,024 discretization points. (**b**) Sample paths of coupled fine and coarse discretization of the asset price process of the Heston model with parameters from Table 9.1 Case I with 16 respectively 4 discretization and (as an approximation to exact path) 1,024 discretization points

"double" precision as infinite or reference precision, written as $p = \infty$. In our mixed precision designs we choose the exponent such that no overflow can occur. This is already fulfilled for 6 exponent bits with a representable range of $\pm 2^{2^{6-1}} \approx \pm 10^9$.

In 2011, Chow et al. have investigated different implementation methodologies for mixed precision architectures on reconfigurable architectures for numerical function comparison [11]. Their approach aims at aggressively using low precisions for performing most of the comparisons, and only re-compute with higher precision if the result is ambiguous. In contrast to a pure high precision implementation, they could achieve a speedup of more than five times with their mixed precision methodology on the same FPGA device.

9.3 Mixed-Precision Idea

We now introduce a novel mixed precision methodology for reconfigurable architectures. First, we consider the same situation as in Sect. 9.2.2, i.e. the aim is compute the quantity $a = \mathbb{E}[H(S)]$, where direct sampling of S is possible. Implementing the necessary calculations on a FPGA (or any computing system) will lead to round-off errors due to finite precision. The naive idea, and almost always used strategy, is to do all calculations with a very high precision. For most practical applications this leads to negligible round-off errors. Since CPU computing systems are optimized for using predefined precision (single and double) this is a feasible approach. On FPGA devices huge cost savings are possible if reduced precision is used.

In analogy to the multilevel idea for SDEs we introduce a hierarchy of different levels now with respect to precision instead of time discretization.

In this case the multiple levels are given by

$$\mathbb{E}\left[H\left(S^{(p_L)}\right)\right] = \mathbb{E}\left[H\left(S^{(p_1)}\right)\right] + \sum_{l=2}^{L} \mathbb{E}\left[H\left(S^{(p_l)}\right) - H\left(S^{(p_{l-1})}\right)\right], \qquad (9.14)$$

where $S^{(p_l)}$ uses the same operations as S, but with reduced precision p_l. Chow et al. have done this for two levels, i.e., $L = 2$ in [12]. For two levels the idea was also presented in [27] using single and double precision.

In the following we will explain the idea of a multilevel algorithm with respect to precision in more detail. If exact sampling of S is possible, then there exists $d \in \mathbb{N}$ and a function

$$\varphi : [0,1)^d \to \mathbb{R}$$

such that for a random variable U which is uniformly distributed on $[0,1)^d$, the distribution of $\varphi(U)$ is the same as the distribution of S. We will, in an informal way, assume that φ is implementable, meaning that φ is given as a sequence of arithmetic operations, case distinctions and "elementary functions" like exp, log, etc. Consider now approximations

$$\varphi^{(p)} : [0,1)^d \to \mathbb{R}$$

of φ with

$$\varphi^{(p)}(x) = \varphi^{(p)}\left(2^{-p}\lfloor 2^p x \rfloor\right)$$

for all $x \in [0,1)^d$. This condition ensures that $\varphi^{(p)}$ depends only on the first p fractional bits of the input. The approximation $\varphi^{(p)}$ could for example be given as the elementary operations building φ executed with p fractional bits applied only to the first p fractional bits of the input. Then, for a suitable sequence of precisions $p_1, \ldots, p_L$, $S^{(p_l)} = \varphi^{(p_l)}(U)$ and $S^{(p_{l-1})} = \varphi^{(p_{l-1})}(U)$ in (9.14) are coupled via U. We need p_l random bits to simulate the coupled random variables $\left(S^{(p_l)}, S^{(p_{l-1})}\right)$: Consider $B_{1,n}^{(l)}, \ldots, B_{p_l,n}^{(l)}$ independent and uniformly distributed on $\{0,1\}$ for $l = 1, \ldots, L$ and $n = 1, \ldots, N_l$. Set

$$Y_n^{(p_l),f} = \varphi^{(p_l)}\left(\sum_{i=1}^{p_l} 2^{-i} B_{i,n}^{(l)}\right)$$

and for $l > 1$

$$Y_n^{(p_l),c} = \varphi^{(p_{l-1})}\left(\sum_{i=1}^{p_{l-1}} 2^{-i} B_{i,n}^{(l)}\right).$$

Hence, $Y_n^{(p_l),f}$ and $Y_n^{(p_l),c}$ are coupled and $Y_n^{(p_l),f}$ and $Y_{n'}^{(p_{l+1}),c}$ for $n \neq n'$ are independent and distributed like $\hat{S}^{(p_l)}$. The multilevel algorithm with respect to precision only is now given as

$$\begin{aligned} \mathcal{A}^{\mathrm{MP}} &= \frac{1}{N_1} \sum_{n=1}^{N_1} H\left(Y_n^{(p_1),f}\right) \\ &\quad + \sum_{l=2}^{L} \frac{1}{N_l} \sum_{n=1}^{N_l} \left[H\left(Y_n^{(p_l),f}\right) - H\left(Y_n^{(p_l),c}\right)\right]. \end{aligned}$$

9.4 Mixed Precision Multilevel

In the following we present the idea of our mixed precision methodology when S is given as a one-dimensional SDE and direct sampling of $H(S)$ is not possible. We develop an extension of the MLMC method that exploits the mixed precision setting by applying a customized precision for each numerical discretization $\hat{S}^{(l)}$. The finest numerical discretization is computed in reference precision. Hence the resulting error introduced through the bias stays the same. The coarser numerical discretization are computed with aggressively reduced precision and therefore at much lower cost. This typically will increase the variance of the corresponding level. Our proposed methodology first chooses the precision for each numerical discretization such that the level variance in reduced precision is close to the level variance in reference precision. We then select the number of MC runs on each level such that the overall computational effort is minimized, while guaranteeing the desired accuracy.

A similar idea has been presented in [12], where Chow et al. have shown that mixed precision simulations can achieve the same overall accuracy as pure reference simulations in a MC simulation, while providing a speedup of $8\times$. They use an auxiliary sample function to determine the mixed precision error in their setting and correct it afterwards. Our presented approach is related to the work of Chow et al., but addresses some more points:

- They only consider two different precision levels, while we consider the general MLMC setting with an arbitrary number of levels.
- Our algorithm determines the appropriate precision at runtime with a heuristic analysis, while Chow et al. precompute this before simulating in a time-consuming static analysis.
- We are solving the SDE numerically, and cannot rely on exact sampling of the random variable as in their simpler setting. This adds another dimension of complexity to the problem, in particular another source of error.

9.4.1 Algorithm

Our novel *mixed precision MLMC algorithm* follows the classic MLMC line. However, in addition to multiple discretization schemes $\hat{S}^{(l)}$ with different number of steps $D_l = M^{l-1}$ we introduce customized precisions p_l to calculate the approximation $\hat{S}^{(l)}$ for each scheme. We denote this as $\hat{S}^{(l,p_l)}$:

$$\hat{S}^{(l,p_l)} = \varphi^{(l,p_l)}(W)$$

where $\varphi^{(l,p_l)}$ is a discretization scheme as described in Sect. 9.2.3.1, but each operation is performed only with p_l fractional bits. The increments of W necessary as input for $\varphi^{(l,p_l)}$ are generated in reference precision and then cut-down to p_l fractional bits. Recently, an approach that already uses reduced precision for the generation of the increments was outlined in [28].

For the following fix a maximal level L and a sequence of precisions $p_1, \ldots, p_L$. A way to choose these precision will be explained in Sect. 9.4.5. Due to the definition of $\hat{S}^{(l,p_l)}$ one can establish the telescopic sum

$$\begin{aligned}\mathbb{E}\left[H\left(\hat{S}^{(L,p_L)}\right)\right] &= \mathbb{E}\left[H\left(\hat{S}^{(1,p_1)}\right)\right] \\ &\quad + \sum_{l=2}^{L} \mathbb{E}\left[H\left(\hat{S}^{(l,p_l)}\right) - H\left(\hat{S}^{(l-1,p_{l-1})}\right)\right] \qquad (9.15)\end{aligned}$$

where $H\left(\hat{S}^{(l,p_l)}\right) = H\left(\varphi^{(l,p_l)}(W)\right)$ and $H\left(\hat{S}^{(l-1,p_{l-1})}\right) = H\left(\varphi^{(l-1,p_{l-1})}(W)\right)$ are coupled via the same Brownian motion. This gives a way to represent the target expectation as a sum of expectations of finer and coarse approximations, but now with two dimensions of discretization: A time discretization and a discretization due to finite precision.

As in classic MLMC, we compute each expectation on the right hand side of Eq. (9.15) with a classic MC estimate. This leads to the Mixed Precision Multilevel (MPML) algorithm

$$\mathcal{A}^{\text{MPML}} = \mathcal{A}^{(1,p_1)} + \sum_{l=2}^{L} \mathcal{A}^{(l,p_l)} \tag{9.16}$$

with

$$\mathcal{A}^{(1,p_1)} = \frac{1}{N_1} \sum_{n=1}^{N_1} H\left(\hat{S}_n^{(1,p_1)}\right) \text{ and}$$

$$\mathcal{A}^{(l,p_l)} = \frac{1}{N_l} \sum_{n=1}^{N_l} \left[H\left(\hat{S}_n^{(l,p_l)}\right) - H\left(\hat{S}_n^{(l-1,p_{l-1})}\right)\right] \text{ for } l = 2, \ldots, L,$$

where $\hat{S}_n^{(l,p_l)}$ are independent copies of $\hat{S}^{(l,p_l)}$ and N_l denotes the number of MC runs on each level. As for the classic MLMC algorithm, it is crucial that $\hat{S}^{(l,p_l)}$ and $\hat{S}^{(l-1,p_{l-1})}$ within one level depend on the same Brownian path.

Analogously as for the MLMC algorithm we obtain,

$$\mathrm{MSE}\left(\mathcal{A}^{\text{MPMC}}\right) = \sum_{l=1}^{L} \frac{V_l^{\text{p}}}{N_l} + \left(b_L^{\text{p}}\right)^2, \tag{9.17}$$

where $b_L^{\text{p}} = \mathbb{E}\left[H\left(\hat{S}^{(L,p_L)}\right)\right] - a$ and

$$V_1^{\text{p}} = \mathbb{V}\text{ar}\left[H\left(\hat{S}^{(1,p_1)}\right)\right] \text{ and}$$

$$V_l^{\text{p}} = \mathbb{V}\text{ar}\left[H\left(\hat{S}^{(l,p_l)}\right) - H\left(\hat{S}^{(l-1,p_{l-1})}\right)\right] \text{ for } l = 2, \ldots, L.$$

The number of simulations N_l for each level has to be chosen in an optimal way, meaning that the MPML estimate has to be calculated with the minimal possible computational costs for a given statistical error ε^2. In the following we show how this optimization problem can be solved considering cost and accuracy depending on N_l, cf. [26].

With $\mathfrak{c}_1$ and $\mathfrak{c}_l$ being the average runtimes for sampling one element on the starting level $H\left(\hat{S}^{(1,p_1)}\right)$ and other levels $\left\{H\left(\hat{S}^{(l,p_l)}\right) - H\left(\hat{S}^{(l-1,p_{l-1})}\right)\right\}$ respectively, the cost for computing the MPML estimate (9.16) is

$$\text{cost}\left(\mathcal{A}^{\text{MPML}}\right) = \sum_{l=1}^{L} N_l \cdot \mathfrak{c}_l. \tag{9.18}$$

On the other hand, the statistical error of the estimate, measured by its variance, can be written as

$$\mathbb{V}\text{ar}\left[\mathcal{A}^{\text{MPML}}\right] = \sum_{l=1}^{L} \frac{1}{N_l} \cdot V_l^{\text{p}}, \tag{9.19}$$

where V_l^{p} are the variances of the particular levels l.

Treating $N_1, \ldots, N_L$ as real numbers, cost $\left(\mathcal{A}^{\text{MPML}}\right)$ can be minimized such that $\mathbb{V}\text{ar}\left[\mathcal{A}^{\text{MPML}}\right] \leq \varepsilon^2$ using the method of Lagrange multipliers. This leads to the following choice of replication numbers

$$N_l = \left\lceil \left(\frac{1}{\varepsilon^2} \sum_{k=1}^{L} \sqrt{V_k^{\text{p}} \mathfrak{c}_k}\right) \cdot \sqrt{V_l^{\text{p}} / \mathfrak{c}_l} \right\rceil. \tag{9.20}$$

This selection leads to

$$\mathbb{MSE}\left(\mathcal{A}^{\text{MPMC}}\right) \leq \varepsilon^2 + \left(b_L^{\text{p}}\right)^2$$

and by setting the precision for the last level to reference precision, i.e., $p_L = \infty$, we obtain the same MSE as for the MLMC algorithm. Note, that the MLMC algorithm is given as a special case of our MPML algorithm by setting $p_1 = \cdots = p_L = \infty$.

The above consideration results in the MPML Algorithm 1. The selection of appropriate precision is given by Algorithm 2 together with the explanation in Sect. 9.4.5.

Algorithm 1 Mixed precision multilevel

Input: ε and L
Output: $\mathcal{A}^{\text{MPML}}$

1: choose $p_1, \ldots, p_L$ using Algorithm 2
2: estimate $V_1^{\text{p}}, \ldots, V_L^{\text{p}}$ using an initial $N_l = 10^4$ samples
3: define optimal $N_l, l = 1, \ldots, L$ using Eq. (9.20)
4: evaluate extra samples at each level as needed for new N_l
5: calculate $\mathcal{A}^{\text{MPML}}$ according to Eq. (9.16)

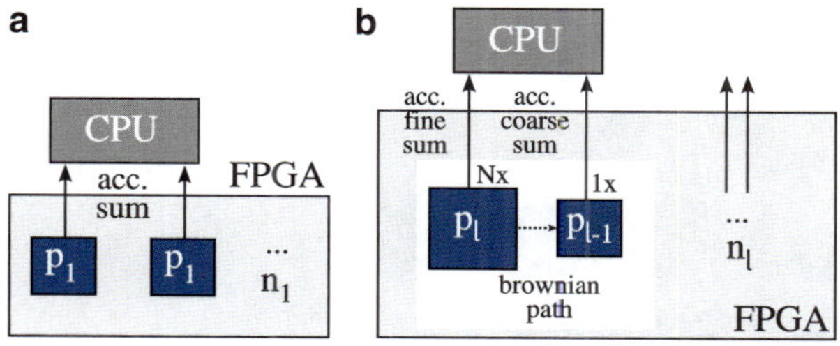

Fig. 9.2 For the first level (**a**), the FPGA contains n_1 accelerators with precision p_1, while for higher levels (**b**) the FPGA contains groups of accelerator with precision p_l and p_{l-1} that exchange their Brownian path. The control and final payoff computation is done on the CPU that receives the accumulated paths

9.4.2 Reconfigurable Architecture

This section depicts the Hardware (HW) architecture template underlying the MPML algorithm introduced in the previous section. We consider a hybrid system that consists of a CPU and an FPGA that is used as a hardware accelerator as illustrated in Fig. 9.2. This setup exploits the benefits of both architectures and is currently considered as one key architecture for the next generation computing systems [23]. The CPU handles the algorithmic control, the simulation in reference precision, and the estimation of the level variances V_l^p. Moreover Step 1 of Algorithm 1, the precision selection with Algorithm 2, is executed on the CPU as well. The FPGA implements the discretized price paths $\hat{S}^{(l,p_\cdot)}$ in reduced precisions p_l and the payoff computation.

There are basically two models on how to schedule the work on the FPGA: Firstly, all the levels could be calculated in parallel with one configuration containing all necessary accelerators, if the number of levels is fixed in advance. Secondly, the levels could be calculated one by one on the whole FPGA serially, with the FPGA being reconfigured between each level. In this chapter we consider the serial model with reconfiguration, while finding optimal configurations for the parallel model is ongoing research.

We have measured that the FPGA reconfiguration times are between 50 and 500 ms for common Xilinx FPGA devices. For maximum levels $L = 5,\ldots,7$, the overall reconfiguration overhead is between 0.5 and 7 s Since we consider small target precisions ε with overall runtimes of multiple minutes, the overhead by reconfiguration is insignificant.

For the first level the FPGA is filled with accelerators with precision p_1 until the FPGA is full, see Fig. 9.2a. The accelerators calculate one path step in each clock cycle. To estimate the terms for higher levels we have to calculate the value $H\left(\hat{S}^{(l,p_l)}\right)$ of the fine path and $H\left(\hat{S}^{(l-1,p_{l-1})}\right)$ of the coarse path. The fine path requires M times more steps, hence we build groups of M accelerators with precision p_l and one with p_{l-1} for optimal resource utilization. They communicate their Brownian path and send both results to the CPU, see Fig. 9.2b.

9.4.3 Showcase Settings

We demonstrate the benefits of our proposed mixed precision methodology by a showcase setting that we describe now. This setup remains fixed for the rest of this chapter. Nevertheless, our approach is applicable to other products and discretization schemes as well.

We use Asian call options in the Heston setting as a showcase because they are practically relevant and no closed-form solutions for pricing are available in general. Their discounted payoff is of the form

$$H(S) = e^{-rT} \max\left(\frac{1}{T}\int_0^T S_t dt - K, 0\right), \tag{9.21}$$

where T is the time to maturity and K is the strike price.

In the following we explain the discretization scheme we used for the Heston model. Since cases with violated Feller condition are relevant for practice, we decided to choose a discretization scheme which works for all choices of parameters. As a first choice, we will present here the approach of [43] which we use because of its good performance and simplicity. At the moment it is not immediately clear how the coupling necessary for multilevel algorithms can be achieved with the methods approximating the Broadie and Kaya approach. Also, the case distinctions required in these approaches will be much more expensive for a FPGA implementation than for a CPU-implementation. The advantage of these methods reported in [31] might therefore vanish. The same might hold true for the scheme reported in [2]. But these topics have to be investigated further.

Now, we will describe the configuration as proposed in [43] using the notation from Sect. 9.2.3.1. Consider two independent Brownian motions W^1 and W^2 and set

$$W^S = \rho W^1 + \sqrt{1-\rho^2} W^2 \text{ and}$$
$$W^\nu = W^1$$

to capture the correlation of the Brownian motions driving the asset prices respectively the volatility process.

The Euler discretization scheme is used to approximate the stochastic volatility process (9.1b) together with full truncation correction to avoid negative volatility values. This gives

$$\hat{\nu}_0^{(l)} = \tilde{\nu}_0^{(l)} = \nu_0$$

and

$$\hat{v}_{t_{m+1}}^{(l)} = \left(\tilde{v}_{t_{m+1}}^{(l)}\right)^{+},$$

where $x^{+} = \max(x, 0)$ and

$$\tilde{v}_{m+1}^{(l)} = \tilde{v}_{m}^{(l)} + \kappa\left(\theta - \left(\tilde{v}_{m}^{(l)}\right)^{+}\right)h^{(l)} + \sigma\sqrt{\left(\tilde{v}_{m}^{(l)}\right)^{+}}\Delta W_{m+1}^{v,(l)}.$$

Instead of discretizing the asset price process directly, we first consider the *log*-transformation, i.e. SDE (9.1a) is transformed with Îto's Lemma to

$$d\bar{S}_t = \left(r - \frac{1}{2}v_t\right)dt + \sqrt{v_t}dW_t^S,$$

where $\bar{S} = \log(S)$. Then, we apply the Euler scheme as well. This gives

$$\bar{S}_0^{(l)} = \log(s_0),$$

$$\bar{S}_{m+1}^{(l)} = \bar{S}_m^{(l)} + \left(r - \frac{1}{2}\hat{v}_m^{(l)}\right)h^{(l)} + \sqrt{\hat{v}_m^{(l)}}\Delta W_{m+1}^{S,(l)}.$$

The integral in the payoff function H from Eq. (9.21) can be calculated exactly for the approximations $\hat{S}^{(l)} = \exp\left(\bar{S}^{(l)}\right)$ by the trapezoidal rule. For the multilevel methods we set the level refinement constant $M = 4$ as in [26].

9.4.4 Cost Model for the FPGA Architecture

This section shows how the cost-model defined in (9.18) emerges from the HW architecture. To calculate the runtime, it is important to know how many accelerators fit on the FPGA. We have therefore synthesized the accelerators for the Maxeler MaxWorkstation system to obtain the HW resource usage data. It contains a Xilinx Virtex 6 FPGA (XC6VSX475T-2) and an Intel i7 CPU connected via PCIe [50].

The critical resource on the FPGA is the number of Lookup Tables (LUTs). Figure 9.3 shows the relative accelerator size λ_p on the FPGA for different precisions p. Accelerators for "single" take 3.27 %, for "double" 9.99 %, of the total amount of LUTs. Custom precision accelerators with $p = 8$ fractional bits take 0.92 %. This means we can fit 10 times more accelerators on the FPGA with this reduced precision instead of "double".

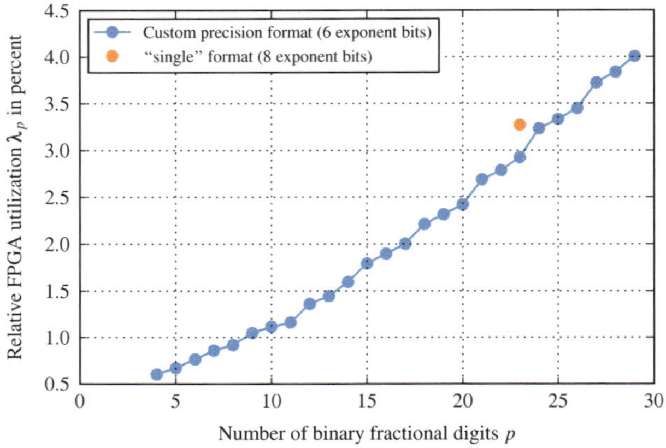

Fig. 9.3 Showing relative device usage λ_p of the MC accelerator on the MaxWorkstation. While CPUs or GPUs only provide the "single" floating-point format, FPGAs can save huge amount of area exploiting reduced precision formats

To avoid routing congestion we assume that we can use up to 80 % of the FPGA area without problems. For interconnect, we attribute an overhead of $\Theta = 0.2\,\% = 600\,\text{LUTs}$ for each accelerator. The number of accelerators for the first level n_{p_1} and higher levels $n_{p_l,p_{l-1}}$ that fit on the FPGA are therefore:

$$n_{p_1} = \left\lfloor \frac{0.8}{\lambda_p + \Theta} \right\rfloor, \; n_{p_l,p_{l-1}} = M \left\lfloor \frac{0.8}{M\lambda_{p_l} + \lambda_{p_{l-1}} + (M+1)\Theta} \right\rfloor.$$

It is important to note that there is no actual dependence on the level l itself in the definition above. The reason is that the accelerators calculate the next step of a discretized solution of the SDE only. Computing on a finer path and therefore using more discretization steps only changes the step size parameter and runtime, but not the form of the increment procedure for each step.

We now calculate the runtime of the MPML algorithm. One accelerator running at frequency $f = 100\,\text{MHz}$ needs $f^{-1} = 10\,\text{ns}$ to calculate one step. The whole FPGA with n accelerators calculating N_l paths with M^{l-1} discretization steps takes $N_l M^{l-1}/fn$ in time. Each FPGA configuration can be run with a unique frequency f_{p_1} respectively $f_{p_l,p_{l-1}}$. Therefore the whole runtime for all levels is:

$$\text{runtime} = \frac{N_1}{f_{p_1} n_{p_1}} + \sum_{l=2}^{L} \frac{N_l M^{l-1}}{f_{p_l,p_{l-1}} n_{p_l,p_{l-1}}} + \text{const.} \tag{9.22}$$

The work on the CPU takes less time and can mostly be interleaved with the work on the FPGA. Therefore it does not increase the overall runtime. As the FPGA path simulation part dominates the computing time, this is also the overall runtime of the algorithm. Assuming the constant in Eq. (9.22) is small compared to the overall runtime we can define the constants $\mathfrak{c}_l$ from Eq. (9.18) as:

$$\mathfrak{c}_1 = \left(f_{p_1} n_{p_1}\right)^{-1}, \quad \mathfrak{c}_l = \frac{M^{l-1}}{f_{p_l,p_{l-1}} n_{p_l,p_{l-1}}}.$$

All accelerators have been synthesized for $f = 100$ MHz. Based on the numbers for λ_p from Fig. 9.3 all the weights $\mathfrak{c}_l$ can be derived.

9.4.5 Heuristic for Precision Selection

As depicted in Sect. 9.4.1, the precision p_l of each numerical approximation $\hat{S}^{(l,p_l)}$ has to be set appropriately. In this section we present our proposed heuristic for precision selection and show that it can be executed at runtime with a low computational overhead.

Substituting the infinite precision simulations with the reduced ones leads to increased level variances V_l^{p}. The goal of our heuristic is to choose p such that the variance V_l^{p} is only slightly enlarged compared to V_l, as higher variances would require more repetitions.

To test this we define the following marker fraction

$$\xi_l\left(p\right) = \frac{\mathrm{Var}\left[H\left(\hat{S}^{(l+1,\infty)}\right) - H\left(\hat{S}^{(l,p)}\right)\right]}{\mathrm{Var}\left[H\left(\hat{S}^{(l+1,\infty)}\right) - H\left(\hat{S}^{(l,\infty)}\right)\right]}, \tag{9.23}$$

where the nominator

$$H\left(\hat{S}^{(l+1,\infty)}\right) - H\left(\hat{S}^{(l,p)}\right) = H\left(\varphi^{(l+1,\infty)}(W)\right) - H\left(\varphi^{(l,p)}(W)\right)$$

and the denominator

$$H\left(\hat{S}^{(l+1,\infty)}\right) - H\left(\hat{S}^{(l,\infty)}\right) = H\left(\varphi^{(l,\infty)}(W)\right) - H\left(\varphi^{(l,\infty)}(W)\right)$$

depend on the same Brownian motion.

For p approaching infinity, $\xi_l\left(p\right)$ converges to 1. We choose p_l such that $\xi_l(p_l) < 1.1$ which ensures that V_l^{p} is close to V_l. In Fig. 9.4 the value of $\xi_l\left(p\right)$ is plotted over p. The threshold value 1.1 is reasonable in the sense that it can indicate whether a sufficient precision has already been reached. The figure indicates that there is a region where $\xi_l\left(p\right)$ is significantly above the threshold and for high p significantly below the threshold. In between there is a region of 2–3 bits where the heuristic is ambiguous. We later empirically show that this has a minor impact on the performance of the MPML algorithm.

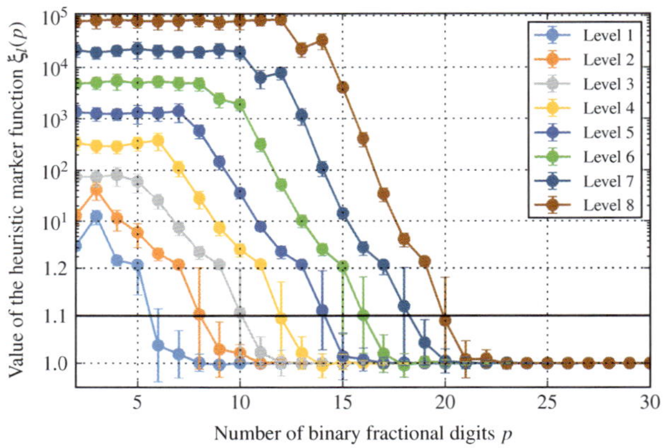

Fig. 9.4 Showing the value of the heuristic $\xi_l(p)$ for different levels for benchmark I. The region up to 1.2 is plotted linearly to highlight the behavior close to the acceptance threshold, while higher values are plotted on a logarithmic scale. The error bars in the plot are the standard deviation of $\xi_l(p)$, when evaluated with 100 paths

---

Algorithm 2 Precision selection

Input: L
Output: $p_1, \ldots, p_L$
  $p_0 \leftarrow 3$ {practically observed}
  for $l = 1$ to $L - 1$ **do**
    $p_l \leftarrow p_{l-1}$
    repeat
      $p_l \leftarrow p_l + 1$
      simulate 100 paths
      estimate $\xi_l(p_l)$ according to Eq. (9.23)
    until $\xi_l(p_l) < 1.1$
  end for
  $p_L \leftarrow \infty$ {i.e., 23 or 52}

Algorithm 2 shows our proposed heuristic, which iteratively selects the precisions. The variances are estimated with only 100 paths, thus the runtime is negligible. The selection procedure has a stochastic nature, thus it might produce different outcomes for independent runs. However, we show that the heuristic works and is stable in Sect. 9.5. It is important to use the same Brownian paths for the estimation of numerator and denominator of ξ_l, since we have observed that otherwise the heuristic fails.

Note that we set the precision p_L for the finest discretization to the reference precision. This ensures that the *bias error* of the MPML estimate is the same as for the classic MLMC method. One could also choose p_L according to the general procedure in Algorithm 2. This will in general yield a higher speedup. We think the impact on the bias in this case is negligible for practical applications, but a thorough analysis is ongoing research.

9.5 Numerical Results

In this section we present the outcome of the numerical performance analysis we have done on the proposed MPML method described in Algorithm 1. For all results we have considered the option pricing problem for Asian call options with the maturity $T = 1$ and the strike $K = 100$ in the Heston model with either benchmark parameter set I or II from Table 9.1.

In particular, we have focused on two points:

- The stability of the heuristic precision selection Algorithm 2,
- and the speedup of our proposed MPML method compared to the classic MLMC algorithm.

9.5.1 Heuristic Stability Analysis

This section shows that the proposed precision selection given by Algorithm 2 is stable in our setting. Since the main aim of the MPML methodology is speeding up the classic MLMC algorithm, *stability* in this context means the ability to be consistently faster than classic MLMC. Note that our algorithm ensures that the resulting error will always be the same as for the classic MLMC method. For testing, we used parameter set I from Table 9.1 and independently run the MPML Algorithm 1 for 50 times with fixed $L = 6$.

Figure 9.5 shows the speedup of our MPML pricer compared to reference precision MLMC for the same number of levels L and statistical error ε^2. It can be seen that the precisions selected by the heuristic may vary within the range of 3 bits for independent runs. However, this small variability in the chosen precision has only minor influence on the overall performance of our MPML algorithm. It is obvious that the speedup does not differ much, and therefore we can conclude the overall stability of the heuristic selection procedure.

9.5.2 Speedup Analysis over Classic MLMC

As a fair reference for the speedup analysis, we have implemented the classic MLMC algorithm for a fixed number of levels L on the same architecture both in "single" and "double" precision. The speedup depends of the specified number of levels L. Note that the speedup is independent of the demanded statistical error ε^2 because the cost of both algorithms depend linear on $1/\varepsilon^2$.

Table 9.1 Benchmark Heston parameters

	S_0	κ	θ	σ	r	v_0	ρ	Feller condition
I	100	3	0.16	0.4	0.02	0.1	-0.8	Fulfilled
II	100	1.5	0.16	0.9	0.02	0.1	-0.8	Violated

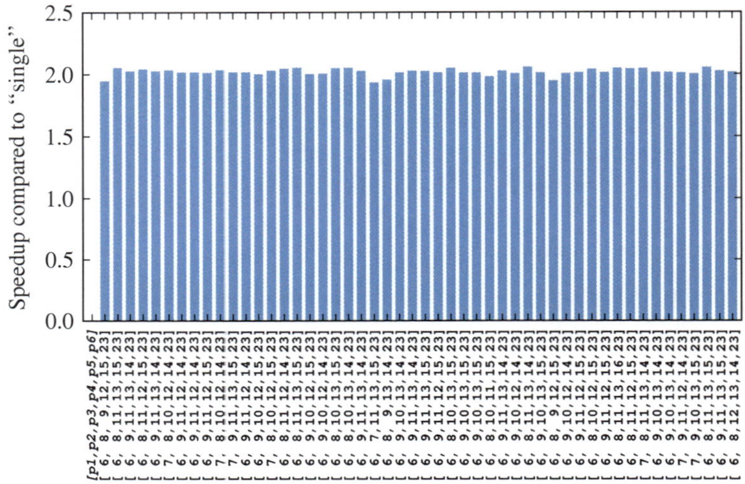

Fig. 9.5 Showing the speedup of MPML compared to MLMC for benchmark I. The target precision is "single" and maximum level $L = 6$. The bottom shows 50 configurations of precision $p_1, \ldots, p_6$ generated by Algorithm 2. For each of the configuration the speedup is drawn compared to MLMC. It is slightly different due to changes in FPGA resources (Fig. 9.3) and variance differences for different bit selections p_l (Fig. 9.6). Although the heuristic sometimes picks lower or higher bits, the speedup is very consistent

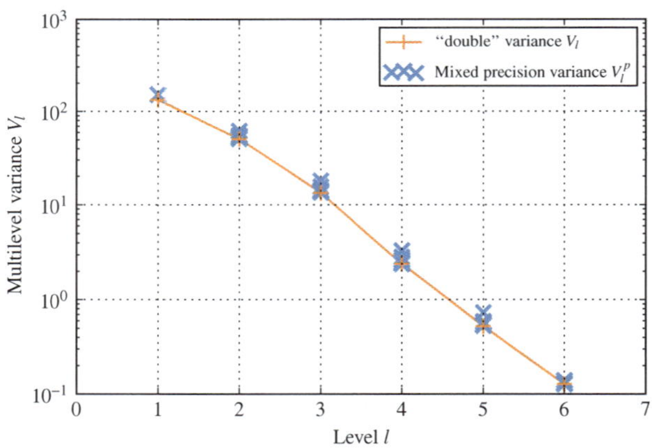

Fig. 9.6 Showing the level variance for Heston benchmark I from Table 9.1. The *orange line* shows the "double" variances V_l of the MLMC algorithm. The *blue crosses* are the mixed precision variance V_l^{p}. They are based on the 50 configurations for $p_1, \ldots, p_L$ in Fig. 9.5 generated by Algorithm 2. Since the mixed precision variances depend on both p_l and p_{l-1}, *multiple blue crosses* are drawn with slightly different variances. The heuristic chooses the bits so that the mixed precision variance is only slightly worse than the "double" variance

As explained in Sect. 9.4.5 the idea of our precision selection heuristic is to choose the precision for each approximation such that V_l^{p} is close to V_l. In Fig. 9.6 we see in orange V_l and in blue V_l^{p} as occurred for independent realizations of our

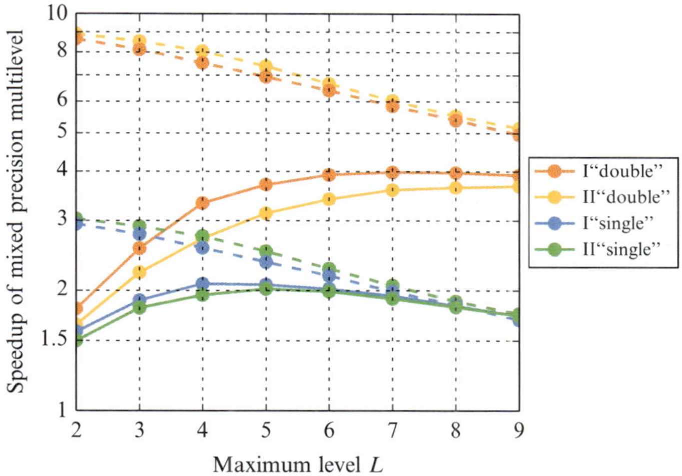

Fig. 9.7 Showing the speedup of MPML compared to MLMC for different maximum level L. The speedup is shown for target precision "single" and "double" and benchmark set I and II. Each point is the average speedup of 50 configurations $p_1, \ldots, p_L$ selected by Algorithm 2, like shown in Fig. 9.5 for $L = 6$. The *solid line* represent speedups when the last precision p_L is set to the target precision, while the *dashed lines* are speedups when p_L is chosen by the heuristic as well

heuristic. The variances in reduced precision are only slightly higher than those in reference precision.

In Fig. 9.7 the solid lines show the observed speedups for benchmarks I and II from Table 9.1 for the following setups:

- MPML $p_L = 23$ vs. "single" precision MLMC
- MPML $p_L = 52$ vs. "double" precision MLMC

One can clearly see that the MPML algorithm is about 1.5–2 times faster than the "single" precision MLMC for both parameter sets from Table 9.1. For "double" precision, the speedup is about 2 for small numbers of the levels and grows towards 4 when moving to higher levels. Moreover, the measured speedup values slightly differ for different parameter sets. This is because of differences in the variance V_l for low levels. Furthermore, it is worth noting that we obtain the results regardless of the Feller condition violation at least for the violated case II. Further investigations are ongoing research.

Additionally, the dashed lines in Fig. 9.7 show the observed speedups when p_L is also selected according to our heuristic. Again the comparison is made in respect to "single" and "double" precision MLMC. We observe a speedup of up to 9× for this setup.

Altogether, we observe a completely stable heuristic behavior, leading to speedups up to more than three when using our MPML methodology over pure reference precision architectures on the same platform.

9.6 Conclusion and Outlook

In this chapter we have presented a novel methodology to exploit reduced precision data format in one of the most advanced algorithms for option pricing. We have formulated a Mixed Precision Multilevel (MPML) algorithm that is aware of the characteristics of custom precision operations. By using reduced precision we lower the overall runtime of the algorithm. Due to the multilevel strategy the final accuracy is not changed.

We have implemented the algorithm on a hybrid CPU/FPGA architecture, which perfectly fits into our mixed precision setting. A key aspect of the algorithm is the choice of appropriate precisions. We have proposed a novel heuristic that selects the precisions at runtime with negligible overhead and demonstrated its effectiveness.

We have showed numerically that our Mixed Precision Multilevel (MPML) algorithm achieves speedups of 3–9$\times$ compared to classic Multilevel Monte Carlo (MLMC) for pricing Asian options in the Heston model. The comparison is made with respect to an already very elaborate algorithm on the same hybrid platform. With our heuristic we are able to determine the precisions for the specific problem under consideration.

In total, our approach is one step towards bridging the gap between financial algorithm design and execution platform refinement. Our idea can also be applied to any numerical method that improves an initial approximation step-by-step by iterative refinements.

Future work includes investigations of the required precision for the finest discretization and of the approach, presented in [28], which already uses reduced precision in the generation of the Brownian increments and detailed studies of energy saving in various setups, for example on the recent Xilinx Zynq-7000 APP or other hybrid CPU/FPGA architectures.

Acknowledgements We gratefully acknowledge the partial financial support from the Center of Mathematical and Computational Modelling $(CM)^2$ of the University of Kaiserslautern. Furthermore we thank Maxeler Technologies Ltd. for providing their technology.

References

1. Alfonsi, A.: On the discretization schemes for the CIR (and Bessel squared) processes. Monte Carlo Methods Appl. **11**(4), 355–384 (2005)
2. Alfonsi, A.: High order discretization schemes for the CIR process: application to affine term structure and Heston models. Math. Comput. **79**(269), 209–237 (2010)
3. Alfonsi, A.: Strong convergence of some drift implicit Euler scheme. Application to the CIR process (2012). arXiv preprint arXiv:1206.3855
4. Andersen, L.B.G.: Efficient simulation of the Heston stochastic volatility model. J. Comput. Financ. **11**(3), 1–42 (2008)
5. Bakshi, G., Cao, C., Chen, Z.: Empirical performance of alternative option pricing models. J. Financ. **52**(5), 2003–2049 (1997)

6. Baldeaux, J., Gnewuch, M.: Optimal randomized multilevel algorithms for infinite-dimensional integration on function spaces with ANOVA-type decomposition. SIAM J. Numer. Anal. **52**(3), 1128–1155 (2014)
7. Berkaoui, A., Bossy, M., Diop, A.: Euler scheme for SDEs with non-Lipschitz diffusion coefficient: strong convergence. ESAIM: Probab. Stat. **12**(1), 1–11 (2008)
8. Black, F., Scholes, M.: The pricing of options and corporate liabilities. J. Politi. Econ. **81**(3), 637–654 (1973)
9. Broadie, M., Kaya, Ö.: Exact simulation of stochastic volatility and other affine jump diffusion processes. Oper. Res. **54**(2), 217–231 (2006)
10. Brugger, C., de Schryver, C., Wehn, N., Omland, S., Hefter, M., Ritter, K., Kostiuk, A., Korn, R.: Mixed precision multilevel Monte Carlo on hybrid computing systems. In: Proceedings of the 2014 IEEE Conference on Computational Intelligence for Financial Engineering Economics (CIFEr), London, pp. 215–222 (2014)
11. Chow, G., Kwok, K.W., Luk, W., Leong, P.: Mixed precision comparison in reconfigurable systems. In: 2011 IEEE 19th Annual International Symposium on Field-Programmable Custom Computing Machines (FCCM), Salt Lake City, pp. 17–24 (2011)
12. Chow, G.C.T., Tse, A.H.T., Jin, Q., Luk, W., Leong, P.H., Thomas, D.B.: A mixed precision Monte Carlo methodology for reconfigurable accelerator systems. In: Proceedings of the ACM/SIGDA International Symposium on Field Programmable Gate Arrays, FPGA'12, New York, pp. 57–66. ACM (2012)
13. Clark, I.J.: Foreign Exchange Option Pricing: A Practitioners Guide. 1st edn. Wiley, Chichester (2011)
14. Cox, J.C., Ingersoll, J., Jonathan, E., Ross, S.A.: A theory of the term structure of interest rates. Econometrica **53**(2), 385–407 (1985)
15. Creutzig, J., Dereich, S., Müller-Gronbach, T., Ritter, K.: Infinite-dimensional quadrature and approximation of distributions. Found. Comput. Math. **9**(4), 391–429 (2009)
16. de Schryver, C., Shcherbakov, I., Kienle, F., Wehn, N., Marxen, H., Kostiuk, A., Korn, R.: An energy efficient FPGA accelerator for Monte Carlo option pricing with the Heston model. In: Proceedings of the 2011 International Conference on Reconfigurable Computing and FPGAs (ReConFig), Cancun, pp. 468–474 (2011)
17. Deelstra, G., Delbaen, F.: Convergence of discretized stochastic (interest rate) processes with stochastic dift term. Appl. Stoch. Models Data Anal. **14**(1), 77–84 (1998)
18. Deelstra, G., Delbaen, F.: An efficient discretization scheme for one dimensional SDEs with a diffusion coefficient function of the form $|x|^\alpha$, $\alpha \in [1/2, 1)$. INRIA Rapport de recherche (5396) (2007)
19. Dereich, S., Neuenkirch, A., Szpruch, L.: An Euler-type method for the strong approximation of the Cox–Ingersoll–Ross process. Proc. R. Soc. A: Math. Phys. Eng. Sci. **468**(2140), 1105–1115 (2012)
20. Dick, J., Gnewuch, M.: Infinite-dimensional integration in weighted Hilbert spaces: anchored decompositions, optimal deterministic algorithms, and higher-order convergence. Found. Comput. Math. **14**(5), 1027–1077 (2014)
21. Dick, J., Gnewuch, M.: Optimal randomized changing dimension algorithms for infinite-dimensional integration on function spaces with ANOVA-type decomposition. J. Approx. Theory 184(0), 111–145 (2014)
22. Duffie, D., Pan, J., Singleton, K.: Transform analysis and asset pricing for affine jump-diffusions. Econometrica **68**(6), 1343–1376 (2000)
23. Duranton, M., Black-Schaffer, D., Bosschere, K.D., Mabe, J.: The HiPEAC Vision for Advanced Computing in Horizon 2020 (2013). Available at http://www.hipeac.net/system/files/hipeac_roadmap1_0.pdf
24. Dyrting, S.: Evaluating the noncentral chi-square distribution for the Cox-Ingersoll-Ross process. Comput. Econ. **24**(1), 35–50 (2004)
25. Engelmann, F.K.B., Oeltz, D.: Calibration of the Heston stochastic local volatility model a finite volume scheme (2011). Available at SSRN: http://dx.doi.org/10.2139/ssrn.1823769

26. Giles, M.B.: Multilevel Monte Carlo path simulation. Oper. Res.-Baltim. **56**(3), 607–617 (2008)
27. Giles, M.B.: Multilevel Monte Carlo methods (2013). arXiv preprint arXiv:1304.5472
28. Giles, M.B.: Multilevel Monte Carlo methods (2015, in preparation). Preprint available at http://people.maths.ox.ac.uk/gilesm/files/acta_draft.pdf
29. Glasserman, P.: Monte Carlo Methods in Financial Engineering. Stochastic Modelling and Applied Probability, 8th edn. Springer, New York (2003)
30. Glasserman, P., Kim, K.-K.: Gamma expansion of the Heston stochastic volatility model. Financ. Stoch. **15**(2), 267–296 (2011)
31. Haastrecht, A.V., Pelsser, A.: Efficient, almost exact simulation of the Heston stochastic volatility model. Int. J. Theor. Appl. Financ. **13**(1), 1–43 (2010)
32. Heinrich, S.: Monte Carlo complexity of global solution of integral equations. J. Complex. **14**(2), 151–175 (1998)
33. Heston, S.L.: A closed-form solution for options with stochastic volatility with applications to bond and currency options. Rev. Financ. Stud. **6**(2), 327 (1993)
34. Hickernell, F.J., Müller-Gronbach, T., Niu, B., Ritter, K.: Multi-level Monte Carlo algorithms for infinite-dimensional integration on $\mathbb{R}^{\mathbb{N}}$. J. Complex. **26**(3), 229–254 (2010)
35. Higham, N.J.: Accuracy and Stability of Numerical Algorithms, 2nd edn. Society for Industrial and Applied Mathematics, Philadelphia (2002). ISBN:978-0898715217
36. Higham, D., Mao, X.: Convergence of Monte Carlo simulations involving the mean-reverting square root process. J. Comput. Financ. **8**(3), 35–61 (2005)
37. Hutzenthaler, A.J.M., Noll, M.: Strong convergence rates and temporal regularity for Cox-Ingersoll-Ross processes and Bessel Processes with accessible boundaries (2014). Online at arXiv: http://arxiv.org/abs/1403.6385
38. Kahl, C., Jäckel, P.: Fast strong approximation Monte-Carlo schemes for stochastic volatility models. J. Quant. Financ. **6**(6), 513–536 (2006)
39. Kahl, C., Schurz, H.: Balanced Milstein methods ordinary for SDEs. Monte Carlo Methods Appl. **12**(2), 143–170 (2006)
40. Kloeden, P.E., Platen, E.: Numerical Solution of Stochastic Differential Equations. Stochastic Modelling and Applied Probability, vol. 12. Springer, New York (2010)
41. Korn, R., Korn, E., Kroisandt, G.: Monte Carlo Methods and Models in Finance and Insurance. CRC, Boca Raton (2010)
42. Kuo, F.Y., Sloan, I.H., Wasilkowski, G.W., Wozniakowski, H.: Liberating the dimension. J. Complex. **26**(5), 422–454 (2010)
43. Lord, R., Koekkoek, R., van Dijk, D.: A comparison of biased simulation schemes for stochastic volatility models. Quant. Financ. **10**(2), 177–194 (2010)
44. Malham, S., Wiese, A.: Chi-square simulation of the CIR process and the Heston model. Int. J. Theor. Appl. Financ. **16**(3), 1350014-1-1350014-38 (2013)
45. Marxen, H.: Aspects of the application of multilevel Monte Carlo methods in the Heston model and in a Lvy process framework. PhD thesis, University of Kaiserslautern (2012)
46. Müller-Gronbach, T., Ritter, K.: Variable subspace sampling and multi-level algorithms. In: Ecuyer, P.L., Owen, A.B. (eds.) Monte Carlo and Quasi-Monte Carlo Methods 2008, pp. 131–156. Springer, Berlin/Heidelberg (2009)
47. Ninomiya, S., Victoir, N.: Weak approximation of stochastic differential equations and application to derivative pricing. Appl. Math. Financ. **15**(2), 107–121 (2008)
48. Niu, B., Hickernell, F.J., Müller-Gronbach, T., Ritter, K.: Deterministic multi-level algorithms for infinite-dimensional integration on $\mathbb{R}^{\mathbb{N}}$. J. Complex. **27**(3–4), 331–351 (2011)
49. Novak, E.: The real number model in numerical analysis. J. Complex. **11**(1), 57–73 (1995)
50. Pell, O., Averbukh, V.: Maximum performance computing with dataflow engines. Comput. Sci. Eng. **14**(4), 98–103 (2012)

Chapter 10
Accelerating Closed-Form Heston Pricers for Calibration

Gongda Liu, Christian Brugger, Christian De Schryver, and Norbert Wehn

Abstract Calibrating models against the markets is a crucial step to obtain meaningful results in the subsequent pricing processes. In general, calibration can be seen as a minimization problem that tries to fit modeled product prices to the observed ones on the market (compare Chap. 2 by Sayer and Wenzel). This means that during the calibration process the modeled prices need to be calculated many times, and therefore the run time of the product pricers have the highest impact on the overall calibration run time. Therefore, in general, only products are used for calibration for which closed-form mathematical pricing formulas are known.

While for the Heston model (semi) closed-form solutions exist for simple products, their evaluation involves complex functions and infinite integrals. So far these integrals can only be solved with time-consuming numerical methods. However, over the time, more and more theoretical and practical subtleties have been revealed for doing this and today a large number of possible approaches are known. Examples are different formulations of closed-formulas and various integration algorithms like quadrature or Fourier methods. Nevertheless, all options only work under specific conditions and depend on the Heston model parameters and the input setting.

In this chapter we present a methodology how to determine the most appropriate calibration method at run time. For a practical setup we study the available popular closed-form solutions and integration algorithms from literature. In total we compare 14 pricing methods, including adaptive quadrature and Fourier methods. For a target accuracy of 10^{-3} we show that static Gauss-Legendre are best on Central Processing Units (CPUs) for the unrestricted parameter set. Further we show that for restricted Carr-Madan formulation the methods are 3.6× faster. We also show that Fourier methods are even better when pricing at least 10 options with the same maturity but different strikes.

G. Liu (✉) • C. Brugger • C. De Schryver • N. Wehn
Microelectronic Systems Design Research Group, University of Kaiserslautern, Kaiserslautern, Germany
e-mail: gliu946@gmail.com; brugger@eit.uni-kl.de; schryver@eit.uni-kl.de; wehn@eit.uni-kl.de

C. De Schryver (ed.), *FPGA Based Accelerators for Financial Applications*,
DOI 10.1007/978-3-319-15407-7_10

10.1 Introduction

Simulating the future market behavior is a basic need for many financial applications such as product pricing, risk assessment and -management, or trading. It is obvious that the meaning of simulated asset prices strongly depends on the correct parameterization of the underlying models. Fitting the models against the market observations is called *model calibration*. For details refer to Chap. 2 by Sayer and Wenzel.

Calibrating their models against the market is one of the most compute-intensive tasks for financial institutes and can occupy a big compute cluster for several hours every night. The result is that the achieved accuracy is in general limited by the available compute time, and the energy consumed for calibration in total is immense. Especially sophisticated models with stochastic or local volatility processes, such as Heston [10], incorporate a number of parameters that need to be adjusted carefully to the current market observations. For those models, usually plain vanilla call and put options for which (semi) closed-form solutions exist are employed in the calibration process (see Sect. 10.3). Due to its popularity in business, we focus on the Heston model in this work.

Because of the high computational effort, many institutes run overnight calibration jobs based on local optimization techniques and gauge their settings once a week with a global optimizer run. However, most of the compute time is not spent in the optimizer itself, but in the evaluation of the cost function (i.e. the vanilla option prices, see Table 10.1). The reason is that although analytical formulas exist, they come with a number of substantial difficulties in general [1]:

- All available formulas contain infinite complex integrals that can only be solved numerically.
- The shapes of the characteristic functions strongly depend on the given parameters, what makes it very hard to pick suitable integration methods a priori.
- The characteristic functions may show discontinuities like jumps or peaks that need to be taken into account.

In addition to the different formulas, we can apply various approximation methods for the quadrature itself (non-adaptive vs. adaptive/truncate vs. domain transformation). However, not every method is optimal for every formula and an appropriate choice even depends on the actual Heston setting. This results in a large parameter space we need to navigate through in order to pick the best available setup for a specific set of inputs.

Table 10.1 Calibration time breakdown

	Time (s)	Percent (%)
Heston prices P^{Heston}	72.1	99.5
Cost function overhead $\sum_i [\ldots]^2$	0.2	0.3
Optimizer overhead (Simplex)	0.1	0.2
Total	72.4	100.0

A number of publications are available that investigate theoretical and practical aspects of implementing the Heston pricing formula for calibration purposes, both for classical quadrature and Fast Fourier Transform (FFT) methods. Unfortunately, to the best of our knowledge, a systematic study for comparing different approaches is not established and it is hard to determine which methods are most suitable under specific market conditions, e.g. for finding the best method to price with a given accuracy on a specific platform. Furthermore, directly comparing the results from literature is not possible in most cases due to disparate settings and target accuracies.

This chapter summarizes our findings first presented at the 7th Workshop on High Performance Computational Finance (WHPCF '14) in November 2014 [5]. We show a comprehensive analysis of available solutions and their computational complexity on CPUs. Fur this purpose, we introduce an objective methodology for determining the fastest pricing method in a calibration process that meets a specific predetermined accuracy. We apply this methodology to in total 14 pricing method and provide throughput results for one well-defined setting with a given accuracy of 10^{-3}. In addition, we quantitatively study the benefits and drawbacks of FFT methods for Heston calibration and compare them to classic quadrature methods.

10.2 Related Work

Pricing vanilla options in the Heston model is the subject matter of many publications. However, most contributions are from the financial mathematics community and focus on theoretical considerations. These aspects are covered in Sect. 10.3. As a result, only a few number of papers deal with numerical accuracies, implementation aspects of executing Heston calibration tasks on CPU or General Purpose Graphics Processor Unit (GPGPU) systems, or include runtimes. This section summarizes the latter category.

In 1999, Carr and Madan have compared their novel FFT approach to classic solutions and could show that it is up to 45 times faster on a standard CPU for one specific setting [6]. They claim that the FFT method should be used whenever a closed-form solution for the underlying characteristic function is available.

Aichinger et al. have presented a method that combines global and local optimizer techniques for Heston calibration in 2011 [1]. They provide runtimes for (multi-core) CPUs and Graphics Processor Units (GPUs). Bernemann et al. have achieved a speedup of around 70 compared to an Intel Xeon E5620 CPU by accelerating accelerating the Gauss-Laguerre integration method on an Nvidia Tesla C1060 GPU [4]. Unfortunately, further descriptions of the employed setting are missing, in particular which error measure has been used.

The most comprehensive study we have seen up to now has been made by Schmelzle in 2010 [17]. He gives a very good overview about available analytical formulas and integration methods. In his work, Schmelzle provides CPU runtimes for FFT based approaches and adaptive quadrature methods, both for Root Mean Squared Errors (RMSEs) down to 10^{-15} (however, we are not sure if this level of accuracy is relevant for practical applications). Nevertheless, he only examines one very specific Heston parameter set that seems to come with a rather smooth characteristic function. In this work he shows that the shape of this function can become very challenging for quadrature methods for a number of parameter constellations. Furthermore, Schmelzle only considers domain transformation for dealing with the infinite integrals, while we investigate the performance of finite upper integration borders.

10.3 Background

This section summarizes available formulas and integration methods from literature, focusing on the Heston model. For a general introduction into calibration and its challenges refer to Chap. 2 by Sayer and Wenzel in this book.

10.3.1 Heston Model

The Heston model is a mathematical model used to price products on the stock market [10]. Nowadays, it is widely used in the financial industry [15]. One main reason is that the Heston model is complex enough to describe important market features, especially volatility clustering [12]. At the same time, closed-form solutions for simple products are available. This is crucial to enable calibrating the model against the market in realistic time.

Under the risk-neutral measure, the Heston model consists of two correlated Stochastic Differential Equations (SDEs). Equation (10.1a) describes the asset price process S that matches the Black-Scholes SDE, except for the non-constant volatility $\sqrt{\nu_t}$ driven by its own SDE (10.1b).

$$dS_t = S_t r dt + S_t \sqrt{\nu_t} dW_t^S, \tag{10.1a}$$

$$d\nu_t = \kappa(\theta - \nu_t)\,dt + \sigma\sqrt{\nu_t} dW_t^{\nu}. \tag{10.1b}$$

The two driving Brownian motions W^S and W^{ν} are correlated with the correlation coefficient ρ to reflect the observable volatility clustering effect of the market.

Today's price of a derivative can be calculated as $P = \mathbb{E}g(S)$ where g is a corresponding discounted payoff function. For some products closed form solutions exist for P as we will see later.

The parameters of the Heston model are the initial variance ν_0, the reversion rate κ, the long-term average volatility θ, the volatility of the volatility σ, and the correlation ρ. The initial asset price S_0 and the riskless rate r are provided from market data.

10.3.2 Calibration

While the initial asset price S_0 and r can be observed directly at the market, all other parameters need to be found with a so-called *calibration* process. For this, products based on the the same underlying S can be used. Calibration is often restricted to the vanilla put and call options that are currently traded, i.e. products for which closed-form solutions exist. Call and put options e.g. give the buyer the right to buy (call) or sell(put) the underlying asset for a predefined price K (strike) and date T (maturity).

Calibration in general is an optimization problem. For the given market situation it tries to find the Heston parameters that describe the currently observed product prices in the most optimal way: Assuming we observe N vanilla call options for an underlying S with maturities T_i, K_i, and market prices C_i^{Market}, the calibration problem can be stated as:

$$\min_{v_0,\kappa,\theta,\sigma,\rho} \sum_{i=1}^{N} \left[C^{\text{Heston}}\left(v_0,\kappa,\theta,\sigma,\rho;K_i,T_i\right) - C_i^{\text{Market}}\right]^2.$$

The result of the calibration process are the five Heston parameters: $v_0, \kappa, \theta, \sigma, \rho$.

When calibrating to market data it is very common to have several options for the same maturity T (e.g. 1 year) with several strike values K. We will see later that the FFT method can benefit from this.

Table 10.1 shows the runtime breakdown of one calibration process with a downhill simplex optimizer on a dual socket Intel Xeon X5660 server. It is important to state that most institutes are interested in calibrating complete books with thousands of positions. We can see that 99.5 % of the calibration time is spent in evaluating the Heston prices itself. That means for a fast calibration an efficient pricer implementation is crucial, while the optimizer itself can be neglected. In the next section we discuss various closed-form solutions that are available to calculate these option prices. The methods are outlined in Fig. 10.1.

10.3.3 Closed-Form Solutions for C^{Heston}

Already with its introduction in 1993, the Heston model was known as one of the few stochastic models that provide closed-form solutions. Heston himself presented a closed-form solution for vanilla call and put options in his paper [10]. We call this the *Original Heston Formula*.

10.3.3.1 Heston Trap Formula

Albrecher et al. have theoretically shown that the original formulation of the characteristic function as given by Heston himself [10] comes with instabilities

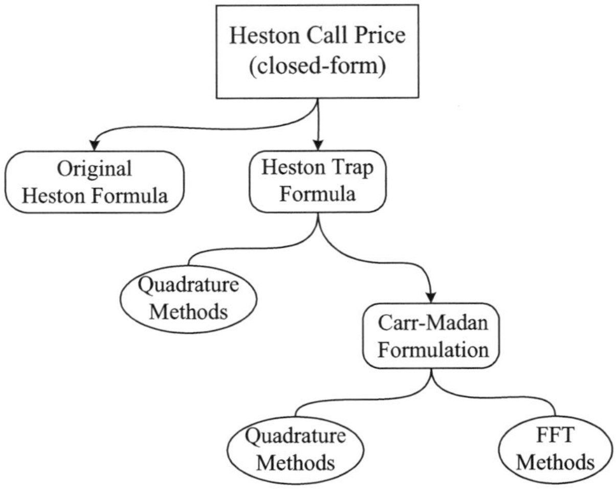

Fig. 10.1 Available closed-form approaches to determine the price of a call option in the Heston model. While the original Heston formula showed instabilities, the Heston trap formula is free of such limitations and can be integrated with quadrature methods. Under certain conditions, the Carr-Madan formulation provides a faster approximation and allows the use of FFT methods

due to a branch cut of the complex logarithm [2]. They have proven that a second version used by Schoutens-Simons-Tistaert [18] or by Gatheral [8] is stable for the complete parameter space. It is a variation of the original Heston formula, which we call *Heston Trap Formula*. The characteristic function in that case looks like:

$$\begin{aligned}
\phi(v,t) &= \exp\left(iv(\log S_0 + rt)\right) \\
&\quad \times \exp\left(\theta\kappa\sigma^{-2}\left((\kappa - \rho\sigma iv - d)t - 2\log\frac{1-ge^{-dt}}{1-g}\right)\right) \\
&\quad \times \exp\left(v_0{}^2\sigma^{-2}(\kappa - \rho\sigma iv - d)\frac{1-e^{-dt}}{1-ge^{-dt}}\right), \\
g &= (\kappa - \rho\sigma iv - d)/(\kappa - \rho\sigma iv + d), \\
d &= \sqrt{(\rho\sigma iv - \kappa)^2 + \sigma^2(iv + v^2)}.
\end{aligned}$$

The Heston call price can be calculated by integrating the characteristic function as following:

$$\begin{aligned}
c(K,T) &= \frac{1}{2}\left(S_0 - e^{-rT}K\right) + \frac{1}{\pi}\int_0^\infty \left(f_1(v) - Kf_0(v)\right)\mathrm{d}v, \\
f_a(v) &= \mathrm{Re}\left[\frac{e^{-iv\log K}\phi(v - ia;T)}{iv}\right].
\end{aligned} \tag{10.2}$$

10.3.3.2 Carr-Madan Formulation

Carr and Madan have introduced a new method to calculate the Heston call price: They Fourier-transform the option price. The huge benefit of this formulation is that the integrated function $h(v)$ does not depend on K. This makes the pricing of options with different strike K but same maturity T very efficient. The expression for c is in this case:

$$\begin{aligned} c(K,T) &= \frac{e^{-\alpha \log K}}{\pi} \int_0^\infty e^{-iv\log K} h(v)\,\mathrm{d}v, \\ h(v) &= \frac{e^{-rT}\phi(v-(\alpha+1)i;T)}{\alpha^2+\alpha-v^2+i(2\alpha+1)v}. \end{aligned} \tag{10.3}$$

In order to make these integrals finite they have introduced the *damping factor* α. α can be chosen with the following different strategies: Lord and Kahl presented a methodology to find an optimal α for each specific setting [14]. This method of searching has to be done before each pricing and for us took much longer than the pricing itself, rendering this strategy useless. In the literature one can find two popular choices for α: $\alpha = 1/4\,\alpha^+$ as suggested in [5] or a fixed $\alpha = 0.75$ like in e.g. [11, 18]. Ng has demonstrated that the first choice leads to bad results in many cases and suggests to use $\alpha = 0.75$ [16]. This is in line with our investigations, therefore we fix $\alpha = 0.75$ for all of our settings.

In addition, the Carr-Madan formulation requires that the characteristic function is finite at the origin [6]. Lee has derived an upper and lower bounds for α for the case $\rho < \kappa/\sigma$ [13]. The case $\rho \geq \kappa/\sigma$ has also been studied in several papers, nicely summarized in Gatheral and Jacquier [9]. They show that many problems may arise in this second region. But due to lack of theoretical proofs it is not clear when the Carr-Madan formulation can be safely applied here. Based on that we have decided to use the Carr-Madan formula only in the first region:

$$\begin{aligned} \rho < \frac{\kappa}{\sigma} \quad &\text{and } \alpha^- < \alpha < \alpha^+ \;\Rightarrow\; \phi((\alpha+1)i;T) < \infty \\ \alpha^\pm &= -\frac{\sigma - 2\kappa\rho \mp \sqrt{\sigma^2 - 4\sigma\kappa\rho + 4\kappa^2}}{2(\rho^2-1)\sigma} - 1 \end{aligned} \tag{10.4}$$

While the main benefit of the Carr-Madan formulation is that the integral can be solved with FFT methods, the integral in Eq. (10.3) can also be evaluated with quadrature methods. We will therefore study both approaches in the following.

Both the Heston trap formula and the Carr-Madan formulation require the numerical evaluation of infinite integrals. In the next section we discuss numerical integration methods. Figure 10.1 summarizes the closed-form solution formulas and available numerical methods to solve them.

10.3.4 Integration Methods

A vast amount of integration methods exists for evaluating the integrals we see in the closed-form Heston call price formulas. However, making one particular choice has a high impact on the accuracy and runtime of the pricing methods. Furthermore, it is hard to tell a priori which methods are suitable for the closed-form solutions due to the complex behavior of the integrals. Issues that may arise are oscillations, high peaks, and long tails as shown in Fig. 10.3. For that reason, we evaluate all the standard approaches in this work, including fixed, adaptive quadrature, and FFT methods as outlined in Fig. 10.2.

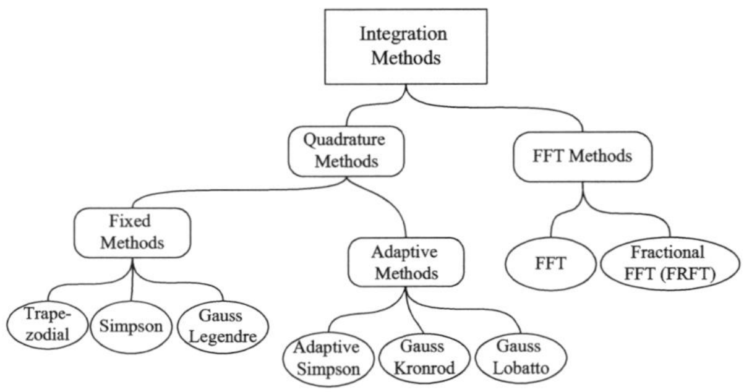

Fig. 10.2 Considered integration methods for the closed-form solutions of the Heston call price. They include fixed and adaptive quadrature methods and FFT methods

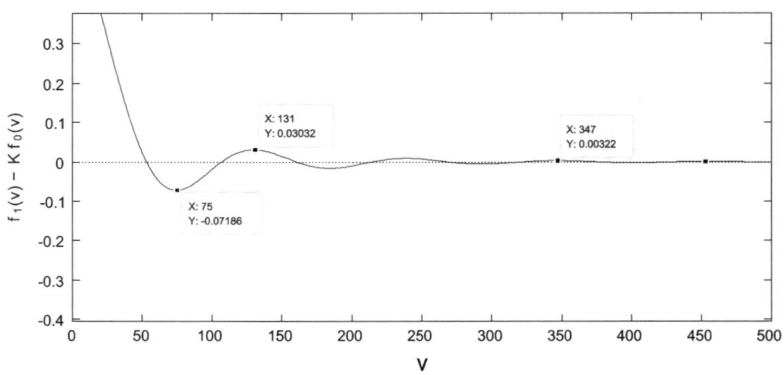

Fig. 10.3 The integrand of the Heston trap pricing formula (10.2). In this case the function has a specifically long tail. The parameters are: $S_0 = 138, K = 150, T = 0.62, r = 0.037, \kappa = 1.17, \theta = 0.068, \sigma = 7.9, \rho = -0.45, v_0 = 0.0027$

10.3.4.1 Fixed Quadrature Methods

Quadrature methods are algorithms for calculating the numerical value of integrals based on interpolating functions. Fixed quadrature methods evaluate the integration domain at predefined points.

The most basic one is the trapezoidal approach. It evaluates the integrand on a uniformly spaced grid and approximates the region under the graph as a trapezoid.

The Simpson method also operates on the a uniform grid, but uses three points effectively approximating the area with polynomials of degree two. This so-called *quadratic integration* may lead to a better approximation of the integrand and furthermore provides and error estimate for each section which is required for adaptive methods.

Gaussian quadrature methods provide the best numerical estimates by picking optimal evaluation points. However, this results in unequally spaced points in general. With this freedom it is possible to approximate polynomials of degrees up to $2n-1$ with only n evaluations of the integrand. The standard Gauss quadrature method uses *Legendre* polynomials. It is a widely used method for numerical integration.

10.3.4.2 Adaptive Quadrature Methods

Adaptive quadrature methods approximate the integrand while adaptively refining subintervals of the integration domain where needed.

In the first step they make an estimation over the full interval with a static quadrature rule based on a few points only (e.g. five). Then they estimate the error for each interval. If the estimated error is larger than the tolerance τ, the interval is subdivided into two halves. For each half the procedure is repeated until the tolerance is reached.

A requirement for adaptive methods is the availability of a good error estimate. The *adaptive Simpson method* is the first approach we are considering here. It shows two important beneficial characteristics: It provides an adequate error estimate and subintervals that can reuse the evaluation points of the previous iteration.

Because of the irregular grid of standard Gauss methods it is not possible to reuse points for subintervals in general. One approach for overcoming this issue is the *Gauss-Kronrod extension*. It extends the Gauss-Legendre method by evaluating additional points. With this solution a more accurate approximation can be computed by re-using existing points. However, one side effect is that for the same points the Gauss-Kronrod extension provides a second quadrature rule of higher order. Nevertheless, the difference with the original one can be taken as a good error estimate for the adaptive method.

The Gauss-Lobatto quadrature is an alternative to the original Gauss-Legendre quadrature. In its formula it includes the points at the border of the integration region. This method can be nested easily at the cost of sacrificing some accuracy.

10.3.4.3 FFT Methods

The call price c in the Carr-Madan formulation in Eq. (10.3) is the Fourier integral of $h(v)$ with respect to $k = \log K$. It lends itself the application of the FFT.

Applying the Simson quadrature rule to the integral in Eq. (10.3) we can rewrite it as:

$$c(k_u) = \frac{e^{-k_u\alpha}}{\pi}\mathrm{Re}\,[X_u],\; k_u = \frac{\pi}{\beta}\left(\frac{2u}{N} - 1\right), \tag{10.5a}$$

$$X_u = \sum_{n=0}^{N-1} x_n e^{-i2\pi u\frac{n}{N}},\; u \in \{0,..,N-1\}, \tag{10.5b}$$

$$x_n = e^{i\pi n}\, h(n\beta)\, \beta\, \frac{3+(-1)^{n+1} - \delta_{n0}}{3}. \tag{10.5c}$$

It is worth noting that Eq. (10.5b) can be efficiently solved by an FFT within $O(N\log N)$ for all u when N is a power of two. Furthermore, once we have the components X_u, prices for several options with different strike K but the same maturity T can be calculated very efficiently through interpolation.

The integration grid is equally spaced out in $logK$ and v. However, in order to evaluate the price of a call option with a specific strike K it has to be interpolated in the $C(k_u)$ vector. In literature linear and cubic spline interpolation schemes are mentioned that we both consider here.

The FFT method effectively integrates the infinite integral in Eq. (10.3) up to $v_{\max} = (N-1)\beta$. The spacing in the resulting strike price vector is given as $\Delta k = 2\pi/(\beta N) = O\left({v_{\max}}^{-1}\right)$. For a given $v_{\max}$ it is therefore not possible to decrease the spacing by increasing N.

The Fractional Fourier Transform (FRFT) methods overcome this limitation by choosing a different expression for k_u [3, 7]. As the name suggests they only evaluate a fraction of the whole $\log K$ space that is chosen by the user.

Due to numerical instabilities for the FFT methods we needed to introduce some checks that the problem is not ill-conditioned. Although by condition (10.4) it is guaranteed that the initial point is bounded, we check whether it lies in a reasonable numeric range. When sampling ϕ it starts with a positive real component and then oscillates around 0. When this oscillation happens too fast, the FFT method will always fail. We therefore introduce the check:

$$\begin{aligned} &\mathrm{Re}\,[\phi(n\beta - (\alpha+1)i;T)] > 0 \quad \forall\, n \in \{0,..,5\}, \beta = 0.02 \\ &\text{and}\;\; \phi((\alpha+1)i;T) < 10^{16}. \end{aligned} \tag{10.6}$$

These additional conditions filter out less than 1 % of the parameters for our setup we introduce later.

10.4 Methodology

Calibration is a time consuming task. Although having closed-form solutions for the model prices, we have seen that they are the bottleneck for calibration from Table 10.1.

We have also seen that all closed-form solutions require the use of numerical integration methods. They only approximate the result and are not exact. In addition, they have parameters like the number of integration points N to increase the accuracy with additional evaluations. Various methods exists with varying complexity and accuracy, with a huge design space.

In this section we present a methodology for finding the fastest pricing method in an automatic fashion. For a fair comparison it is crucial that we compare the runtime of methods with the same accuracy. In our point of view it does e.g. not make sense to compare an adaptive Gauss method that is able to reach $\varepsilon = 10^{-8}$ with a FRFT method that only achieves $\varepsilon = 10^{-2}$. A big part of our methodology will be on how to tune the parameters of our methods such that they fulfill the required target accuracy without exceeding it.

Additionally, we think it is crucial to test the various methods for many different pricing parameters, i.e. the Heston parameters $r, S_0, v_0, \kappa, \theta, \sigma, \rho$ and the option parameters maturity T and moneyness $M := S_0/K$. In Fig. 10.3 we have seen that the parameters can have a huge impact on the shape of the curve. That means: If we would tune our methods to one specific curve, they might perform vary badly for different parameters. In order to make our method robust and objective we instead sample our parameters from a distribution. This allows us to generate thousands of pricing settings and to derive conclusions for the average case.

To make the result as meaningful as possible the distributions should be identical to how likely they are picked by the optimizer during a real calibration processes.

We define our error ε to be the RMSE of the method under consideration $\hat{c}$ compared to the true prices c. Our methods for approximating the true prices $\hat{c}$ are no fixed, but parameterized. The fixed quadrature methods e.g. have two parameters that influence the runtime and accuracy: The maximum integration bound $v_{\max}$ and the number of discretization points N. To distinguish between all these estimators we use $\hat{c}^N_{v_{\max}}$. We assume that for high N and $v_{\max}$ our estimators are unbiased, meaning that we captured all errors:

$$\varepsilon^2 = \text{MSE} = \text{e}\left(\hat{c}^N_{v_{\max}} - c\right)^2 \xrightarrow[v_{\max}\to\infty]{N\to\infty} 0.$$

By splitting up the errors as following we can influence the error on the parameters independently:

$$
\begin{aligned}
\mathrm{e}\left(\hat{c}^{N}_{v_{\max}} - c\right)^2 &= \mathrm{e}\left(\hat{c}^{N}_{v_{\max}} - \hat{c}^{N=\infty}_{v_{\max}} + \hat{c}^{N=\infty}_{v_{\max}} - c\right)^2 \\
&= \underbrace{\mathrm{e}\left(\hat{c}^{N}_{v_{\max}} - \hat{c}^{N=\infty}_{v_{\max}}\right)^2}_{\text{Discretization Error}} + \underbrace{\mathrm{e}\left(\hat{c}^{N=\infty}_{v_{\max}} - c\right)^2}_{\text{Range Error}} \\
&= \left(\Delta_{\text{discr.}}\right)^2 + \left(\Delta_{\text{range}}\right)^2
\end{aligned}
$$

While the *discretization error* $\Delta_{\text{discr.}}$ depends here on the step size $h = v_{\max}/N$ and also in general on both N and $v_{\max}$, the *range error* Δ_{range} only depends on $v_{\max}$. This has two important consequences: First, we can find a proper choice for $v_{\max}$ through $\Delta_{\text{discr.}}$ independently of N, and based on that estimate a proper choice for N through Δ_{range}. Second, as $\hat{c}^{N=\infty}_{v_{\max}}$ is independent of the integration scheme, it is sufficient to estimate one $v_{\max}$ for all quadrature methods operating on the same integrand.

Our adaptive quadrature methods have as parameters $v_{\max}$ and the tolerance τ while our FFT methods depend on $v_{\max}$, N, and the selected interpolation method. For our adaptive methods the derivation is analog to the fixed method, and we call the error due to the tolerance *tolerance error*. For the FFT methods we have an additional *interpolation error*.

To find the fastest method that has a RMSE of ε under the considered setting we therefore follow the following procedure:

1. Find the optimum $v^*_{\max}$ such that $\Delta_{\text{range}} = \varepsilon/\sqrt{2}$ for each integral separately.
2. For this $v^*_{\max}$ find optimum N^* or τ^* such that $\Delta_{\text{range}} = \varepsilon/\sqrt{2}$ for each method and integral.
3. Measure the runtime of all methods with the found optimal parameters $v^*_{\max}, N^*, \tau^*$ and pick the fastest.

In the next section we will apply this novel methodology to a well defined setting for all of our 14 numerical methods.

10.5 Evaluation Setting

In our work we model the distribution of the parameters with nine marginal distributions. We have extracted those distributions from our experience with market data that we have seen over the time. The probability density functions of the distributions are shown in Fig. 10.4. They have the following exact form:

$$
\begin{aligned}
&r \sim 0.05\ \text{Beta}(\alpha = 1, \beta = 3),\ \rho \sim 2\ \text{Beta}(2,5) - 1, \\
&S_0 \sim \text{unif}(50, 150),\ v_0 \sim 9.999\ \text{Beta}(1,3) + 0.001, \\
&\kappa \sim 9.9\ \text{Beta}(1,3) + 0.1,\ \theta \sim 4.999\ \text{Beta}(1,3) + 0.001, \\
&\sigma \sim 9.9\ \text{Beta}(1,3) + 0.1,\ T \sim 4.8\ \text{Beta}(1,3) + 0.2, \\
&M/T \sim \text{unif}(0.8, 1.2).
\end{aligned}
$$

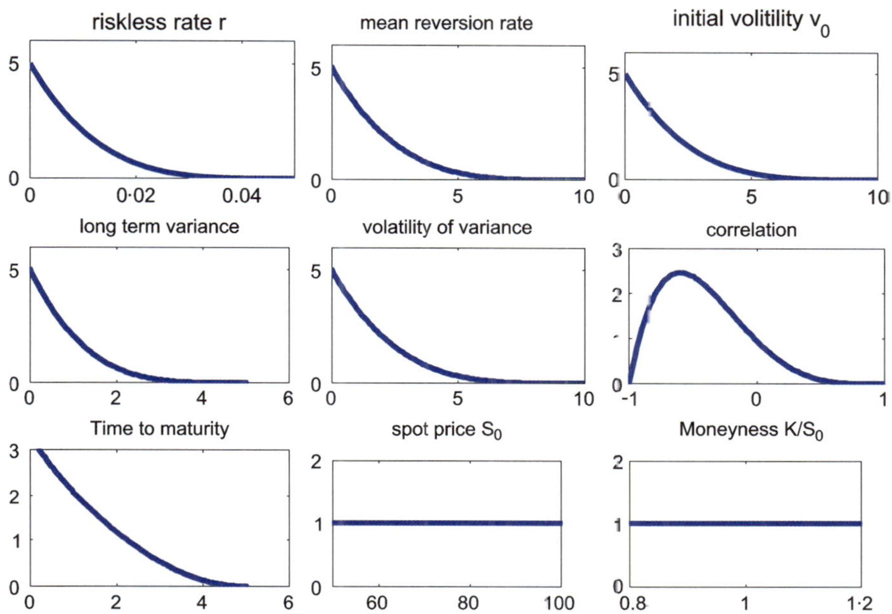

Fig. 10.4 Showing marginal x density functions of the parameter as defined in our setup. We sample from these distributions to calculate RMSE errors and assure that our methods work for many parameters

For example, we have researched the market data and observed that most maturities are within 1 year. However, there exist still some long term maturities T from 6 years up to 20 years. Since for practitioners the most interesting strike prices lie within a range of 20 % difference from the current spot price, we have set the moneyness $M = S_0/K = [0.8, 1.2]$. With the help of the calibration results, we have derived a meaningful parameter range for the Heston model. Based on that we have built appropriate Random Number Generators (RNGs) by using the suitable beta distributions from Fig. 10.4.

We have set our target accuracy to $\varepsilon = 10^{-3}$. Although this might sound very coarse from a scientific point of view it is a sufficient choice for most practical calibration settings (in many cases it doesn't make any sense to calibrate more accurate than 1 cent).

As a reference method c to obtain the correct prices we have integrated the Heston Trap formula with an adaptive Gauss-Kronrod method. For that, we have chosen a tolerance $\tau = 10^{-6}$ and upper integration bound $v_{\max} = 10{,}000$. We have ensured that the prices converge for this setting and do not change for higher τ or $v_{\max}$.

For our FRFT methods we have generated our target grid for K from 80 % to 120 % of the spot price S_0, making it possible to interpolate in this regime. For the

Fourier methods all the plots have been made with cubic interpolation, we will later comment on linear interpolation.

We have implemented all of our methods in Matlab. For the quadrature formulas and FFT we have use the efficient library functions of Matlab itself. We have evaluated each RMSE based on 10,000 parameter samples from the distribution and used the same parameters for all methods when selecting the optimal parameters of the methods $v^*_{\max}, N^*, \tau^*$. However, for the final runtime and accuracy test we have drawn a second independent set of parameters. We have run our test on an Intel Core i5-460M CPU. The runtimes are for one CPU core only.

10.6 Results

In this section we present the results of applying our methodology for finding the fastest integration method to our setting for the 14 pricing methods summarized in Sect. 10.3.

10.6.1 Optimal Integration Bounds v^*_{max}

The first step of our methodology is to estimate the optimal integration bound $v^*_{\max}$ for all of our integrals. In the introduction we have introduced the two pricing formulas (10.2) and (10.3). For the method used to evaluate the range error $\hat{C}^{N=\infty}_{v_{\max}}$ we have employed the adaptive Gauss-Kronrod with a tolerance of $\tau = 10^{-6}$. We have ensured that this method does not have any discretization error by checking wether the prices change for higher tolerance τ. The results are shown in Fig. 10.5.

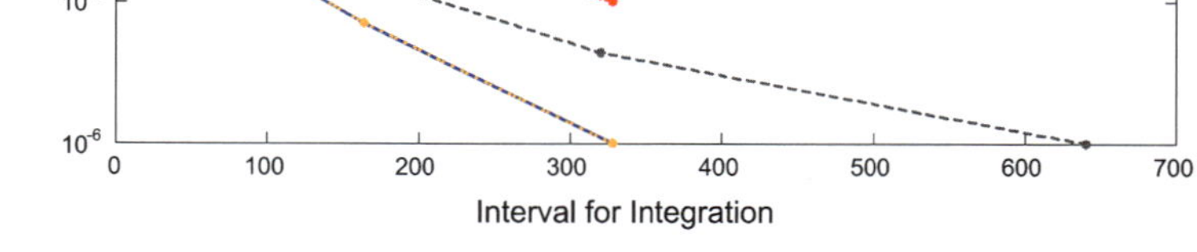

Fig. 10.5 Range error Δ_{range} over the integration upper bound $v_{\max}$ for different integrals. The points where the curves cross the target accuracy $\varepsilon/\sqrt{2}$ define the optimal integration range $v^*_{\max}$

We can see that the Heston trap integral leads to higher errors and needs a higher integration bound. The reason is that the Carr-Madan integral can only be applied for some of the parameters, (approximately 80 % for our setting). These parameters fulfilling the conditions (10.4) seem to converge earlier than when pricing the full set.

We have repeated this study for the FFT and FRFT method. As they are both based on the Carr-Madan formula (10.3) we would expect that the curves are identical to the above test. While this is true for the FRFT method with $N = 32{,}768$, we see a higher error for the FFT method with the same N. The reason is that although we can eliminate the discretization error $\Delta_{\text{discr.}}$ by choosing a big N we have seen before that increasing N will not effect the spacing in the resulting result vector. The effect is that we have a significant interpolation error in this cases and for this reason the curve is higher.

10.6.2 *Optimal Settings N^* and τ^**

From Fig. 10.5 we can read of $v^*_{\max}$ for the different integrals. With these intervals we can tune the discretization error of our methods.

10.6.2.1 Heston Trap Formula

The results for the Heston trap formula (10.2) are shown in Fig. 10.6. We can see that the static quadrature methods have significant difficulties to reach the target accuracy. Comparing trapezoidal to Simpson we can see almost no difference. Although the Simpson method should provide a better theoretical convergence, we cannot observe this in our setting. Only the Gauss-Legendre method reaches the pre-given accuracy. This is due to the shape of the curve and the large integration range of $v_{\max} = 188$. In Fig. 10.3 we see that the integrand has a peak at the beginning and then quickly decays. The fixed methods are just not capable of capturing enough details of this peak since their grid is spaced out evenly over the whole integration range.

The adaptive quadrature methods can reach the accuracy with ease. It is interesting to see that while Gauss-Kronrod methods reach our target already with a tolerance of $\tau = 1$, the adaptive Simpson rule requires a accuracy of $\tau = 10^{-5}$. However, since both are using very different methods to estimate the error this is not surprising. What we should take away from this is that although the name suggests it the tolerance of the integration has little to do with the RMSE of the pricing method. E.g. setting a τ to your target precision, as some paper do, is very risky and might not lead to the desired results.

10.6.2.2 Carr-Madan Formulation

The results of integrating the Carr-Madan formula (10.3) are shown in Fig. 10.7. This time the static quadrature methods can reach the target accuracy easily. This is mainly due to the drastically decreased integration interval of $v_{\max} = 73$.

In addition, the adaptive methods are able to reach the target without a lot of refinements. We can tell this from the almost flat decay for all the methods.

For the Fourier methods we can see that they also easily reach a very low accuracy. At the beginning the FRFT method is clearly better. However, in order to be able to apply the FFT, N needs to be restricted to $2^m, m \in \mathbb{N}$. Although the FRFT is almost reaching the target accuracy for $N = 1{,}024$ we have to take the next higher $N = 2{,}048$ for our final method. We will later see that this might be one reason why the FRFT methods are slower in our specific setup compared to FFT methods.

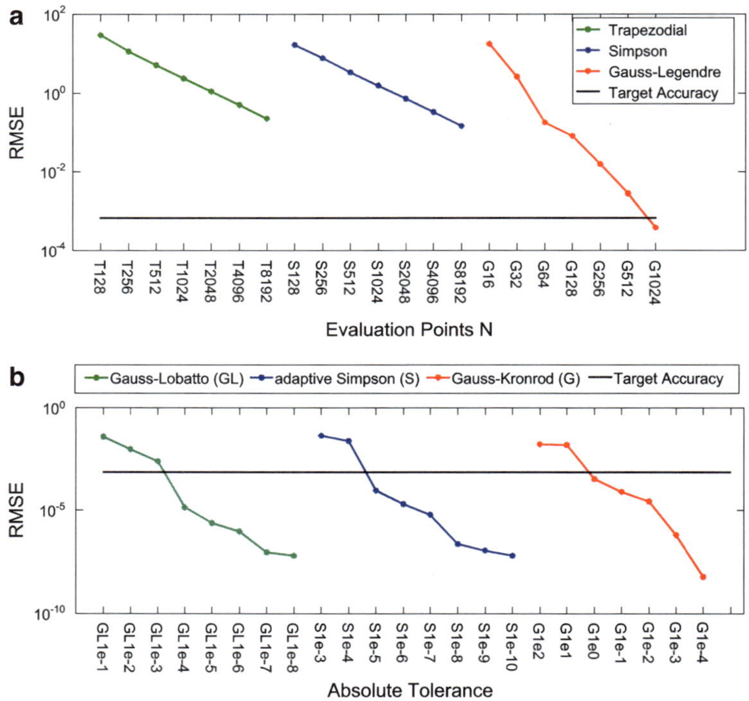

Fig. 10.6 Heston trap formula (10.2) integrated with quadrature methods. The points where the curves cross the target accuracy $\varepsilon/\sqrt{2}$ define the optimal setting N^* or τ^* for the specific integration methods. (**a**) Static quadrature. (**b**) Adaptive quadrature

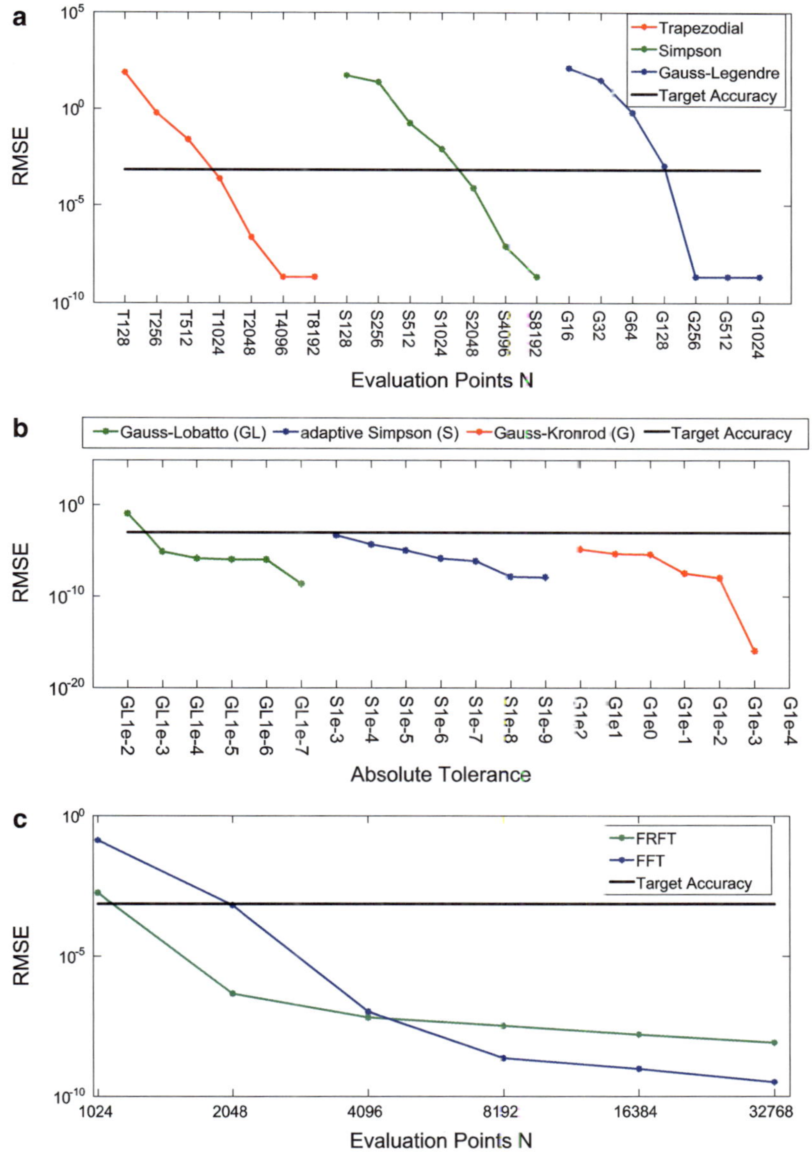

Fig. 10.7 Carr-Madan formula (10.3) integrated with quadrature and FFT methods. The points where the curves cross the target accuracy $\varepsilon/\sqrt{2}$ define the optimal setting N^* or τ^* for the specific integration methods. (**a**) Static quadrature. (**b**) Adaptive quadrature. (**c**) FFT methods

Table 10.2 Runtime and accuracy of pricing methods

Method	Setting	Time (ms)	Accuracy ($\times 10^{-3}$)
Heston Trap ($\nu_{\max} = 188$):			
Trapezodial	Any	Not able to reach	
Simpson	Any	Target precision	
Gauss-Legendre	1,024	1.2	2.9
ad. Gauss-Lobatto	10^{-4}	8.5	2.9
ad. Simpson	10^{-5}	2.7	2.9
ad. Gauss-Kronrod	10^{-3}	5.1	2.9
Carr-Madan ($\nu_{\max} = 73$):			
Trapezodial	1,024	1.2	0.25
Simpson	2,048	2.4	0.25
Gauss-Legendre	128	0.28	1.00
ad. Gauss-Lobatto	10^{-4}	8.5	0.25
ad. Simpson	10^{-5}	2.7	0.30
ad. Gauss-Kronrod	10^{-3}	5.1	1.00
FRFT - Grid	2,048	5.4	
Interpolation	Linear	<0.01	–
	Cubic	<0.01	0.10
Carr-Madan ($\nu_{\max} = 132$):			
FFT - Grid	2,048	2.5	
Interpolation	Linear	<0.01	–
	Cubic	< 0.01	1.00

10.6.3 Runtime Evaluation

Based on the optimal settings $\nu^*_{\max}, N^*, \tau^*$ we can now compare the 14 methods regarding their CPU runtime. Table 10.2 shows the setup we have chosen and the average time to price one option. We have also checked the accuracy of our methods on a second sample of our parameters. We can see that all methods are well within one order of magnitude of our target accuracy of $\varepsilon = 10^{-3}$. This shows the effectiveness of our methodology.

For the Heston trap formula, the static Gauss-Legendre formula is the fastest method with 1.2 ms and is able to price the full parameter range.

For the restricted parameter range (10.4) of the Carr-Madan formulation the situation is a bit more complex. For a single option, the fixed Gauss method takes a clear lead with 0.28 ms, far in front of all the adaptive methods. While the fixed methods had it quite hard for the Heston trap formula, they show their strength in the Carr-Madan case of being simple and efficient.

For the Fourier methods we can split the overall runtime into two parts: One to generate the price grid for all the strikes and a second for looking up prices in this grid by interpolation. When doing calibration it is common that we have many options with the same maturity T in many cases. However, for them we only have to generate the grid once and then only interpolate for each of the strikes K. From the numbers we can see that FFT methods are faster than FRFT methods for our setting. Linear interpolation showed no benefit in runtime, while being less accurate. That is why we have chosen cubic interpolation for our final FFT setting. Furthermore, we can conclude that the FFT method is the fastest method when pricing 10 or more options for different strikes K but the same maturity T.

It is apparent that the Carr-Madan methods are at least 3.6× faster than the fastest Heston trap method. This suggests to use a hybrid pricer that favors the Carr-Madan method as far as possible and employs the Heston trap method as a backup when needed. For our setup this would be the case in only 20 % of all scenarios.

10.6.4 A Unified IP Block for Heston Calibration on FPGAs

Regarding an appropriate hardware architecture for acceleration the (semi) closed-form call price formula, there's no need to build three individual methods for solving the Heston model. The reason is that the Heston trap formula and the Carr Mandan formula share some common parts. By reusing the shared part, a unify Intellectual Property (IP) block can be built for this case.

Let us take the integration function, for instance. The integration function units can be divided into two parts: One is the exponential part, another one is the non-exponential part. The exponential part contains the characteristic function part and the strike information part. In the non-exponential part, the residuary equations are located. In particular, the split is the following:

- Exponential part:
  - FFT: $\phi(v,t)$,e^{-ibv},e^{-rT}
  - Carr Mandan: $\phi(v,t)$,$e^{-ivlog(k)}$,e^{-rT},$e^{-\alpha log(k)}$
  - Heston Trap Quadrature: $\phi(v,t)$,$e^{-ivlog(k)}$,e^{-rT}
- Non-Exponential part:
  - FFT: $\frac{1}{\alpha^2+\alpha-v^2+i(2\alpha+1)v}$,$\beta \cdot W_{FFT}$
  - Carr Mandan: $\frac{1}{\alpha^2+\alpha-v^2+i(2\alpha+1)v}$, W_{Carr}
  - Heston Trap Quadrature: $\frac{1}{iv}$, W_{Trap}

where the W means the weighting matrix of the selected integration method. One advantage of this categorization is the shorter critical path and less resource utilization. Another benefit is the that shared resource part and replaceable block can be easily selected out. Based on shared resource methodology generates it the following: Exponential part:

- Shared part above all: characterized function $\phi(v,t)$
- Shared part of Carr Mandan and FFT : e^{-rT}
- Shared part of Carr Mandan and Heston Trap Quadrature: $e^{-ivlog(k)}$

Individual exponential block

- FFT : e^{-ibv}
- Carr Mandan: $e^{-\alpha log(k)}$
- Heston Trap Quadrature: none

Non-exponential part:

- Divisor part
  - FFT and Carr Mandan: complex divisor part $\alpha^2 + \alpha - v^2 + i(2\alpha + 1)v$
  - Heston Trap Quadrature: simple divisor part $\frac{1}{iv}$
- Weight Matrix block
  - FFT : $\beta \cdot W_{FFT}$
  - Carr Mandan: W_{Carr}
  - Heston Trap Quadrature: W_{Trap}

Figure 10.8 shows how each individual modules and shared blocks are placed in the scheduling table. The number of each methods are shown above the block, which means the block is shared with the corresponding methods. The exponential and non-exponential paths are computed separately. In the end, this two paths are multiplied and generate the integrate function unit $f_a\bar{(}v)$.

For example, if the system wants to execute the FFT method, corresponding blocks will be activated and react to the corresponding inputs of the hybrid pricer unit at certain locations. None common blocks and red blocks will remain deactivated. Both the exponential part and non-exponential part can execute simultaneously.

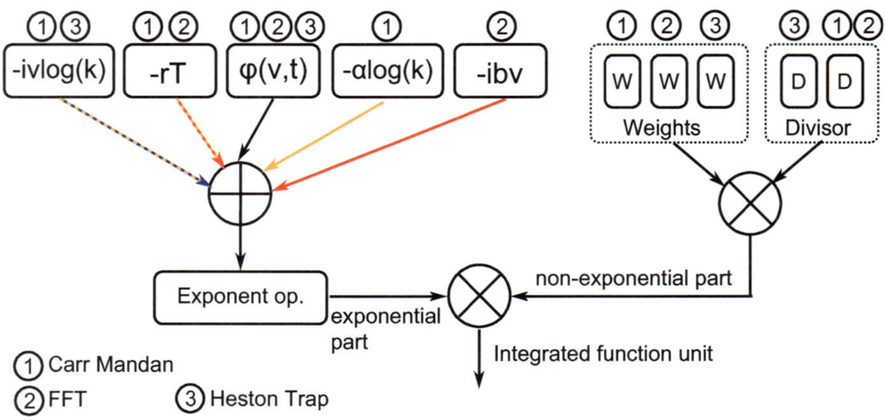

Fig. 10.8 Pricer unit scheduling

10.7 Conclusion

Calibration is a very computationally demanding problem for financial institutions. In this work we have shown that the actual challenge of calibration is not the optimization but evaluation the closed-form Heston pricing formula itself.

We have seen that there are actually two closed-form formulas, the Heston trap and the Carr-Madan formulation. Both formulations require numerical integration methods. We have identified 14 standard methods as possible pricing candidates.

Our main contribution is a methodology to find the fastest pricing methods within the Heston model that meets the desired target accuracy. With this novel approach we tune the parameters of the pricing methods independently by guaranteeing a given accuracy. The independent study of the errors makes the method robust and efficient. With our methodology it is now possible to adjust methods close to a specific accuracy. We have demonstrated this for fixed and adaptive quadrature and FFT methods.

We have applied our methodology for a practical setup with a target accuracy of $\varepsilon = 10^{-3}$. For our setup we have shown that Carr-Madan methods are $3.6\times$ faster than using the Heston trap formulation, although they can only be applied in around 80 % of the cases. We have figured out that fixed Gauss-Legendre methods are best for both the Carr-Madan and Heston trap formulation. When pricing multiple options simultaneously, FFT methods are best when pricing at least 10 options with the same maturity, but different strikes.

By applying our methodology to their setup, practitioners can now speed up their pricing systems by adjusting their systems to the accuracy they really need.

Acknowledgements We gratefully acknowledge the partial financial support from the Center of Mathematical and Computational Modelling $(CM)^2$ of the University of Kaiserslautern, from the German Federal Ministry of Education and Research under grant number 01LY1202D, and from the Deutsche Forschungsgemeinschaft (DFG) within the RTG GRK 1932 "Stochastic Models for Innovations in the Engineering Sciences", project area P2. The authors alone are responsible for the content of this work.

References

1. Aichinger, M., Binder, A., Fürst, J., Kletzmayr, C.: A fast and stable Heston model calibration on the GPU. In: Guarracino, M., Vivien, F., Träff, J., Cannatoro, M., Danelutto, M., Hast, A., Perla, F., Knüpfer, A., Di Martino, B., Alexander, M. (eds.) Euro-Par 2010 Parallel Processing Workshops, Ischia. Volume 6586 of Lecture Notes in Computer Science, pp. 431–438. Springer, Berlin/Heidelberg (2011)
2. Albrecher, H., Mayer, P., Schoutens, W., Tistaert, J.: The little Heston trap. Technical report 05, K.U. Leuven Section of Statistics, Sept 2006. Issue 05
3. Bailey, D.H., Swarztrauber, P.N.: The fractional Fourier transform and applications. SIAM Rev. **33**(3), 389–404 (1991)
4. Bernemann, A., Schreyer, R., Spanderen, K.: Accelerating exotic Option pricing and model calibration using GPUs. Available at SSRN 1753596 (2011)

5. Brugger, C., Liu, G., de Schryver, C., Wehn, N.: A systematic methodology for analyzing closed-form Heston pricer regarding their accuracy and runtime. In: Proceedings of the 7th Workshop on High Performance Computational Finance, WHPCF '14, Piscataway, pp. 9–16 (2014). IEEE
6. Carr, P., Madan, D.: Option valuation using the fast Fourier transform. J. Comput. Finance **2**(4), 61–73 (1999)
7. Chourdakis, K.: Option pricing using the fractional FFT. J. Comput. Finance **8**(2), 1–18 (2004)
8. Gatheral, J.: The Volatility Surface: A Practitioner's Guide, vol. 357. Wiley, Hoboken (2006)
9. Gatheral, J., Jacquier, A.: Convergence of Heston to SVI. Quant. Finance **11**(8), 1129–1132 (2011)
10. Heston, S.L.: A closed-form solution for options with stochastic volatility with applications to bond and currency options. Rev. Financ. Stud. **6**(2), 327 (1993)
11. Janek, A., Kluge, T., Weron, R., Wystup, U.: FX smile in the Heston model. In: Čižek, P., Hardle, W., Weron, R. (eds.) Statistical Tools for Finance and Insurance, pp. 133–162. Springer, Berlin/Heidelberg/New York (2011)
12. Korn, R., Korn, E., Kroisandt, G.: Monte Carlo Methods and Models in Finance and Insurance. CRC, Boca Raton (2010)
13. Lee, R.W.: Option pricing by transform methods: extensions, unification and error control. J. Comput. Finance **7**(3), 51–86 (2004)
14. Lord, R., Kahl, C.: Optimal Fourier inversion in semi-analytical option pricing. Technical report, Tinbergen Institute Discussion Paper (2006)
15. Lord, R., Koekkoek, R., van Dijk, D.: A comparison of biased simulation schemes for stochastic volatility models. Quant. Finance **10**(2), 177–194 (2010)
16. Ng, M.W.: Option pricing via the FFT and its application to calibration. PhD thesis, Master Thesis, Applied Institute of Mathematics, TU Delft (2005). [Online]. Available: http://ta.twi.tudelft.nl/mf/users/oosterle/oosterlee/ng.pdf. Last access: 31 Dec 2014
17. Schmelzle, M.: Option pricing formulae using Fourier transform: theory and application. Technical report (2010)
18. Schoutens, W., Simons, E., Tistaert, J.: A perfect calibration! Now what? Wilmott Mag. March. **576**, 66–78 (2004)

Chapter 11
Maxeler Data-Flow in Computational Finance

Tobias Becker, Oskar Mencer, Stephen Weston, and Georgi Gaydadjiev

Abstract Computational finance is an area that includes many algorithms in trading and analytics that are both computationally very complex and performance critical. As financial institutions intend to perform a steadily increasing number of computations and obtain the results as quickly as possible, computer systems are expected to satisfy these growing performance demands. However, recent years have brought the end of "free" processors speed-ups, and single-thread performance is no longer the driving force behind automatic performance gains enjoyed by the industry for many decades. Nowadays, high-performance computing systems have to increasingly rely on parallel programming models where the original application has to be modified to exploit many parallel cores. This requires considerable redesign efforts and yet, the desired performance improvements are not guaranteed. Some financial applications may also reach practical physical limits imposed by the space and power provisions available in the data centre. A solution to the above problems can be the use of custom accelerators implemented in reconfigurable hardware. Reconfigurable implementations can deliver both high computational throughput and low compute latency in addition to superior energy efficiency. However, porting applications for such devices requires a special skill set in hardware design, complicating their practical adoption. Maxeler Technologies offers conveniently programmable, high-performance computing systems and a software toolchain that exploit the sheer computational power of reconfigurable devices while abstracting the programming into a high-level data-flow model. Our vision is to empower domain experts with the necessary means to create highly customised, efficient hardware/software implementations for their specific applications. This approach enables vertical optimisations across the different layers of abstraction that are typically not exposed to an application designer. The final result is a productive application development process that often delivers speed-ups by orders of magnitudes over traditional CPU implementations.

T. Becker (✉) • O. Mencer • S. Weston • G. Gaydadjiev
Maxeler Technologies, London, UK
e-mail: tbecker@maxeler.com

C. De Schryver (ed.), *FPGA Based Accelerators for Financial Applications*,
DOI 10.1007/978-3-319-15407-7_11

11.1 Introduction

Computer technology has become an essential driver for the financial industry in almost all its areas. Advances in hardware and software technology, numerical methods, financial models and algorithms have made computers a key technology that is crucial to all financial institutions. High-performance computing systems are used to price financial products, to calculate risk, or to carry out electronic trades automatically by following sophisticated pre-programmed trading strategies. Often, the available compute power determines the types of problems that can be practically solved. Being able to solve a more complex problem or to obtain the results faster than other institutions directly translates into a competitive advantage.

Conventional computer architectures used in many areas of everyday life including mobile devices, desktop computers, and high-performance computer (HPC) systems generally follow the basic concepts of general-purpose processing [6]. Such processors perform calculations by executing a sequence of instructions that can either carry out arithmetic, control or IO operations. This model of execution is extremely flexible; however, it is also inherently sequential. Over many decades, the performance of processors has been improved by increasing the clock rates, but also by extending the basic processor architecture with complex structures to deal with issues like control divergence, main memory access penalties and to recover low-level binary instructions parallelism. Many micro-architectural innovations such as caches, branch prediction, out-of-order execution, and Single Instruction Multiple Data (SIMD) extensions were developed to alleviate the fundamental drawbacks of the general-purpose processor paradigm. This has led to modern processor architectures where only a tiny part of the chip area is used to perform useful calculations at very high speeds while the rest of the device is used for auxiliary functions such as caching of instructions and data. With the end of clock frequency improvements offered by the CMOS technology scaling, additional performance can now be only obtained through exploiting parallelism. Multi-threaded implementations on multiple cores or SIMD extensions are just two examples. However, the individual cores (or threads) still rely on a fundamentally sequential computing principle, i.e. performing a sequence of instructions. In addition, legacy applications have to be re-written, analysed and optimised in order to achieve satisfactory performance levels. Attempting to compute larger and larger problems by simply scaling over existing processor technology is no longer practical for many current and future HPC applications [11]. Even if performance requirements can be met by using a large number of machines, the cost, area and power requirements may exceed practical limits.

These limitations have led to an increased interest in special-purpose computing where an algorithm or parts of it are targeted onto a customised architecture, leading to both increased performance and power efficiency. A special-purpose architecture can be customised and tailored to the unique requirements of the application, resulting in a combination of increased performance, reduced power, smaller area and lower economical cost as compared to its general-purpose counterparts.

Nowadays, special-purpose units are added to many processors to perform specific, frequently used tasks such as encryption or video decoding. However, these special-purpose units are available only for a limited set of common functionalities, and they are fixed during the design of the processor. HPC can also benefit from special-purpose processing, but due to the vast space of possible applications with different characteristics, pre-fabricated (and hence fixed) accelerators are not practical. Instead, a flexible computing substrate that can be customised on demand by the designer according to the application requirements is required. Reconfigurable devices, such as Field-Programmable Gate Arrays (FPGAs), offer such a substrate technology, and significant speed-ups over both multi-core processors and GPUs have been reported for a range of applications [2, 12]. However, the downside of targeting an FPGA is often a complex, low-level programming model that requires special knowledge in hardware design.

Maxeler Technologies is pioneering a novel approach of data-flow oriented supercomputers. Maxeler computing systems are a type of special-purpose system that can be customised to the unique requirements of an application. At the heart of a Maxeler system are one or several Data-Flow Engines (DFEs) that combine a large and powerful reconfigurable device with significant amounts of DRAM memory. DFEs are programmed using a simple data-flow model that enables domain experts to optimise both their algorithms and the underlying architecture simultaneously, cutting through the typical layer approach of custom libraries, standard libraries, operating systems, and hardware organisation. This approach has led to orders-of-magnitude higher performance, lower power consumption and significantly lower data-centre space as compared to traditional approaches. A wide range of applications ranging from 3D finite-difference partial differential equation solvers [9] to Monte Carlo simulations have been successfully accelerated in commercial products [13]. In addition, speed-ups can also enable completely new computational models that were previously not feasible under hard timing constraints. For example, computing a 24-h forecast is not practical if the computation takes 48 h to complete. If, however, the same computation can be achieved in 2 h, then running 24-h forecasts becomes a realistic scenario.

11.2 The Data-Flow Paradigm

Maxeler's data-flow oriented computing paradigm fundamentally differs from conventional processors which are control-flow centric. This approach is illustrated in Fig. 11.1 and it represents an evolution of data-flow and systolic array concepts [3, 8]. A conventional processor operates by reading and decoding an instruction, loading the required data, performing an operation on the data, and returning the result to memory. This process is iterative in nature and requires complex control mechanisms that manage the operation of the processor. The data-flow execution model is greatly simplified in comparison. Data is streamed from memory

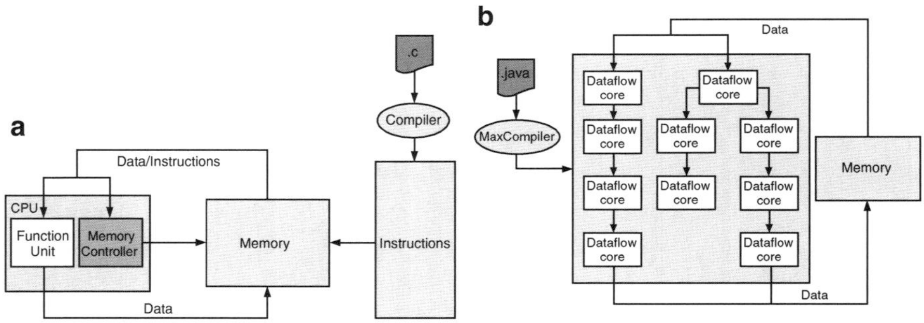

Control-flow system Data-flow system

Fig. 11.1 A conventional control-flow oriented processor (**a**) as compared to a Maxeler's DFE (**b**)

into the chip where arithmetic operations are performed by chains of functional units (data-flow cores) statically interconnected in a structure corresponding to the implemented functionality. It should be noted that the data-flow structure preforms computations entirely without instructions. Data flows from one functional unit directly to the next without the need of complex control mechanisms. Data simply arrives when it is needed and the final results are streamed back into memory. Each data-flow core performs only a simple operation such as addition or multiplication and hence, thousands of operations can fit into the given chip area.

Unlike the control-flow based processor where operations are computed on a time-shared functional unit ("computing in time"), the complete data-flow computation is laid out in space over the entire chip ("computing in space"). Dependencies in the data flow are resolved statically at compile time, and because there is no data-dependent behaviour present at run time, the entire data-flow engine can be deeply pipelined. Every stage of the pipeline computes in parallel with the data-flow architecture maintaining overall throughput of one result per clock cycle. A simple analogy of this approach is an assembly line in a car factory. The most efficient way to realise large scale productions (computations) is a specialised assembly line (pipeline) where parts (data) move from storage (memory) to a chain of dedicated workstations (custom functional units) where there are assembled together (data is processed) and moved forward in lock-step to produce complete cars (final results) at the end. There are no instructions and hence, instruction decoding logic is not required. Also, a static data-flow model does not require control-flow techniques such as branch prediction and out-of-order execution for obvious reasons. General purpose caches are equally not necessary and data is always kept on chip with the minimum amount of buffering memory for intermediate results. By eliminating these extraneous functions, all the chip's resources can be dedicated to perform useful computations instead of managing the execution.

Maxeler realises this data-flow oriented computing approach by mapping an application described in its data-flow model onto a DFE. DFEs are highly efficient for large scale computations with a static execution model due to the elimination of sequential execution and control, and the optimisation of memory access to a simple feed-forward model. However, DFEs are inefficient for small-scale computations with control-dominated dynamic behaviour. The key to effective data-flow computing systems is therefore the combination of DFEs with a conventional processor. The DFE carries out the compute-intensive part of the application while host-processors are tasked with control-intensive tasks and also with setting up and controlling the computation on the DFE. Depending on the nature of the problem, one can also adopt a combined processing approach where the processor computes the less demanding part of the application while the DFE will target the performance-critical part. This results in a co-design approach where we develop a conventional processor application together with a customised DFE implementation. In the following, we first cover Maxeler data-flow systems, followed by programming principles and application optimisations.

11.3 Maxeler Data-Flow Systems

At the centre of Maxeler's data-flow systems sits its proprietary DFE hardware. In state-of-the-art MAX4 generation systems, DFEs are based on large Altera Stratix-V FPGAs that provide the reconfigurable computing substrate for data-flow cores. This device is surrounded by large amounts of DRAM memory (currently between 48–96 GB), providing a very high memory capacity enabling large computational problems. This is called Large Memory (*LMem*). In addition, the FPGA itself also provides embedded on-chip memories which are spread throughout the chip's fabric and can be used to hold local values of the computation. These embedded memories are called Fast Memory (*FMem*) as they can be accessed with a total bandwidth of several terabytes/second. This is an important factor for the efficiency of DFE computations because data can be kept locally were it is needed and accessed with very high speeds. This is in contrast to CPU caches where data is kept on a speculative basis, and replicated several times, with only the smallest L1 cache providing very high speed to the computational unit.

As previously mentioned, DFEs are not intended to fully replace conventional CPUs; instead, they are integrated into an HPC-system consisting of CPUs, DFEs, storage, and networking. Various system architectures are possible and the overall balance of components can be tailored to the requirements of the user application. As a key feature, DFEs always contain large amounts of DRAM to facilitate the previously described model of data-flow processing. Various configurations between DFEs and CPU, as well as between multiple DFEs are possible. In the following, we give a brief overview of the current Maxeler MPC-C, MPC-X and MPC-N series.

Maxeler MPC-C systems couple x86 server-grade CPUs with up to four DFE cards (see Fig. 11.2). Each DFE card contains 48 GB of DRAM as LMem, and each DFE card is connected to the CPUs via a PCI Express (PCIe) bus. DFE cards are also directly connected to each other through a dedicated high-speed, low-latency link called MaxRing. This provides fast communication between neighbouring

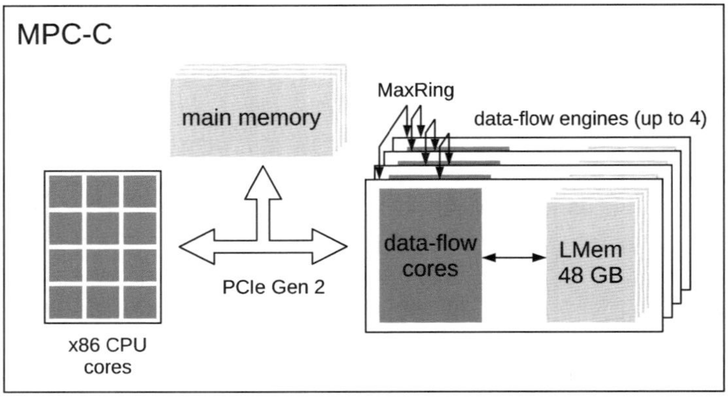

Fig. 11.2 MPC-C series architecture. A single node contains both x86 CPUs and four data-flow engines connected via PCIe and MaxRing

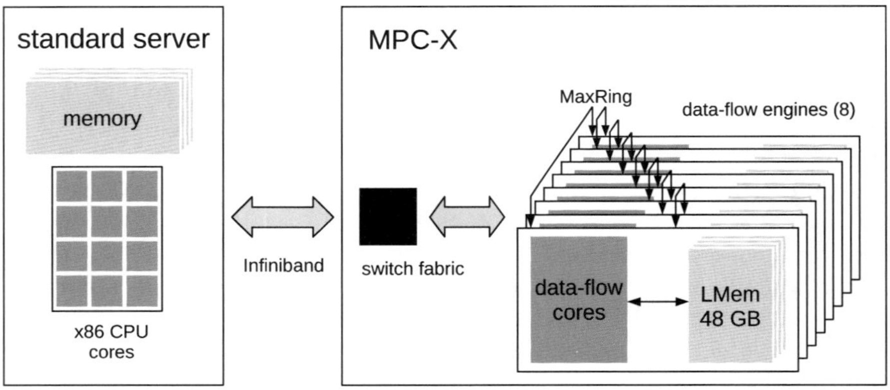

Fig. 11.3 MPC-X series architecture with eight data-flow engines inside a node. Multiple MPC-X nodes and CPU nodes are connected through an Infiniband network, and the number of DFEs used by each CPU can be varied dynamically

DFEs, enabling larger applications to scale across multiple DFEs without the PCIe link becoming a communication bottleneck. The system also includes storage and networking, and it is integrated into a dense 1U industry standard rack unit. Such a system supports simple stand-alone deployment of DFE technology, tightly coupled with high-end CPUs. The architecture is beneficial for high-performance applications that run on a fixed number of CPU cores and continuously use one or multiple DFEs.

The MPC-X series enable a more heterogeneous system architecture supporting dynamic balancing of CPU and DFE resources. MPC-X series systems are pure DFE nodes without any CPUs (see Fig. 11.3). An MPC-X system combines 8 DFE cards in a 1U chassis directly connected through MaxRing. DFEs are also linked through Infiniband to a cluster of CPU nodes. The system can allocate large numbers of

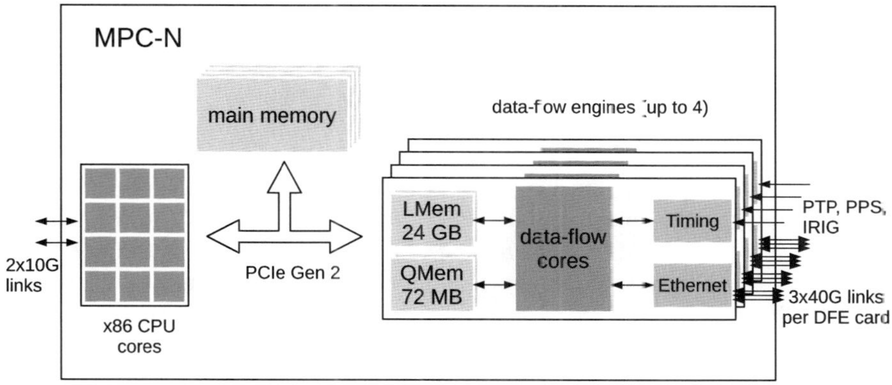

Fig. 11.4 MPC-N series architecture with four data-flow engines inside a node. Each DFE card also provides three 40G Ethernet connections directly to the DFE

DFEs dynamically, providing good scalability and flexibility for applications with changing behaviour, e.g. when the computation has several stages that vary in their characteristics. The ratio of CPU servers to MPC-X nodes can be tuned to user application requirements.

Maxeler's MPC-N series systems (see Fig. 11.4) are a network oriented platform that provides Ethernet connections directly to the data-flow engines, supporting ultra low-latency line-rate processing of multiple 10–40 Gbit data streams. A single MPC-N node contains up to 4 DFE cards similar to the MPC-C series architecture. However, each DFE card also supports up the 3 QSFP+ 40 Gbit Ethernet connections where each 40 Gbit port can be split into 4 × 10 Gbit ports. Providing fast Ethernet connections directly to the DFE enables network processing with minimal latency. The memory architecture in DFE also differs from the two previous system architectures: in addition to 24 GB DRAM available as LMem, the DFE also integrates 72 MB of QDR SRAM (QMem) supporting very low latency off-chip data access. The system contains additional 10 Gbit connections to the CPU. MPC-N series systems are well suited for a range of networking applications including gateways, aggregators, or endpoints.

Maxeler systems are provided with a compilation and simulation environment (called MaxCompiler) for application development, and the MaxelerOS system management environment. MaxelerOS coordinates the use of DFE resources at run time, and manages the scheduling and data movement within Maxeler systems. MaxCompiler provides a high-level programming environment to express data-flow structures, and produces the necessary binaries for CPU and DFE binaries.

11.4 Data-Flow Programming Principles

In the following, we outline the data-flow oriented programming model that is used in Maxeler systems. As described in the previous section, Maxeler data-flow systems are based on a combination of DFEs and CPUs. The basic logical architecture of such a system is illustrated in Fig. 11.5. The CPU is responsible for setting up and controlling the computation on the DFE. The DFE contains one or multiple data-flow kernels that perform the accelerated arithmetic and logical computations. Each DFE also contains a manager that is responsible for the connections between kernels, DFE memory, and the various interconnects such as PCIe, Infiniband and MaxRing.

Separating computation and communication into kernels and managers is beneficial because it allows the data-path inside the kernels to be deeply pipelined without any synchronisation issues. When developing the kernel, a designer would simply focus on achieving high degrees of pipelining and parallelism without worrying about scheduling or synchronisation. The scheduling of operations inside the kernel will be performed automatically by the compiler. The manager code describes how kernels are connected to memory and other IO interfaces, and the necessary synchronisation logic will also be generated by the compiler.

Developing an application for a DFE-based system therefore includes three parts:

1. A CPU application typically written in C/C++, Matlab, Python or FORTRAN;
2. One or multiple data-flow kernels written in extended Java[1];
3. A manager configuration, also written in extended Java.[1]

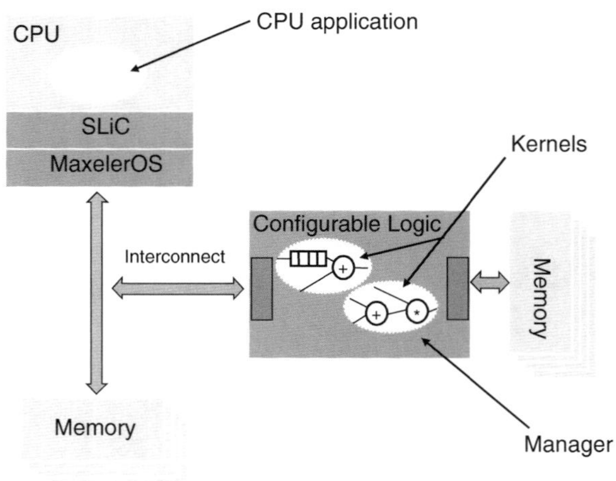

Fig. 11.5 Logical architecture of a data-flow computing system with one CPU and one DFE

[1]Maxeler provides extensions to the Java language, referred to as MaxJ.

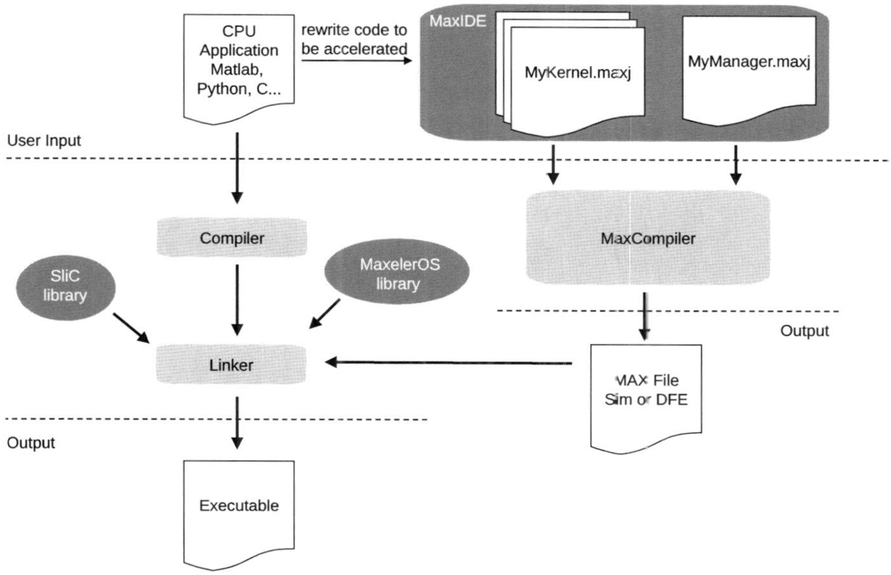

Fig. 11.6 Compiling a data-flow application with MaxCompiler

The compilation flow of a Maxeler data-flow design is illustrated in Fig. 11.6. The design typically starts with a CPU application where a performance-critical part needs to be accelerated. This part of the application will be targeted on a DFE. Designing a DFE application involves describing one or multiple kernels and a manager in MaxJ. MaxJ is Java-based meta-language that describes data-flow. It is important to note that executing the MaxJ program will not perform the computations described within the program. Instead, it will trigger the generation of a configuration file for the DFE (the so-called .max file). The computation will later be performed by loading the .max configuration file into the DFE and streaming the data through it. Before we can do this, we need to modify the CPU application to invoke the DFE. To simplify this process, MaxCompiler will generate the necessary function prototypes and header files. The CPU code is then compiled as usual and is linked with the .max file and Maxeler's Simple Live CPU (SLiC) interface library. The result of this is a single executable file that contains all the binary code to run on both the conventional CPUs and the DFEs in a system.

Let us focus on the principles of data-flow programming in MaxJ. As mentioned previously, MaxJ is a meta-language that describes data-flow computing structures; it uses Java syntax but is in principle different from regular Java programming (or other imperative programming paradigms that describe computations by changing state). The most important principle in MaxJ is that we describe a fixed spatial data-flow structure that can perform computations by simply streaming through data, and not a sequence of instructions to be executed on a traditional processor.

```
for (i = 0; i < numDataElements; i++) {
    float x = input[i];
    float y = x * x + 3 * x + 17;
    output[i] = y;
}
```

Fig. 11.7 C code of a simple computation inside a loop

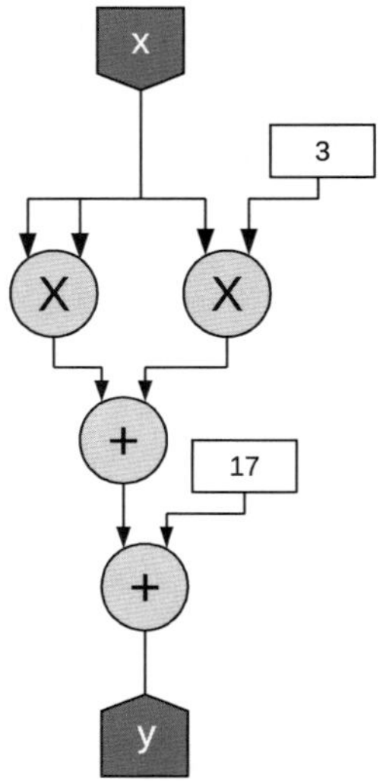

Fig. 11.8 A data-flow implementation for the computation inside the loop body

To illustrate these principles, we show how a simple loop computation can be transformed into a data-flow description using MaxJ. Let us assume we want to calculate $y = x^2 + 3x + 17$ over a data set. Even though there is nothing inherently sequential in this computation, a conventional C program would require a for loop. This is illustrated in Fig. 11.7. The calculation is repeated for the number of data elements in a loop. Within the loop body, all operations also run sequentially.

In contrast, a data-flow implementation would focus on identifying the core part of the computation and creating a data path for it. Figure 11.8 illustrates such a data-flow implementation. The same computation that is described inside the loop body can be performed by a fixed data path that contains two multipliers and two adders. It is one of the key features of data-flow computing having several operators present at the same time and running concurrently, instead of using a time-shared functional unit inside a processor. A practical data-flow implementation can have thousands of operators in a data path all running concurrently. Another important principle is the absence of control and instructions. The data path is fixed and the computation is performed by streaming data from memory directly into the data path.

Figure 11.9 depicts the MaxJ kernel description that can generate the data-path shown in Fig. 11.8. The MaxJ descriptions begins by extending the `kernel` class (line 1). The kernel class is part of the Maxeler Java extensions and the user develops their own kernels by using inheritance. Next, we define a constructor for `SimpleCalc` class (line 2). It is important to remember that this MaxJ program

```
class SimpleCalc extends Kernel {
    SimpleCalc() {
        DFEVar x = io.input("x", dfeFloat(8,24));
        DFEVar y = x * x + 3 * x + 17;
        io.output("y", y, dfeFloat(8,24));
    }
}
```

Fig. 11.9 An MaxJ description that generates the data-flow implementation shown in Fig. 11.8

```
int x = 10;
DFEVar y;
DFEVar z;
y = x; // ok, assign constant to run-time variable
x = y; // compiler error, cannot read run-time variable into compile-time Java variable
z = y; // ok, both handle run-time data
```

Fig. 11.10 DFEVars handle run-time data, Java constants are evaluated only at compile time

will only run once to build the DFE configuration; the constructor will facilitate building the data-flow implementation. To create the streaming inputs and outputs for the kernel, the methods `io.input` (line 3) and `io.output` (line 5) are used. Streaming inputs and outputs replace the `for` loop in the original C code that iterates over data. The input method takes two arguments: the name on the input that will be used by the manager to connect the kernel, and the data type of the input. In this case, we use a standard single precision floating point format (8-bit exponent and a 24-bit mantissa), but MaxJ also supports custom data types that can be defined by the user. This is useful when optimising the numerical behaviour and performance, which will be covered later. The output method uses three arguments: the name of the output to be used by the manager, the variable to connect to the output, and the data format. The computation itself is expressed in a very similar way as in the original C code (line 4).

In MaxJ the `DFEVar` object is used to handle run-time data. Since MaxJ describes a data-flow graph rather than a procedure, we have to distinguish between run-time values and compile-time values. Regular Java variable such as `int` will be evaluated and fixed at compile time. Such variables can be used as constants for improved code readability, or to control the build of the data-flow graph. The values of DFEVars are known only at run time when data is streamed through the kernel. This means assigning a Java variable to a DFEVar will result in a constant. However, it is not possible to read a DFEVar and assign it to a Java variable (Fig. 11.10). This principle means that we can use Java variables and control constructs to shape the structure of our data-flow graph. Let us consider an example of a nested loop as shown in Fig. 11.11. We observe that the outer `for` loop performs an iteration over data, while the inner `for` loop describes a computation with a cyclic dependency of `v` from one loop iteration to another.

This example can be effectively transformed into a data-flow description as illustrated in Fig. 11.12. Again, the outer loop is replaced by streaming inputs and outputs. The inner loop is described with the same for `for` loop statement in Java, but the compilation of this loop will result in an unrolled implementation of the loop body in space, as depicted in Fig. 11.13. Unlike the original loop in C, the `for`

```
for (i = 0; i < numDataElements; i++) {
    float d = input[i];
    float v = 2.91 - 2.0 * d;
    for (iteration = 0; iteration < 4; iteration++) {
        v = v * (2.0 - d * v);
    }
    output[i] = v;
}
```

Fig. 11.11 C code of a nested loop with dependency

```
class Loop extends Kernel {
    Loop() {
        DFEVar d = io.input("d", dfeFloat(8,24));
        DFEVar v = 2.91 - 2.0 * d;
        for (int iteration = 0; iteration < 4; iteration += 1) {
            v = v * (2.0 - d * v);
        }
        io.output("output", v, dfeFloat(8,24));
    }
}
```

Fig. 11.12 A MaxJ implementation of the inner loop will be statically evaluated resulting in spatial replication

loop in MaxJ does not carry out four iterations at run time. Instead, the compiler can resolve the dependency of `v` from one loop iteration to another and construct an unrolled, acyclic data path where the calculation inside the loop body is replicated four times, and each `v` is connected to the result from the previous iteration.

The previous example has shown how a Java `for` loop can be used to control the replication of statements inside the loop body into an unrolled data path. Likewise, it is possible to use Java conditionals such as `if` or `case` to control the construction of the data-flow graph. The Java `if` condition is evaluated at compile time, and the block of code inside the conditional statement will be added into the data-flow graph only if the condition is evaluated as true.

However, we cannot use a Java conditional on DFEVars because their value will be only known at run time. As previously mentioned, run-time dependent behaviour is undesirable as it is against the principles of static data-flow computing. If a data-dependent decision needs to be made then this can be expressed using the ternary operator `? :` (see Fig. 11.14). This example results in data-dependent control, but in the data-path, both `y1` and `y2` will be computed concurrently. At the output we simply select one of the two results, depending on the value of `a`. This switching will be very fast and will not delay or stall the stream processing. However, it also means the we require resources for both computations on the DFE chip even though only one of the two outputs will be used at any time. This makes this type of control effective for fast, small-scale switching. For switching between larger blocks of computation, it might be more effective to implement separate DFE kernels and handle the switching and control from the CPU host.

Figure 11.14 also illustrates that custom number formats other than conventional single or double-precision floating point can be used. In this example, we use a 9-bit exponent and a 31-bit mantissa, which offers better scaling and precision than single precision (8, 24 bit) but less than double precision (11, 53 bit). Likewise, it is

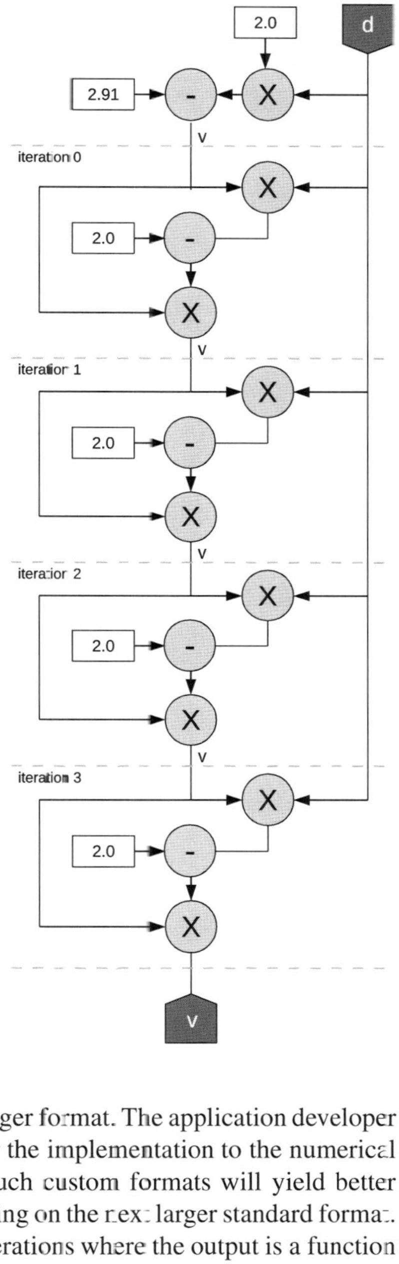

Fig. 11.13 The result of the MaxJ loop is an unrolled and pipelined data path

possible to use any arbitrary fixed-point or integer format. The application developer can use such custom number formats to tailor the implementation to the numerical requirements of the application, and using such custom formats will yield better resource utilisation and performance than relying on the next larger standard format.

All previous examples have considered operations where the output is a function of inputs with the same array index within the stream, e.g.:

$$z_i = 5x_i + y_i, \quad z_{i+1} = 5x_{i+1} + y_{i+1}, \quad \dots \tag{11.1}$$

```
DFEVar x = io.input("x", dfeFloat(9,31));
DFEVar a = io.input("y", dfeFloat(9,31));

DFEVar y1 = x * 5;
DFEVar y2 = x - 7;

DFEVar y = a > 3 ? y1 : y2;

io.output("y", y, dfeFloat(9,31));
```

Fig. 11.14 Data-dependent control with the ternary operator, and use of a custom number format

```
DFEVar x = io.input("x", dfeFloat(8,24));
DFEVar prev = stream.offset(x,-1);
DFEVar next = stream.offset(x,1);
DFEVar y = (prev + x + next)/3;
io.output("y", y, dfeFloat(8,24));
```

Fig. 11.15 Using stream offsets to access values with relative offsets in the stream

However, in some cases we need to access values that are ahead or behind the current element in the data stream. For example, in a moving average filter we need to compute:

$$y_i = \frac{x_{i-1} + x_i + x_{i+1}}{3} \tag{11.2}$$

In data-flow computing, `x` is a stream rather than an indexed array, and we need a way of accessing elements of the same stream with other indices than the current one. This can be achieved with the `stream.offset` method that accesses values with a relative offset from the current value in the stream. In the moving average example, we need the previous value (-1) and the next value ($+1$) (Fig. 11.15).

Figure 11.16 illustrates how a DFE application interacts with the CPU host application. On the right side, we see the moving average kernel `MAVKernel` from our last example. As previously mentioned, we also create a manager to describe the connectivity between the kernel and the available DFE interfaces. In Fig. 11.16, the kernel is connected directly to the CPU, and all of the communication will be facilitated via PCIe. The manager also makes visible to the CPU application all the names of the kernel streaming inputs and outputs. Compiling the manager and kernel will produce a .max file that can be included in the host application code. In the host application, running the moving average calculation will be performed with a simple function call to `MAVKernel()`. In this example, the host application is written in C but MaxCompiler can also generate bindings for a variety of other languages such as MATLAB or Python.

MaxelerOS and the SLiC library provide a software layer that facilitates the execution and control of the DFE applications. The SLiC Application Programming Interface (API) is used to invoke the DFE and process data on it. In the example in Fig. 11.16 we use a simple SLiC interface and the simple function call `MAVKernel()` will carry out all DFE control functions such as loading the binary configuration file and streaming data in and out over PCIe. More advanced SLiC interfaces are also available that provide the user with additional control over the

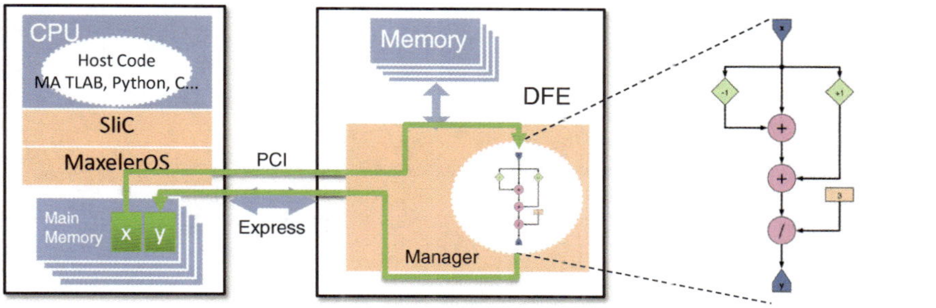

Host Code (.c)

```
#include "MAVKernel.max"

float x[n], y[n];

MAVKernel(x, y, n);
```

Manager (.maxj)

```
Manager m = new Manager();
Kernel k = new MAVKernel();
m.setKernel(k);
m.setIO(
   link("x", CPU),
   link("y", CPU));
m.build();
```

MAVKernel (.maxj)

```
DFEVar x = io.input("x", dfeFloat(8, 24));

DFEVar prev = stream.offset(x, -1);
DFEVar next = stream.offset(x, 1);

DFEVar y = (prev+x+next) / 3;

io.output("y", result, dfeFloat(8, 24));
```

Fig. 11.16 Interaction between host code, manager and kernel in a data-flow application

DFE behaviour. For example, in many cases it is beneficial to transfer the data to DFE memory (LMem) first and then start the computation. This is one of many performance optimisations, which we will briefly cover in the next section.

11.5 Development Process and Design Optimisation

In the previous section we have introduced the principles of data-flow programming. We now outline how to develop data-flow applications in practice, and how to improve their performance. In traditional software design, a developer usually targets a given platform and optimises the application based on available libraries that reflect the capabilities and architectural characteristics of the targeted platform. Developing a data-flow implementation fundamentally differs in that we *codesign* the application and architecture. Instead of mapping a problem to pre-existing APIs and data-types, we enable domain experts, e.g. physicists, mathematicians, and engineers to create a solution all the way from the formulation of the computational problem down to design of the best possible data-flow architecture. A developer would therefore optimise the scientific algorithm to match the capabilities of the data-flow architecture while at the same time optimising the data-flow structure to match the requirements of the algorithm. Another key difference to traditional software design is the implementation and optimisation cycle. In software design, a developer would typically implement a design, go on to profile and evaluate the performance of the current implementation, and then tweak the implementation.

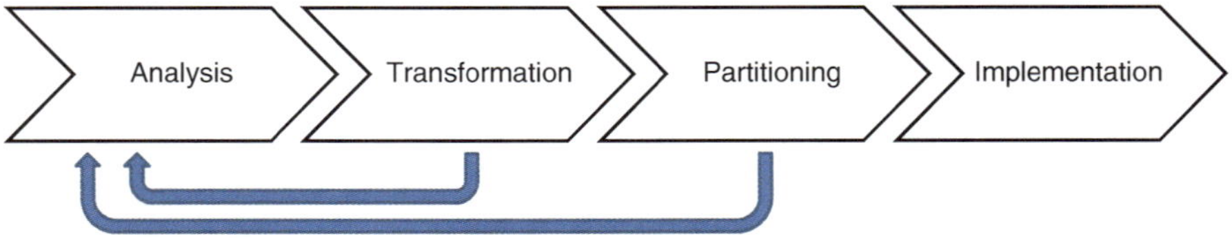

Fig. 11.17 Process for developing and optimising data-flow applications

In data-flow design, we adopt a different approach where the design is optimised before it is implemented: The behaviour inside a DFE is very predictable and we can therefore plan and precisely predict the performance of a possible solution without even implementing it. This means the design will be analysed and optimised with simple spreadsheet calculations before we create the final implementation.

This development process is illustrated in Fig. 11.17. The first step consists of an application analysis phase. The purpose of this step it to establish an understanding of the application, the data set, the algorithms used, and the potential performance-critical parts. Since we will codesign an algorithm and its data-flow architecture, this analysis should cover all parts of the computational problem, from the mathematical formulation and algorithm to the architecture and implementation details. Typical considerations are the type and regularity of the computation, the ratio between computation and memory accesses, the ratio of computation to disk IO or network communication, and the balance between recomputation and storage of pre-computed results. All these aspects can have a significant impact on the performance of the final implementation. If, for instance, an application is limited by the speed at which data can be read from disk, then optimising the throughput of the compute kernel beyond that limit will have no benefit.

The second step involves algorithmic transformations. A designer could attempt to choose a different algorithm to solve the problem, or transform the code, data access patterns or number representations. A typical example of an algorithmic transformations is to change the number format: Choosing a smaller number representation can support more IO bandwidth, and higher computational performance, but the numerical effects of the algorithm have to be well understood. The reconfigurable technology used inside the DFEs support far greater flexibility in the available number formats than all conventional processors. Instead of choosing from single or double precision floating point, a design can exploit a custom format with arbitrary bit-widths of its exponent and mantissa. Another common optimisation is the reordering of data-access patterns to support better data flow. The impact of algorithmic transformations has to be evaluated through iterative analysis of the design.

The third step is to partition the application between the CPU and the DFE. This partitioning covers program code as well as data. For the program code, we can chose whether the code should run on the CPU or the DFE. Large scale applications typically involve multiple DFEs and this also involves partitioning DFE code over multiple DFEs. Furthermore, it is often beneficial to follow a co-processing approach where the CPU and DFE work on different parts of the computation at the same time. For instance, the CPU can perform lightweight pre-calculations or more control-intensive parts of the application. For this purpose, the SLiC library provides non-blocking functions to control the DFEs. Another consideration is the partitioning of data. The example in Fig. 11.17 showed DFE data being streamed from main CPU memory. For processing larger data sets, it is usually beneficial to locate the data in the large DFE memory (LMem). Coefficients or frequently accessed values can be kept inside the DFE reconfigurable substrate in fast memory (FMem).

A high-level performance model is used to evaluate the design as it undergoes various transformations, code and data partitionings. The process of analysis and optimisation is repeated iteratively as additional possibilities are explored. Only when the design is fully optimised, the designer will proceed to step four: the implementation of the design.

11.6 Financial application examples

Maxeler data-flow technology has been deployed in a number of areas including finance [7, 13], oil and gas exploration [4, 10], atmospheric modelling [5], and DNA sequence alignment [1]. The range of applications includes Monte Carlo, finite difference, and irregular tree-based partial differential equations, to name a few. Maxeler provides a number of products and solutions in the financial domain, including financial analytics and trading applications, particularly for low-latency/high frequency electronic trading on organised exchanges.

11.6.1 Maxeler RiskAnalytics Platform

Maxeler FinancialAnalytics is a financial valuation and risk management platform designed from the ground-up, where the core analytic algorithms are accelerated on Maxeler data-flow systems. The purpose of the platform is to go beyond simply providing highly efficient computational finance capabilities, but rather the aim is to provide a complete, vertically-integrated application stack that provides a platform containing all the necessary components for streamlined front-to-back portfolio risk management, including:

- Front-end, pre-trade valuation and risk checking;
- Exchange-based, electronic trade execution, portfolio valuation and risk management;
- Front-end trade booking, portfolio management, model and risk reporting and analysis;
- Post-trade model and risk metric selection and verification;

- Rapid and flexible transaction analysis and reporting;
- Application layer in software for quick and flexible functional reconfiguration;
- Large memory to enable rapid and flexible in-memory portfolio risk analysis;
- Regulatory reporting for Basel III, EMIR, Dodd-Frank, Volker-rule, Solvency II, etc.;
- Adaptive load balancing;
- Database integration.

All core FinancialAnalytics components have been implemented in both software and on Maxeler DFE-based systems, requiring integration of the DFE technology with expertise of quantitative analysts with extensive investment banking experience. The platform has been designed in a modular fashion to maximise flexibility and performance. Each module realises a core analytics component, such as curve bootstrapping or Monte Carlo path generation. To support flexible hardware/software co-processing and to enable ease of integration with existing systems, each module is available as both a CPU and DFE library component. As outlined in Sect. 11.5, achieving an efficient implementation depends on the overall system composition, architecture and application structure. Making use of pre-existing CPU and DFE library components greatly simplifies this process. In the following, we show the practical use of Maxeler's RiskAnalytics library in several commercial use cases.

First, let us consider *interest rate swap pricing*. An interest rate swap is a financial derivative with high liquidity that is commonly used for hedging. Such a swap involves exchanging interest rate cashflows based on a specified notional amount from one interest rate to another, e.g. exchanging fixed interest-rate flows for floating interest-rate flows. Figure 11.18 illustrates a typical module configuration for pricing interest rate swaps, involving bootstrapping the Overnight Index Swap (OIS) curve and the London Interbank Offered Rate (LIBOR) curve, followed by generating swap cashflow schedules, valuing swaps and calculating swap portfolio risk. Each stage is available as either a CPU or a DFE library component and can be accessed via number of convenient APIs. The implementation provides construction of and access to all intermediate and final objects.

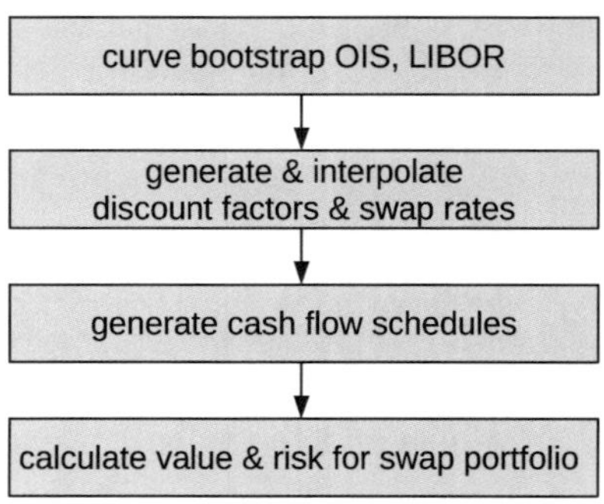

Fig. 11.18 A typical swap pricing pipeline

Table 11.1 Possible configurations for swap pricing pipeline

Application characteristic	OIS	LIBOR	Cashflow	Pricing
Many curves, few swaps	DFE	DFE	CPU	CPU
Few curves, many swaps	CPU	CPU	DFE	DFE

Depending on the characteristics of the swap pricing application, DFE acceleration can be beneficial at one or more stages of the computation. Table 11.1 illustrates two possible module configurations where the performance-critical DFE acceleration can be carried out at different stages of the pipeline. Modular design of Maxeler's FinancialAnalytics allows the user application to dynamically load balance between CPUs and DFEs, and to target heavy compute load to DFEs, leaving CPUs to support application logic and lighter compute loads. DFE functionality can be switched in real time by using MaxelerOS SLiC API functions. Fully pipelined, a Maxeler DFE-equipped 1U MPC-X node can value a portfolio of 10-year interest rate swaps at a rate of over 2 billion per second – including bootstrapping of the underlying interest rate curves.

A second example of the application of DFE technology in finance is the calculation of *value-at-risk* (VaR), a measure widely used to evaluate the risk of loss on a portfolio over a given time period. VaR defines the loss amount that a portfolio is not expected to exceed for a specified level of confidence over a given time frame. VaR can be calculated in a number of ways (e.g. using fixed historical scenarios, or using arbitrarily specified scenarios, a delta-based approach, or using Monte Carlo generated scenarios). Irrespective of the method chosen, the VaR computation involves evaluating many possible market scenarios, a technique that is computationally very demanding. Regardless of the chosen approach, the computation of VaR using conventional technology is frequently slow and often inaccurate, as well as being unstable in the tail of the loss distribution, resulting in uncertainty in risk attribution and difficulty in optimising against portfolio VaR targets. This is illustrated in Fig. 11.19, where the tail of the loss distribution for a mixed portfolio of interest rate swaps exhibits a step-wise profile, making it extremely difficult to accurately manage portfolio VaR.

Mitigating these problems requires massively increased number of scenarios, in order to provide higher resolution in the tail of the loss distribution, in order to significantly improve stability for risk attribution and/or provide greater visibility of the impact of market and portfolio changes. This is clearly illustrated when comparing Figs. 11.19 and 11.20. In the second case, the number of Monte Carlo scenarios is increased by a factor of 50, resulting in far greater granularity in the tail of the loss distribution leading to improved accuracy of portfolio risk management. Fully pipelined, a Maxeler DFE-equipped 1U MPC-X node can compute full revaluation VaR on a portfolio of 250,000 10-year interest rate swaps (equivalent to a rate of over 2 billion swaps per second) – including bootstrapping of the underlying interest rate curves, as well as scenario construction.

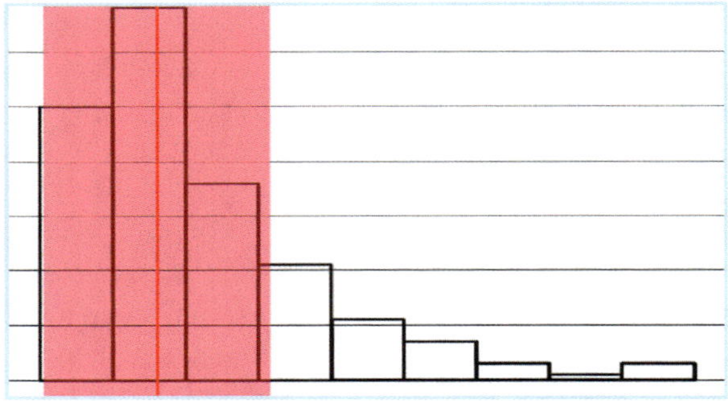

Fig. 11.19 Value-at-Risk with 10,000 scenarios

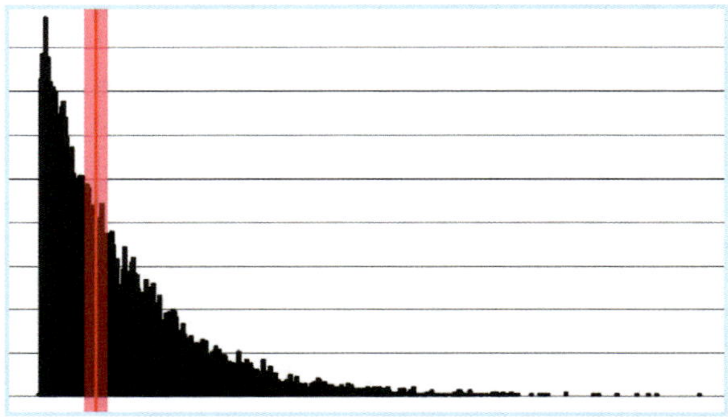

Fig. 11.20 Value-at-Risk with 500,000 scenarios

Increasing the number of Monte Carlo scenarios as suggested above obviously increases the computational requirements, but with DFE-acceleration, the extra scenarios can be easily and practically achieved. When the accuracy of computation is increased, several new approaches to VaR can become feasible:

- Pre-horizon cashflow generation and dynamic portfolio hedging;
- Sensitivity metrics for enhanced risk explain and attribution;
- Stable and efficient portfolio optimisation.

A third application example is *exotic interest rate pricing*. A user might wish to price an exotic product such as a Bermudan swaption, which is an option to enter into an interest rate swap on any one of a number of predetermined dates. One of the industry standard approaches to this pricing problem is to use the LIBOR market model (LMM) which employs a high-dimensional Monte Carlo model with complex dynamics and a large state space. Pricing involves a multi-stage algorithm with forward and backward cross-sectional (Longstaff-Schwartz) computations across the full path space. Here, the challenge is to manage large-path data sets, typically several gigabytes, across multiple stages. Figure 11.21 illustrates the FinancialAnalytics DFE implementation, including cashflow generation and Longstaff-Schwartz backward regression. By closely coordinating between multiple DFE stages and DRAM memory, 6,666 quarterly 30-year Bermudan swaptions can be priced per second on a Maxeler 1U MPC-X node. This represents an

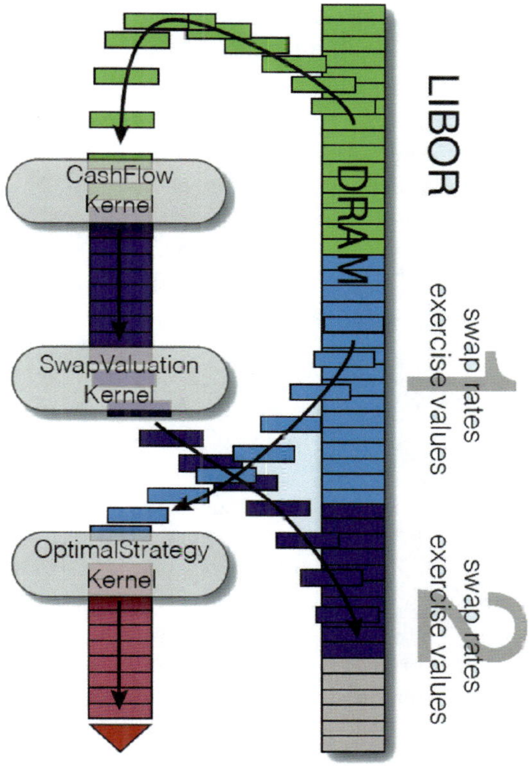

Fig. 11.21 Bermudan swaptions computation on a DFE

Table 11.2 Comparison of CPU and DFE-node performance (instruments priced per second) for various instruments

Instrument	Conventional 1U CPU-Node	Maxeler 1U MPC-X Node	Comparison
European swaptions	848,000	35,544,000	42×
American options	38,400,000	720,000,000	19×
European options	32,000,000	7,080,000,000	221×
Bermudan swaptions	296	6,666	23×
CDS	432,000	13,904,000	32×
CDS bootstrap	14,000	872,000	62×

23× improvement over a 1U CPU node. Table 11.2 provides a comparison of different instruments priced per second for a range of instrument types supported in RiskAnalytics. As it can be seen, a single 1U MPC-X node can replace between 19 and 221 conventional CPU-based units. The power efficiency advantage due to the data-flow nature of the implementation also ranges between one an two orders of magnitude.

11.6.2 Ultra Low-Latency Trading

In addition to high performance computational capabilities, Maxeler also provides products for ultra low-latency trading, leveraging the benefits of data-flow technology through dedicate network oriented systems. The goal is to enable latency-sensitive traders to deploy fast and deterministic trading technology and develop more complex strategies under real-time constraints and execute them faster than the competition. A key concern when deploying specialised technology is not only to achieve lowest possible latencies but also to support rapid algorithm development. A further important feature is the ability to make this technology accessible to existing, front-office, strategy-development teams and keep the strategy and algorithm knowledge in-house. Maxeler's unique offering is that it provides the capability to bring together in hardware low-latency execution, pre and post-trade portfolio risk management, as well as providing the software for simple in-house programming to deliver decision support at a speed that matches market needs.

Maxeler MPC-N series systems provide the basis for a low-latency trading platform. An essential feature of these systems is direct connectivity of the DFE card to 3 QSFP+ ports supporting 12×10 Gbit or 3×40 Gbit Ethernet links, combined with a precision timing interface. A full TCP/IP stack in hardware is also available, and industry-standard trading interfaces for CME, Eurex, NYSE and NASDAQ are supported. This allows creating a programmable low-latency platform entirely within the DFE. Figure 11.22 depicts a high-frequency *top-of-order-book* application based on the low-latency platform. Top of the book refers to the highest bid and the lowest ask in the order book, with the bid being lower than the ask (otherwise this would quickly be resolved through a trade). These values indicate the prevalent market and they can be exploited in user-defined algorithms. In the case of the Chicago Mercantile Exchange, the Maxeler platform receives CME's

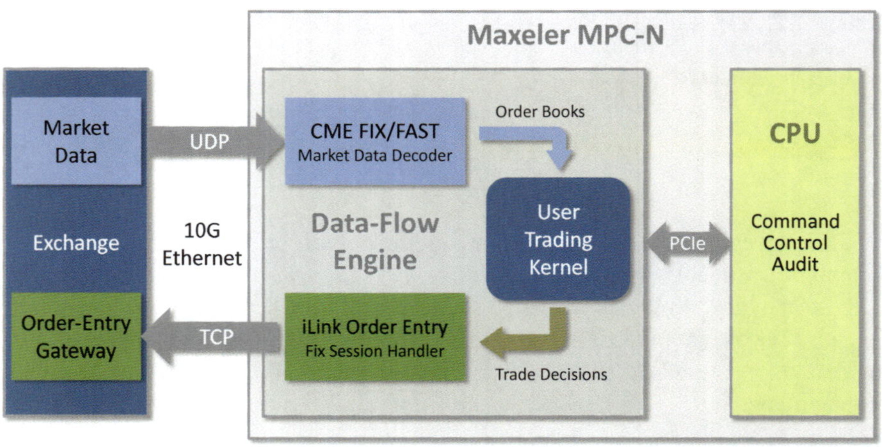

Fig. 11.22 Low-latency trading platform based on MPC-N for top-of-book application

market data via UDP, then decodes the FAST market data messages at line rate, before finally re-constructing the full Level 2 order book. As an example of how this is used in practice, a user-defined trading kernel can be inserted into a DFE to reconstruct the full order book, monitor trading strategies, compute pre-trade risk, before finally issuing FIX-formatted orders for execution when a target variable such as volatility reaches a certain pre-definable level. Efficient user development of such trading kernels is supported by the high-level data-flow programming approach that is described in Sect. 11.4. The output of the kernel is trade decisions, and individual orders are transmitted through a FIX session over TCP/IP to the CME order entry gateway. The application also receives order execution acknowledgements which are passed to the CPU software for post-trade, position risk management. This platform supports a highly deterministic wire-to-wire turnaround time between market data arriving and the order being executed over TCP in under 2.0 μs.

11.7 Conclusion

Cutting-edge applications in computational finance require powerful computational systems, but scaling over current CPU technology is becoming increasingly problematic. Maxeler has pioneered a new vertically-integrated, data-flow oriented approach that can deliver orders-of-magnitude improvement in performance, data-centre space and power consumption for a wide range of applications. DFEs realise a highly efficient computational model for the compute-intensive parts of an algorithm. In addition, they can be balanced with other types of resources such as CPUs and storage according to the requirements of the application. Maxeler supports a high-level programming model that allows application experts to harness the computational power of data-flow systems and optimise their application all the way from the formulation of the algorithm down to the design of the best possible data-flow architecture for its solution. This data-flow technology is key to many finance applications where a more complex model, more frequent re-computation, or lower latency often directly translate into monetisable, competitive advantage. A number of DFE-based products for analytics and trading are available from Maxeler, and we described several practical application scenarios that could not be achieved with conventional CPU technology.

References

1. Arram, J., Luk, W., Jiang, P.: Ramethy: reconfigurable acceleration of bisulfite sequence alignment. In: Proceedings of the International Symposium on Field-Programmable Gate Arrays (FPGA), Monterey, pp. 250–259. ACM (2015)
2. Chau, T.C.P., Niu, X., Eele, A., Maciejowski, J., Cheung, P.Y.K., Luk, W.: Mapping adaptive particle filters to heterogeneous reconfigurable systems. ACM Trans. Reconfigurable Technol. Syst. **7**(4), 36:1–36:17 (2014)

3. Dennis, J.B.: Data flow supercomputers. Computer **13**(11), 48–56 (1980)
4. Fu, H., Gan, L., Clapp, R.G., Ruan, H., Pell, O., Mencer, O., Flynn, M.J., Huang, X., Yang, G.: Scaling reverse time migration performance through reconfigurable dataflow engines. IEEE Micro **34**(1), 30–40 (2014)
5. Gan, L., Fu, H., Yang, C., Luk, W., Xue, W., Mencer, O., Huang, X., Yang, G.: A highly-efficient and green data flow engine for solving euler atmospheric equations. In: 24th International Conference on Field Programmable Logic and Applications (FPL), Munich, 2–4 Sept 2014, pp. 1–6. IEEE (2014)
6. Godfrey, M., Hendry, D.: The computer as von Neumann planned it. IEEE Ann. Hist. Comput. **15**(1), 11–21 (1993)
7. Jin, Q., Dong, D., Tse, A.H.T., Chow, G.C.T., Thomas, D.B., Luk, W., Weston, S.: Multi-level customisation framework for curve based Monte Carlo financial simulations. In: Reconfigurable Computing: Architectures, Tools and Applications – 8th International Symposium (ARC), Hong Kong, pp. 187–201. Springer (2012)
8. Kung, H.T.: Why systolic architectures? Computer **15**(1), 37–46 (1982)
9. Lindtjorn, O., Clapp, R., Pell, O., Fu, H., Flynn, M., Mencer, O.: Beyond traditional microprocessors for geoscience high-performance computing applications. IEEE Micro **31**(2), 41–49 (2011)
10. Pell, O., Bower, J., Dimond, R., Mencer, O., Flynn, M.J.: Finite-difference wave propagation modeling on special-purpose dataflow machines. IEEE Trans. Parallel Distrib. Syst. **24**(5), 906–915 (2013)
11. Pell, O., Mencer, O.: Surviving the end of frequency scaling with reconfigurable dataflow computing. SIGARCH Comput. Archit. News **39**(4), 60–65 (2011)
12. Thomas, D.B., Luk, W.: Multiplierless algorithm for multivariate gaussian random number generation in FPGAs. IEEE Trans. VLSI Syst. **21**(12), 2193–2205 (2013)
13. Weston, S., Spooner, J., Racaniére, S., Mencer, O.: Rapid computation of value and risk for derivatives portfolios. Concurr. Comput. Pract. Exp. **24**(8), 880–894 (2012)

List of Abbreviations

$(CM)^2$	Center for Mathematical and Computational Modelling.
ACP	Accelerator Coherency Port.
AGP	Accelerated Graphics Port.
AJD	Affine Jump Diffusion.
ALU	Arithmetic Logic Unit.
API	Application Programming Interface.
ASIC	Application Specific Integrated Circuit.
ATA	AT Attachment.
ATM	At the Money.
AVX	Advanced Vector Extensions.
AXI	Advanced eXtensible Interface.
BAR	Base Address Register.
BIOS	Basic Input/Output System.
BLAST	Basic Local Alignment Search Tool.
BM	Brownian Motion.
BS	Black-Scholes.
CAN	Controller Area Network.
CAPEX	Capital Expenses.
CDR	Clock Data Recovery.
CI	Confidence Interval.
CPU	Central Processing Unit.
CRUD	Create, Read, Update and Delete.
DAL	Database Abstraction Layer.

C. De Schryver (ed.), *FPGA Based Accelerators for Financial Applications*,
DOI 10.1007/978-3-319-15407-7

DMA	Direct Memory Access.
DRAM	Dynamic Random-Access Memory.
DSL	Domain-Specific Language.
DSP	Digital Signal Processor.
EISA	Extended Industry Standard Architecture.
EMS	Euler-Maruyama scheme.
FF	Flip-Flop.
FFT	Fast Fourier Transform.
FIFO	First in, First Out.
FLOPS	Floating-Point Operations per Second.
FPGA	Field Programmable Gate Array.
FRFT	Fractional Fourier Transform.
GARCH	Generalized Autoregressive Conditional Heteroskedasticity GBM Geometric Brownian Motion.
GNU	GNU's Not Unix.
GPGPU	General Purpose Graphics Processor Unit.
GPIO	General-Purpose Input/Output.
GPU	Graphics Processor Unit.
GSL	GNU Scientific Library.
HDL	Hardware Description Language.
HFT	High-Frequency Trading.
HLS	High-Level Synthesis.
HP	High Performance.
HPC	High Performance Computing.
HPRC	High Performance Reconfigurable Computing.
HTTP	Hypertext Transfer Protocol.
HW	Hardware.
HW/SW	Hardware/Software.
i.i.d.	Independent and Identically Distributed.
I^2C	Inter-Integrated Circuit.
ICDF	Inverse Cumulative Distribution Function.
II	Initiation Interval.
ILP	Integer Linear Programming.
IP	Intellectual Property.
ISA	Industry Standard Architecture.

IT	Information Technology.
ITM	In the Money.
LS	Longstaff-Schwartz.
LUT	Lookup Table.
MC	Monte Carlo.
MCMC	Markov Chain Monte Carlo.
MGT	Multi-Gigabit Transceiver.
MLMC	Multilevel Monte Carlo.
MMU	Memory Management Unit.
MPEG	Moving Picture Experts Group.
MPML	Mixed Precision Multilevel.
MSE	Mean Squared Error.
MSVC	Microsoft Visual C++.
MT	Mersenne Twister.
NAG	Numerical Algorithms Group.
NRE	Non-recurring Engineering.
OCM	On-Chip Memory.
OPEX	Operating Expenses.
OS	Operating System.
OTC	Over-the-Counter.
OTM	Out of the Money.
PC	Personal Computer.
PCI	Peripheral Component Interconnect.
PCI-X	Peripheral Component Interconnect Extended.
PCIe	Peripheral Component Interconnect Express.
PDE	Partial Differential Equation.
PL	Programmable Logic.
PLL	Phase Lock Loop.
PS	Programmable Systems.
QE	Quadratic Exponential.
ReST	Representional State Transfer.
RMSE	Root Mean Squared Error.
RN	Random Number.

RNG	Random Number Generator.
RTL	Register-Transfer Level.
RV	Random Variable.
SCU	Snoop Control Unit.
SD	Secure Digital.
SDE	Stochastic Differential Equation.
SerDes	Serializer/Deserializer.
SIMD	Single Instruction Multiple Data.
SoC	System on Chip.
SV	Stochastic Volatility.
SWIP	Scottish Widows Investment Partnership.
TCO	Total Cost of Ownership.
TLP	Transaction Layer Packet.
TTM	Time to Market.
UART	Universal Asynchronous Receiver/Transmitter.
URI	Uniform Resource Identifier.
USB	Universal Serial Bus.
WWW	World Wide Web.
XML	eXtensible Markup Language.

List of Symbols

Options and Markets

H	payoff function.
K	strike price of the option.
M	moneyness of the option.
S_0	current price of the asset (asset spot price).
S	continuous time asset price process.
T	time to maturity or time to expiration.
W^S	Wiener process for the asset price simulation process.
W^v	Wiener process for the volatility simulation process.
W	Wiener process resp. Brownian motion.
X	price of a financial derivative.
Φ	cumulative distribution function of the standard normal distribution.
α	variance process in the SABR model.
β	distribution parameter in the SABR model.
$\hat{S}$	discrete time asset price process.
$\hat{v}$	discrete time volatility process in the Heston model.
κ	mean reversion rate in the Hull-White model.
κ	mean reversion rate of the volatility in the Heston model.
$\Re\mathfrak{e}$	real part of a complex number.
μ	long term average price in the Black Scholes model.
v_0	current volatility.
v	volatility parameter in the SABR model.
v	continuous time volatility process in the Heston model.
ρ	correlation between two Brownian motions in Hull-White model.
ρ	correlation between two Brownian motions in the SABR model.
σ	volatility of the asset price in the Black-Scholes model.
σ	volatility in the Hull-White model.
σ	volatility of the volatility in the Heston model.
θ	long term average volatility in the Heston model.

C. De Schryver (ed.), *FPGA Based Accelerators for Financial Applications*,
DOI 10.1007/978-3-319-15407-7

φ characteristic function of the logarithmic stock price.
ρ correlation between two Brownian motions in Heston model.
a fair price of a european (possibly path-dependent) option.
c fair price of a call option.
p fair price of a put option.
r risk-free interest rate.
American exercise feature exercisable at any time until maturity, cf. European exercise feature.
at-the-money strike equals spot.
call option giving the buyer the right to buy an asset at maturity for the strike price.
cap series of caplets.
caplet call on the forward interest rate.
European exercise feature exercisable only at maturity, cf. American exercise feature.
floor series of floorlets.
floorlet put on the forward interest rate.
implied volatility value of the volatility parameter in a pricing formula equating model and market price.
in-the-money intrinsic value is positive.
maturity expiration time of a derivative.
out-of-the-money intrinsic value is negative.
put option giving the buyer the right to sell an asset at maturity for the strike price.
strike fixed price at which the owner of the option can trade the underlying asset at maturity.
swaption option on the swap rate.
Vega sensitivity of a product price with respect to the volatility.

Monte Carlo Simulations

L total number of levels in a multilevel Monte Carlo simulation.
M the multilevel constant.
N number executed random experiments.
P physical probability measure.
Q equivalent (risk-neutral) probability measure.
X random variable.
$\mathcal{A}$ Monte Carlo estimator.
$\mathbb{E}$ expectation value.
μ true expectation value of a random variable X.
σ standard deviation.
l current level in a multilevel Monte Carlo simulation.

Stochastic Processes and SDEs

D number of discretization steps in discretized process
X stochastic process.
g functional applied to a stochastic process.
h step width of an equidistantly discretized process.
t time variable.

Calibration process

ω weight assigned to a particular market price in calibration.
calibration process of fitting model parameters to a set of market prices.
objective function the function to be minimized in an optimization problem.
penalty term stabilizing functional used in the calibration procedure.

Parameter sets

$\mathcal{M}$ model parameters.
$\mathcal{O}$ observable market parameters.
$\mathcal{P}$ product parameters.

Equity and interest models

Bates jump diffusion equity model of Bates.
Black-Scholes equity model of Black and Scholes.
Black '76 interest rate model of Black.
Heston stochastic volatility equity model of Heston.
Hull-White interest rate model of Hull and White.
Merton jump diffusion equity model of Merton.
SABR stochastic volatility interest rate model.

Interest rates

discount factor price of a zero bond paying one unit of money at a future time.
forward rate expected interest rate to be paid between to future points in time.
instantaneous forward rate expected interest rate to be paid for an infinitesimal small time step in the future.
zero rate interest rate to be paid from today until a future point in time.

Product prices

ask price lowest price the seller is willing to accept.
bid price highest price the buyer is willing to pay.
bid-ask spread difference between bid and ask price.
market price price at which a financial derivative is traded on the market.
model price price of a financial derivative as implied from its model, market and product parameters.

Printed by Printforce, the Netherlands